MEMS and Nanotechnology-Based Sensors and Devices for Communications, Medical and Aerospace Applications

A.R. Jha, Ph.D.

CRC Press
Taylor & Francis Group
Boca Raton London New York

CRC Press is an imprint of the
Taylor & Francis Group, an **informa** business
AN AUERBACH BOOK

CRC Press
Taylor & Francis Group
6000 Broken Sound Parkway NW, Suite 300
Boca Raton, FL 33487-2742

First issued in paperback 2019

© 2008 by Taylor & Francis Group, LLC
CRC Press is an imprint of Taylor & Francis Group, an Informa business

No claim to original U.S. Government works

ISBN-13: 978-0-8493-8069-3 (hbk)
ISBN-13: 978-0-367-38753-2 (pbk)

Library of Congress Cataloging-in-Publication Data

Jha, A. R.
    MEMS and nanotechnology-based sensors and devices for communications, medical and aerospace applications / A.R. Jha.
        p. cm.
    "An Auerbach title."
    Includes bibliographical references and index.
    ISBN-13: 978-0-8493-8069-3 (alk. paper)
    ISBN-10: 0-8493-8069-3 (alk. paper)
    1. Microelectromechanical systems. 2. Detectors. 3. Telecommunication--Equipment and supplies. 4. Medical instruments and apparatus. 5. Aerospace engineering--Equipment and supplies. 6. Aeronautics--Equipment and supplies. I. Title.

TK7875.J485 2008
620'.5--dc22   2007040799

Visit the Taylor & Francis Web site at
http://www.taylorandfrancis.com

and the CRC Press Web site at
http://www.crcpress.com

Dedication

This book is dedicated to my beloved parents who always encouraged me to pursue advanced research and development studies in the fields of science and latest technology for the benefits to mankind.

Contents

Foreword

One of the remarkable offshoots of the 50-year-old and yet exponentially growing silicon chip revolution has been the microelectromechanical systems (MEMS) technology. These miniaturized electromechanical devices are built on silicon substrates using well-established chip process technologies. Within these MEMS, we typically see integration, on a single silicon substrate, of not just electronic devices as on the chips, but also mechanical elements, sensors, and actuators. In addition to the commonly present materials in silicon integrated circuits (ICs), other materials such as ceramics and most recently carbon nanotube (CNT) arrays are also being incorporated into MEMS. The resulting microsystems have shown, for a variety of applications, unprecedented levels of miniaturization, reliability, and new capabilities.

Various end-user applications are based on devices that get incorporated in hybrid (MEMS and non-MEMS) systems or stand-alone MEMS. Examples include electro-optical, acoustic, and infrared (IR) sensors, RF/mm wave phase shifters, switches, tunable filters, resonators, and gyros for automobile control and safety. In addition to automobiles, the use of these subsystems in unmanned aerial vehicles (UAVs) is becoming increasingly common these days. An even faster growing application is in medical applications. These involve added technology synergies from nanobiotechnology materials. The resulting applications are concerned with advanced diagnostic tools as well as targeted deliveries of drugs to cancer cells.

This comprehensive book on MEMS and nanotechnology (NT) provides a wide-lens perspective of the field. It starts from basic design principles and current technologies for three-dimensional (3-D) MEMS with physical dimensions in the millimeter range. Fabrication processes incorporate modified silicon chip process technologies such as stress-controlled polysilicon thin films, as well as CNT arrays, and smart metal films. These are packaged in conventional as well as multichip modules (MCM) as practiced in the current state of the art. Structural material requirements for operating under harsh thermal, humid, and vibration environments are covered. The fabrication technology demands particular attention to sealing, bonding, passivation, etc. To assure compatibility, one needs to focus on coefficient of thermal expansion (CTE) match, and right choices from the available array of semiconductor technology materials from a diverse group consisting of single crystal and polycrystalline silicon, silicon oxide, silicon nitride, ceramics, glasses, polyimides, aluminum, chromium, titanium nitride, nickel, gold, etc.

There are two subsequent chapters dedicated to actuation mechanisms and integration with sensors and devices. Potential actuation mechanisms are based on electrostatic (ES), piezoelectric, electrothermal, electromagnetic, and electrochemical phenomena, which are discussed in great detail. The design trade-offs for reliability (e.g., cantilever beams), force-generating capacity, response time, and drive voltage/power requirements are covered in some depth. The mechanisms are applications specific and they must be individually customized for medical, auto, communications, and aerospace industries. A parallel-plate (PP) actuator is described in detail as a case study, because it is optimal for delivering a large output force over small displacements as needed in the case of hard-disk drives. Electrostatic (ES) rotary microactuators are also covered, as are bent-beam actuators.

Other chapters are dedicated to design, development, and performance aspects of

- RF-MEMS switches operating at microwave and mm-wave frequencies for wireless communications and electronically steerable antennas
- MEMS phase shifters, both absorption and reflection types, again for electronically steerable-phased array antennas, which offer significant reliability advantage over conventional electronics-based subsystems
- Micropumps (MPs) and microfluidic devices for chemical detection, microbiology, clinical analysis, biological detection, and heat transfer in microelectronic devices, with emphasis on process variables, reliability, and critical design parameters
- Other sensors and devices, such as varactors (tunable capacitors), accelerometers, wideband-tunable filters, and strain sensors are described.

The last chapter deals with a variety of emerging NT materials that enable special purposes of MEMS. These include quantum dots of dye-doped silica nanoparticles, zinc selenium nanocrystals, and fluorescent nanoparticles for clinical applications. Another example involves barium hexaferrite nanocomposite coatings or CNT arrays in high-temperature nanocomposites for high-performance aircraft.

This stimulating book should appeal to a wide audience consisting of design engineers, research scientists, and program managers. It contains a wealth of synergistic information and insights relating to the innovative field of MEMS and NTs. My friend, A.R. Jha, one of the most prolific writers, has covered a wide range of emerging technical fields, not yet included in the established literature. I am proud to write this foreword in support of his most recent book.

Ashok K. Sinha
Retired Senior Vice President, Applied Materials, Inc.

Preface

Recent advancements in nanotechnology (NT) materials and growth of micro/nanotechnology have opened the door for potential applications of microelectromechanical systems (MEMS)- and NT-based sensors and devices. Such sensors and devices are best suited for communications, medical diagnosis, commercial, military, aerospace, and satellite applications. This book comes at a time when the future and well-being of Western industrial nations in the twenty-first century's global economy increasingly depend on the quality and depth of the technological innovations they can commercialize at a rapid pace. Integration of MEMS and NTs will not only improve the overall sensor or device performance including the reliability but it will also significantly reduce the sensor weight, size, power consumption, and production costs. Advancements in MEMS and NTs offer unlimited opportunities in the design and development of various electro-optics sensors, lasers, RF/mm-wave components such as phase shifters, switches, tunable filters and micromechanical (MM) resonators, acoustic and infrared (IR) sensors, photonic devices, accelerometers, gyros, automobile-based control and safety devices, unmanned aerial vehicles (UAVs) sensors capable of providing the intelligence, reconnaissance, and surveillance missions of battlefields. It is important to point out that the latest MEMS- and NT-based sensors and devices are essential to enhance tactical UAV capabilities such as surveillance, reconnaissance, detection, identification, classification, and tracking of battlefield targets. UAVs using miniaturized thermal high-resolution IR sensors incorporating NT and NT-based composite materials will significantly improve the UAV close-range attack capability and covert surveillance and reconnaissance missions in the battlefields. The defensive capability of the UAV can be converted into offensive capability incorporating NT-based tracking sensors miniaturized IR missiles and the UAV will be recognized as an unmanned combat air vehicle (UCAV).

NT can play a key role in the diagnosis and treatment of a disease. NT scientists and nanobiotechnology engineers are deeply involved in research and development activities to search for new therapies, advanced diagnostic tools, and better understanding of the cell and disease symptoms. Nanobiotechnology clinical scientists believe that correct dose of drug will be directly delivered to cancer cells to provide immediate and effective comfort to the patients.

High-resolution IR sensors and thermal high-resolution IR sensors are best suited for commercial, industrial, military, and space applications. Cryogenically cooled MEMS-based photonic, electro-optic, microwave, mm-wave, and IR devices will have significantly improved the system performance, most suitable for space surveillance and reconnaissance, premises security, missile warning, and medical diagnosis and treatment. Cryogenic cooling has demonstrated significant performance improvement of sensors deployed by the UAVs, IR search and tracking systems, radar and missile warning receivers, satellite tracking systems, and imaging sensors.

This book has been written specially for MEMS and NT design engineers, research scientists, professors, project managers, educators, clinical researchers, and program managers deeply engaged in the design, development, and research of MEMS- and NT-based sensors and devices best suited for commercial, industrial, military, medical, aerospace, and space applications. The book will be found most useful to those who wish to broaden their knowledge in the field of MEMS and NTs. The author has made every attempt to provide well-organized material using conventional nomenclature, and a consistent set of symbols and identical units for rapid comprehension by the readers with limited knowledge in the field concerned. The latest performance parameters and experimental data on MEMS- and NT-based sensors and devices are provided in this book, which are taken from various references with due credits to authors and sources. The references provided include significant contributing sources. This book consists of nine chapters, each dedicated to a specific topic and application.

Chapter 1 summarizes the current design and development activities and technological advancements in the field of MEMS and NT. Current research and development activities contributed to rapid maturity of micro/nanotechnology, which has potential applications in MM systems, (MEMS) sensors, and devices most suitable for base stations, satellite communications, cellular and mobile phones, medical diagnosis, and aerospace sensors. Design and development activities of three-dimensional (3-D) MEMS devices with physical dimensions in the millimeter range are briefly discussed. Manufacturing of MEMS components using the very large-scale integration (VLSI) process, fabrication procedure involving a modified integrated circuit (IC) technology, and incorporating a stress-controlled polysilicon process are identified. Integration of advanced technologies such as micromultiple chip module (MCM) and micro-optic (MO) in future nano/micro-spacecraft is briefly discussed. Note a nano/micro-spacecraft system can integrate several functions such as guidance, navigation, and various control functions on a chip, which are most suited for UAVs with reconnaissance and surveillance capabilities. NT-based smart materials such as nanowires (NWs), nanoparticles, carbon nanotubes (CNTs), nanocrystals, and nanostructures are briefly discussed with particular emphasis on their applications in commercial, industrial, military, and aerospace applications.

MEMS- and NT-based sensors will be most attractive for aerospace, military, and space applications, where weight, size, power consumption, and reliability are the critical design requirements.

Chapter 2 identifies and describes various actuation mechanisms, their performance capabilities, limitations, and potential applications. Integration of MEMS technology with next-generation sensors and devices is briefly mentioned. Potential actuation mechanisms such as ES, piezoelectric, electrothermal, electromagnetic, electrodynamic, and electrochemical are discussed in great detail with emphasis on cantilever beam reliability, force-generating capacity, response time, design complexity, and drive voltage and power requirements. An RF-MEMS switch design is briefly described to understand the critical roles played by each element of the actuation mechanism deployed. Design requirements for a freely movable microstructural flexible membrane known as an armature or beam are identified. The mechanical displacement is provided by the microactuator using the force generated by an appropriate actuation mechanism. Plots of mechanical force generated by each actuation mechanism as a function of air gap, spring constant of the beam, beam geometrical parameters, and mechanical properties of materials involved are provided for the benefits of MEMS designers, students, and research scientists. Design requirements for electrodes, cantilever beams, contact surfaces, and passivation layers are identified. Pull-in and sticking problems generally experienced by some microactuators, which can affect the actuator reliability, are briefly mentioned. Actuation mechanisms most suited for communications, medical diagnosis, aerospace, and auto-safety applications are identified. Actuation mechanisms most suitable for RF-MEMS switches, mm-wave phase shifters, micropumps (MPs), health-monitoring devices, and chemical and biological threat-detecting sensors are recommended with emphasis on reliability, cost, actuation voltage, and design aspects.

Chapter 3 describes the latest version of actuation mechanisms capable of providing higher actuation forces to achieve displacements in rectangular dimensions. Performance capabilities of actuation mechanisms described in this chapter are not possible from the mechanisms discussed in Chapter 2. Current microactuators such as the ES rotary microactuator, bent-beam electrothermal microactuator, vertical comb array microactuator (VCAM), and electrochemical microactuator using CNTs are described with major emphasis on enhanced performance capabilities, operational benefits, design simplicity, fabrication aspects, and reliability. Optimum design configurations of electrodes capable of providing uniform and reliable actuation force for the actuators are identified with emphasis on enhanced force-generating capability and higher tracking accuracy over a wide bandwidth. A parallel-plate (PP) actuator configuration is described in great detail, because it is best suited for applications demanding a large output force over small displacements as in the case of hard-disk drives. The design aspects of ES rotary microactuator, which is considered most ideal for the dual-stage servomechanism, are briefly discussed. Design requirements for the critical elements of a rotary microactuator are identified. Computed values of static displacements and power consumption as a function of actuator beam geometrical dimensions and actuation voltages are summarized. Computerized data on ES force generated by the conventional and tilted configurations of rotary microactuator as a function of clearance and tilt angle is provided in a tabular format. Major drawbacks of the bent-beam actuator, namely, the long response time and thermal time constant, which is much higher than the electrical and mechanical time constants, are identified. Plots of generating force as a function of beam dimensions, air gap, and actuation voltages for various actuators are provided. Normalized torque plots as a function of normalized angular displacement and normalized capacitive gap are presented for a rotary microactuator with optimum shaped electrodes. Calculations and plots of displacement and mechanical resonance frequency as a function of piggyback microactuator dimensions, number of electrodes, and structural material parameters are provided for the benefits of the MEMS engineers, graduate students, and research scientists. Performance comparison data on vertical and lateral comb array microactuators is summarized with emphasis on drive voltage and electrical power requirements.

Chapter 4 deals with the packaging, structural material, and fabrication requirements vital for the design, development, and testing of MEMS- and NT-based sensors and devices. In addition, this chapter briefly discusses structural material requirements best suited for MEMS devices capable of operating under harsh thermal, humid, and vibration environments. Thermal, structural, mechanical, electrical, optical, and RF/microwave properties of the materials required in the fabrication, packaging, and bonding for MEMS sensors and devices are summarized. Important characteristics of soft and hard substrates, piezoelectric, and ferromagnetic materials needed for the fabrication of MEMS devices are discussed with emphasis on reliability and device longevity under harsh operating environments. Sealing material requirements are identified with particular emphasis on coefficient of thermal expansion (CTE) match, mechanical integrity, and MEMS device reliability over extended periods. Traditional semiconductors, alloys, and metal materials such as single crystal, polysilicon, silicon oxide, silicon nitride, ceramics, glassed, polymers, polyimides, aluminum, chromium, nickel, gold, and controlled expansion (CE) alloys are briefly discussed. Important properties of zero-level and first-level packaging, sealing, bonding, contact pads, passivation layers, and low-loss electroplating materials are summarized with emphasis on retaining optimum performance over extended periods and under severe operating environments. Properties of alumina, quartz, and fused silica best suited for mm-wave MEMS switches and phase shifters are discussed with major emphasis on dimensional stability, heat dissipation capability, insertion loss, isolation, structural integrity, and strength-to-weight ratio.

Chapter 5 exclusively focuses on the design and development aspects and performance capabilities and limitations of RF-MEMS switches operating at microwave and mm-wave frequencies most suited for wireless communications applications and electronically steerable antennas. Two types of RF-MEMS switch configurations, namely, series and shunt configurations, are discussed in great detail. Advantages of MEMS-based direct-contact switches over the conventional semiconductor RF switches are summarized with emphasis on reliability, power consumption, and control voltage requirements. Critical elements for both the RF-MEMS switches and the RF/microwave phase shifters are described. Performance capabilities and limitations of RF-MEMS shunt and series switches are summarized. Techniques to reduce insertion loss and to enhance isolation, switching speed, power-handling capability, and reliability are outlined. Effects of packaging environments on the functionality and performance of RF-MEMS switches are summarized with emphasis on reliability. Techniques to eliminate failure mechanisms in RF switches are recommended. Benefits of low actuation mechanisms are summarized. In the case of RF-MEMS switches, the nonlinear effects such as signal distortions and intermodulation products generated by the upstate bridge capacitance are mentioned.

Chapter 6 is dedicated to the MEMS phase shifters operating at microwave and mm-wave frequencies. Two categories of phase shifters, namely, absorption types and reflection types, are discussed. It is important to mention that the phase shifter is the most critical component of electronically steerable-phased array antennas, which are widely deployed in electronic warfare (EW) systems, missile tracking radar, forward looking radar used by fighter/bomber aircraft, covert communications systems, and space-based surveillance and reconnaissance sensors. Phase shifters using conventional field-effect transistors (FETs) and PIN-diodes suffer from high insertion loss, excessive power consumption, and poor reliability. However, integration of MEMS switches, MEMS tuning capacitors, and air gap, 3 dB couplers using coplanar waveguide (CPW) technology will lead to design and development of microwave and mm-wave phase shifters with low insertion loss, high isolation, negligible power consumption, enhanced reliability, and significantly reduced intermodulation products and signal distortions. MEMS-based phase shifters are best suited for reconnaissance satellite, missile seeker receivers, and UAVs where weight, size, power consumption, and reliability are the principal design requirements. Optimum design parameters such as Bragg frequency, center conductor width of CPW transmission line, spacing between the MEMS bridges and electrode geometry and dimensions are specified to achieve minimum insertion loss per bit, low-voltage standing wave ratios (VSWRs), reduced phase errors, and high isolation over wideband operations. Optimum design configurations for 2-bit, 3-bit, and 4-bit MEMS-based phase shifters capable of operating wideband in mm-wave regions are identified. Performance capabilities and limitations for MEMS-based true-time-delay (TTD) phase shifters operating in X-band, K-band, V-band, and W-band are summarized.

Chapter 7 concentrates on the design requirements, performance capabilities, and limitations of MPs and microfluidic devices and their potential applications in various disciplines. Note that microfluidics is an important branch of MEMS technology best suited for chemical detection, microbiology, clinical analysis, biological detection, and heat transfer in microelectronic devices. The MP is the critical component of a self-contained microfluidic system or sensor. Performance capabilities of passive MP designs, namely, floating-wall check valves or cantilever-beam "flapper valves," widely known as pneumatic valves, are summarized with emphasis on reliability. Design aspects of MPs using fixed valves involving no moving parts are described with emphasis on flow rate, pressure, and reliability over extended periods.

Performance parameters of various types of MPs such as piezoelectric valve-free (PEVF) MP, electrohydrodynamic (EHD) ion-drag MP, and ferrofluidic magnetic (FM) MP are summarized with emphasis on flow rate, pressure head, and reliability over extended durations. Design aspects for an MP using ES actuation mechanism are identified with emphasis on cost and performance. MPs using ES actuation methods are best suited for drug delivery applications. MPs with fixed-valve designs most ideal for transport of particle-laden fluids are described in great detail. Design, fabrication, and testing of fixed-valve pumps are briefly discussed. Low-order linear model capable of identifying design parameters for optimum performance and predicting resonance behavior is described. Benefits of low-order linear modeling and dynamic nonlinear modeling of MPs are briefly discussed. Numerical computations for resonance frequency, spring constant, pressure gradient, and equivalent mass for a piezoelectric valve-free MP are provided for the benefit of the readers. Curves illustrating the flow rates as a function of natural frequency and differential pressure are provided for the benefit of MEMS-based MP designers.

Chapter 8 describes the performance capabilities of selected MEMS and NT-based sensors and devices best suited for commercial, industrial, health monitoring, military, and aerospace applications. Discussions will be limited to sensors and devices not described previously. Performance capabilities of unique MEMS- and NT-based sensors and devices such as MEMS varactors or tunable capacitors, wideband-tunable filters, accelerometers, NT-based tower actuators using multiwall carbon nanotubes (MWCNTs), biosensors incorporating CNTs, strain sensors using smart materials, and MEMS sensors using smart materials to monitor health of structures, weapon systems, and battlefield environments are summarized. Potential applications and benefits of micro-heat pipes, photovoltaic cells, NT-based radar-absorbing materials, photonic detectors, lithium-ion microbatteries (MBs), and microminiaturized deformable mirrors are discussed identifying their unique performance capabilities. Applications of NWs, nanotubes, nanocrystals, nanorods and nanoparticles in commercial, military, medical, and aerospace disciplines are identified. Timoshenko and Euler–Bernoulli equations are used to compare the computational accuracy of the resonance frequency associated with beams used by various micro-resonators. Potential applications of MEMS varactors in wideband-tunable filters, phase shifters, RF synthesizers, and reconfigurable RF amplifiers are discussed with emphasis on reliability and stable RF performance over extended periods.

Chapter 9 summarizes materials and their important properties critical in the design and development of MEMS- and NT-based sensors, photonic components, and the new generation of MEMS devices for possible applications in aerospace, automobile, clinical research, cancer diagnosis/treatment, and drug delivery. Applications of photonic bandgap devices and photonic bandgap fibers are identified. High temperature stability, frequency stability, and lowest phase noise of optoelectronic-oscillators (OEOs) operating at 10 GHz and beyond are specified. Potential applications of quantum dots dye-doped silica nanoparticles and zinc selenium nanocrystals are identified for clinical research, cancer therapy, and biosensor technology. Benefits of fluorescent nanoparticles capable of providing high-resolution images superior to those obtainable from current magnetic resonance imaging (MRI) and computerized tomography technologies are identified for cancer research and diagnosis. Potential applications of barium hexaferrite nanocomposite coatings vital in the development of stealth technology are briefly discussed with particular application in high-performance fighter or reconnaissance aircraft to avoid detection by enemy radar. Important thermal, mechanical, and electrical properties of bulk and microscale materials best suited for fabrication and packaging of MEMS- and NT-based sensors and microsystems are summarized. Materials and their properties widely used in the design of MEMS- and NT-based sensors and devices such as acoustic sensors; CNT-based transistors; multijunction, high-efficiency photovoltaic cells using organic thin-films; and smart sensors to monitor weapon health, battlefield environments, and chemical/biological/toxic agents are briefly summarized. Applications of CNT arrays in multifunctional, high-temperature nanocomposites best suited for rocket motors and warheads, high-current density electron emitters with cold cathodes, MEMS biosensors, electrochemical actuators, and electrodes for MBs are discussed in great detail with emphasis on reliability and performance parameters not possible with other technologies.

I wish to thank Catherine Giacari, project coordinator in the Editorial Project Development department at CRC Press/Taylor & Francis, who has been very patient in accommodating last-minute additions and changes to the text. Her suggestions have helped the author to prepare the manuscript with remarkable coherency and minimal effort. I wish to thank my wife Urmila Jha, daughter Sarita Jha, son Lt. Sanjay Jha, and my son-in-law Anu Manglani, who inspired me to complete the book on time under a tight schedule. Finally, I am very grateful to my wife, who has been very patient and supportive throughout the preparation of this book.

Author

A.R. Jha, BSc (engineering), MS (electronics), MS (mechanics), and PhD (electronics), technical director, Jha Technical Consulting Service, has design, development, and research experience of more than 35 years in the fields of radar, electronic warfare, lasers, phased array antennas, solid-state mm-wave components, cellular/mobile/satellite communication, receivers, MEMS devices, and IR sensors. Dr. Jha has published more than 75 technical papers (including eight invited papers) and authored eight high-technology books and holds a U.S. patent on a mm-wave terrestrial communication antenna.

Chapter 1

Highlights and Chronological Developmental History of MEMS Devices Involving Nanotechnology

1.1 Introduction

Recent advancements and rapid growth of micro/nanotechnology (MNT) have opened doors for potential applications of microelectromechanical systems (MEMS) devices in various fields. This book comes at a time when the future and well-being of Western industrial nations in the twenty-first century's global economy increasingly depend on the quality and depth of technological innovations they can commercialize at a rapid pace. This chapter in particular focuses on the highlights and chronological developmental history of MEMS devices incorporating nanotechnology (NT). Potential applications of MEMS devices in commercial, industrial, military, and space systems are identified. Benefits from MEMS devices in medical diagnosis and surgical procedures, satellite communications, cellular and mobile phones, perimeter security, airborne surveillance, low-cost thermal imaging, and nondestructive testing of commercial and industrial structures are briefly mentioned. Advancement of MEMS technology offers unlimited opportunities in the design and development of various radio frequency (RF), mm-wave, acoustic wave, infrared (IR), photonic, and electro-optic-based MEMS devices for various applications. It is important to mention that MEMS devices incorporating NT have revolutionized cellular and mobile communications, satellite communications, and optical communication industries, which in turn could open exciting opportunities for the growth of MEMS-based IR and photonic devices. Furthermore, RF-MEMS technology offers superior RF passive and active devices such as switches, phase shifters, tunable capacitors known as varactors, switching capacitors, coplanar waveguide (CPW) transmission lines, memory-on-chips, printed circuit smart antennas, and digital RF memories (DRFMs)—critical elements of electronic warfare (EW) systems.

It is important to mention that MEMS devices are best suited for applications, where low-power consumption, minimum weight and size, high reliability, and reconfigurability are the principal requirements. It is important to mention that communications satellites will be the principal beneficiaries from the MEMS technology [1] because they offer global wireless connectivity, as illustrated in Figure 1.1.

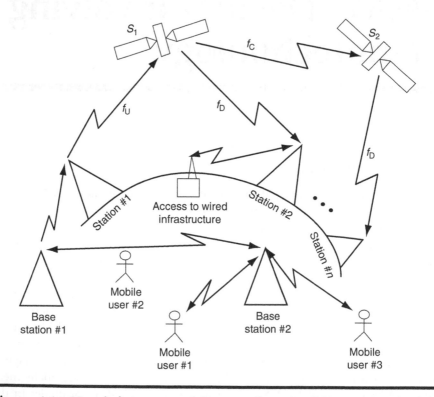

Figure 1.1 RF wireless connectivity paradigm involving communications satellites, base stations, mobile users, and access to wired infrastructures. (From De Los Santos, H.J. et al., *IEEE Microw. Mag.*, 53, 2004. With permission.)

1.2   What Is MEMS?

MEMS is a micrometer-sized device with three-dimensional (3-D) properties, namely sensors, pumps, and valves all capable of sensing and manipulating physical parameters. Some MEMS may not have any moving parts and can be made from plastic, glass, and dielectric materials. The operating principle of MEMS is based on high-frequency devices involving resonant structures and microelectronics. High-frequency elements involve resonating beam cantilevers, diaphragms, and other resonant structures. Production of MEMS devices involves micromachining and various specific processes such as deposition, etching, lithographic, and chemical. MEMS components can be manufactured using very large-scale integration (VLSI) processing techniques. MEMS fabrication involves a modified integrated circuit (IC) technology involving a stress-controlled polysilicon process.

1.2.1   Frequently Used Terms in Nanotechnology

The term NT or MNT encompasses the synthesis and integration of materials and processing devices of submillimeter to submicrometer size. The term micro-engineering deals with machining, assembling, integration, and packaging of miniature two-dimensional (2-D) or 3-D devices and electronic circuits (RF/analog/digital) achieved through micromachining. Molecular electronics (ME) plays a key role in the design and development of RF-MEMS or optical-MEMS devices. The principal advantage of ME is that the molecules are natural nanometer-scale structures. Molecular nanotechnology (MN) offers high level of miniaturization, which leads to significant reduction in weight, size, and power consumption, while maintaining ultrahigh reliability. Nanoelectronics (NE) deals with tens of billionths of a meter. Nanostructures with dimensions up to 100 μm are best suited for designing MEMS devices. Nanotubes conduct large currents without displaying harmful resistance heating. A nanotube has molecules that are 10,000 times smaller than MEMS. Nanotubes can be used in fabrication of field-effect transistors (FETs). Nanowires (NWs) can be used to produce high-speed logic gates, which are best suited for computers, where ultrahigh speed, small size, and minimum power consumption are the principal requirements. It is critical to point out that the molecular form of pure carbon is called a buckytube. Furthermore, a buckytube has a strength-to-weight ratio 600 times greater and can be used in composite structures best suited for aircraft and missiles.

Micro-optical-electromechanical systems (MOEMS), also known as optical-MEMS, indicate that the MEMS could be an optical-based device comprising of optics, microelectronics, and NT circuits merged. This can integrate a MEMS deformable mirror with micro-optics and complementary metal-oxide semiconductor (CMOS) electronics to achieve adaptive optics capability, which yields maximum photon excursion and modulates electronics. It is important to point out that an MOEMS device enables small-scale reflective and diffractive elements controlled by micromachined parts. This device produces the switching and modulation functions without signal loss or cross talk. Unlike electronics, it does not require switching from optics to electronics and vice versa. MOEMS technology plays a key role in the design of optical attenuators and other optical components, integrating both MEMS and optical technologies. It is desirable to mention that tilting mirrors for opto-microelectronics are best suited for space sensor applications.

1.2.2 2005 MEMS Industry Overview and Sales Projections for MEMS Devices

Current market surveys indicate potential sales for MEMS-based devices in wireless and communications sectors. The 2005 MEMS Industry Overview and Forecast report [2] covers the worldwide sales of MEMS devices. It contains forecasts in unit's shipments and revenues by product category and by industry through calendar year 2009. It also includes a ranking of the top 20 MEMS suppliers and a detailed review of new products and developments. This report provides market growth for the MEMS devices over the next few years in terms of unit shipments at a compound annual growth rate (CAGR) approaching 20 percent to nearly 6 billion units in 2009. The report reveals funding increase for MEMS companies by 44 percent in 2004 versus 2003. Year-over-year total MEMS revenues were up to 32 percent from 2003 to 2004. The latest market survey shows microfluidic devices accounted for nearly 69 percent of total unit shipments and 23 percent of total revenues in 2004. The communications market is forecast to experience the highest CAGR for the unit shipments and revenues for MEMS devices deployed by cell phones, mobile phones, inertial sensors, and optical networking sectors. This report further states that the top 20 suppliers of MEMS devices accounted for 62 percent of the revenues in 2004.

1.3 Potential Applications of MEMS Devices in Commercial and Space Systems

In this section, potential applications of MEMS technology and devices will be identified for commercial, industrial, military, and space systems. Integration of MEMS technology will significantly improve the RF performance of high-efficiency power amplifiers, mm-wave circuits, analog ICs, high-speed digital circuits, printed circuit smart antennas, and null-steering retrodirective antennas capable of carrying high data rate under jamming environments. The MEMS technology allows continuous monitoring of patient's medical parameters needed for diagnosis and therapy. This technology also allows design and development of implantable, self-powering, contactless MEMS devices, which are best suited for treating neurological and cardiac disorders.

Integration of MEMS technology has demonstrated significant advantages particularly in the design and development of photonic, IR, electro-optic, RF, and mm-wave devices. It is important to mention that RF and mm-wave MEMS devices offer wide bandwidth, high data rate and reliability, while realizing significant reduction in weight, size, and power consumption. Therefore, MEMS devices are best suited for both wireless and satellite communications. It is important to mention that the broad frequency bandwidths available beyond 50 GHz require the development of single-chip transceiver topologies and techniques based on MEMS technology. Current research and development activities on 80–100 GHz chips will focus exclusively on a single-chip transceiver design capable of transmitting data at a rate exceeding 100 Gbps, which is not possible without MEMS technology. This 100 GHz Ethernet chip could be digitally reconfigured for other applications such as scientific instrumentation and biomedical imaging, where high resolution and accuracy are of critical importance. In summary, MEMS devices are best suited for wireless, commercial, and satellite communications, microwave vehicular systems, RF-monolithic microwave integrated circuit (MMIC) devices, biological, and medical applications, precision control systems for military sensors, active and quasi-optic arrays, light-wave technology devices, superconducting sensors, microwave digital processing and space-based surveillance, and reconnaissance sensors, where high resolution, ultrahigh reliability, minimum weight, size, and power consumption are the principal requirements. Potential applications of RF-MEMS, mm-wave-MEMS, and optical-MEMS in various commercial, industrial, and military applications will be identified and described.

1.3.1 MEMS for Wireless, Base Stations, Satellites, and Nanosatellites

MEMS devices and components incorporating NT have revolutionized the wireless, commercial communications, and satellite communications sectors. MEMS-based RF switches, phase shifters, power amplifiers, and printed circuit antennas are best suited for wireless and space communications systems because NT conserves energy and minimum power consumption. Therefore, as a technology that conserves energy and minimizes electrical power consumption, RF-MEMS switches and phase shifters are critical to increasing satellite communications while minimizing the launch cost. MEMS technology offers cost-effective design for large satellites employing electronically steerable-phased array antennas and for switching matrices to provide pocket-switched communications in constellation and nanosatellites. A nanosatellite is a very small and lightweight spacecraft containing microelectronic equipment, RF components, and payload with launching mass not exceeding 10 kg or 22 lb. Nanosatellites are most attractive due to their cheaper, lightweight construction, versatility of launching means, and applicability to national security missions, either as single units or as clusters working cooperatively. As a cluster comprising of three or five or ten nanosatellites, they may function as smart, distributed apertures for deducing the geolocation of a downed pilot in enemy territory. As a constellation of tens of units, they are most attractive for Internet-in-the-sky applications and continuous reconnaissance of a specific geographical area or battlefront area during a covert military operation.

1.3.1.1 RF-MEMS Amplifier-Switched Filter Bank Capabilities

Recent advances in MEMS technology indicate that a MEMS-based switched filter bank offers several advantages, namely, lowest current consumption while maintaining high linearity over a multi-octave bandwidth. It is critical to point out that with low insertion loss and low stray capacitance, an RF-MEMS switch offers maximum design flexibility in configuration of the switch matrix. A switch with hybrid radial-binary tree configuration shown in Figure 1.2 provides optimum RF performance [2] with minimum cost and complexity. Typical performance parameters of MEMS-based switch filter banks are summarized in Table 1.1.

The above RF assembly uses a complex programmable logic device, which can determine which MEMS switches must be toggled based on the new discrete state, thereby minimizing the switch current requirements. This unit can be designed to provide low distortion amplification of the input signal, while minimizing the switch current. A passive version of this MEMS-based amplifier switch filter bank will

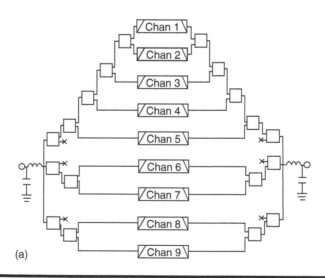

(a)

Figure 1.2 Communications systems showing (a) RF communications system comprising of hybrid radial-binary tree configuration using MEMS technology and

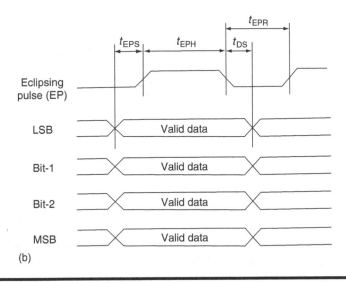

Figure 1.2 (continued) **(b) logic timing of the eclipsing pulse of the system. (From Gawel, R., *Electron. Des.*, 43, 2005.)**

provide a fail safe operation, even under prime power failure. This assembly design concept could be extended to higher frequency operation with various channeling configurations for either passive or active operation. No other technology can meet or beat the performance available from the MEMS-based amplifier switch filter bank.

1.3.1.2 Passive RF-MEMS Components

Active RF-MEMS sensors sometimes do require passive MEMS devices such as couplers, power dividers, matching circuits, and amplitude leveling elements

Table 1.1 Typical Performance Parameters of an RF-MEMS Amplifier-Switch Filter Bank Assembly Comprising of Nine RF Bands

Parameter	Parameter Value
Frequency band (MHz)	25–500
Gain, G (dB)	15 (minimum)
Input voltage standing wave ratio (VSWR)	1.5:1
Isolation (dB)	80 (minimum)
Power consumption	8 mA at 3.3 Vdc
Assembly size (in.)	$4.65 \times 3.00 \times 0.40$

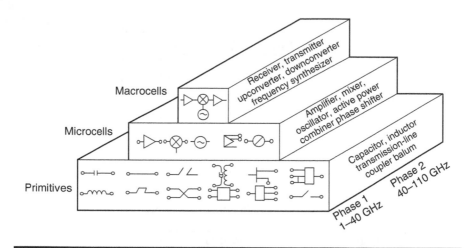

Figure 1.3 MMIC technology cell library indicating the deployment of MEMS-based passive and active components.

(Figure 1.3). These RF-MEMS passive components could use multilayer structures to save cost and design complexity. CPW transmission lines are fabricated on a polyamide, which acts as an insulating dielectric layer. CPW transmission lines are most suitable for fabrication of RF-MEMS and optical-MEMS devices because they offer low insertion loss, minimum dispersion at mm-wave frequencies, and small overlapping at various impedance levels. Typical variation of CPW transmission-line impedance as a function of frequency and overlapping is summarized in Table 1.2.

The fact that RF-MEMS technology offers superior passive devices such as switches, switchable capacitors, tunable capacitors (high-Q varactors), microinductors, transmission lines, and resonators makes this technology most attractive for

Table 1.2 Impedance Variation in CPW Transmission Lines as a Function of Frequency and Overlapping (Ohms, Ω)

Frequency (GHz)	Overlapping (μm)			
	1	*4*	*10*	*20*
5	23	20	17	13
10	22	18	15	10.5
20	20	16	12	8
30	14	11	6	7

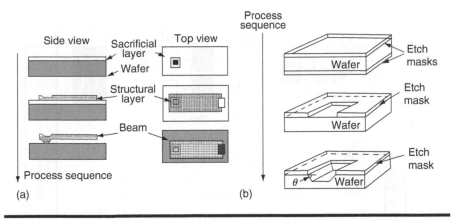

Figure 1.4 Fabrication process sequence for MEMS-based cavities with various configurations (a) surface micromachined cavity and (b) bulk micromachined cavity, using MEMS technology.

wireless appliances operating in the home/ground/mobile/space spheres, such as handsets, base stations, and satellites. As mentioned previously, this technology offers low-power consumption, high reliability, compact packaging, and reconfigurability. Critical issues such as system-level justification for RF-MEMS, device requirements, high-volume manufacturing, packaging, reliability, and performance capabilities will be examined in greater details. It is important to mention that the surface micromachined technique is best suited for RF-MEMS devices used in wireless systems, namely, switchable capacitors, tunable capacitors, ohmic switches, beams, and cantilevers, whereas the bulk micromachining method is used for cavities, as illustrated in Figure 1.4. Potential structural materials, sacrificial materials, and etchant techniques used in the fabrication of RF-MEMS devices are described in Section 1.4.1.

1.3.2 RF-MEMS Technology for Base Station Requirements

An infrastructure vendor is fully equipped to provide a variety of base station configurations (Figure 1.5) to meet network operator needs. In addition to various air interface standards, such as global system for mobile (GSM) communications, universal mobile telecommunications system (UMTS), and code division multiple access (CDMA), there are also different frequency bands of operations and various RF power levels as illustrated in Figure 1.6. It is important to mention that weight, size, and power consumption for RF-MEMS devices are not very stringent compared to those for devices to be used in communications satellites. However, weight,

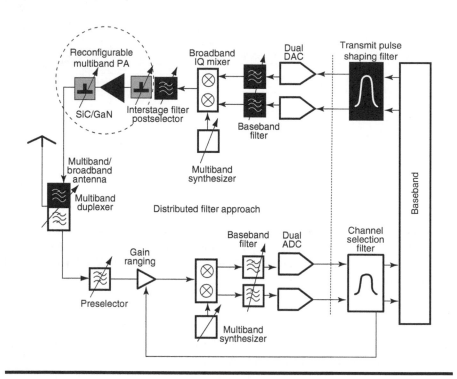

Figure 1.5   Architecture of a communications base station employing RF passive and active MEMS devices.

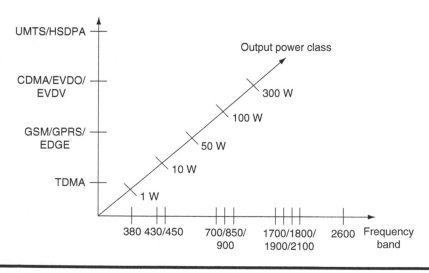

Figure 1.6   RF communications base station variants illustrating power level requirements and operating frequency bands.

size, and power consumption for cell phones and mobile phones are most critical and RF-MEMS devices must meet these requirements, besides circuit reuse capability. It is desirable to mention that RF-MEMS devices show significant improvement in linearity, insertion loss, and reliability, which are not base station invariants including four air interfaces, eleven frequency bands, and five RF power levels leading to 220 ($4 \times 11 \times 5$) different products as shown in Figure 1.6. Assuming 85 percent of the combinations or products are dropped, the remaining 15 percent still implies 33 variants. Upon examination of the remaining variants, it is evident that the greatest variants are in the frequency bands. The variants along the "standards" axis may be limited to two big standards for the future only, which will be UMTS and CDMA, including their evolutions: high-speed downlink packet access (HSDPA) and evolution data only (EVDO). Note that UMTS and CDMA belong to the same family of code multiplex systems. In summary, the axis "frequency band" is the greatest contributor to the base station variants.

1.4 MEMS Technology for Military Systems Applications

RF-MEMS and optical-MEMS devices can play key roles in the design and development of military sensors to meet specific mission requirements. Because of their low-power consumption, compact size, and ultrahigh reliability, MEMS are most suitable in development of sensors for battlefield tanks, surveillance for hostile territory, drone electronics, covert communications, unmanned air vehicles (UAVs), and missiles to shoot down hostile targets. The success of future battlefield conflicts is contingent on rapid integration of MEMS-based microwave and mm-wave devices, according to the deputy director of the Tactical Technology Office of the U.S. Defense Department agency, Defense Advanced Research Projects Agency (DARPA). The battlegrounds of Iraq and Afghanistan have witnessed faster and precision missiles to detonate their weapons on their targets, which is possible through rapid integration of advanced technologies such as MEMS technology or NT in offensive weapon systems. With the integration of these technologies, it is possible for one of the missiles loitering overhead to survey the battlefield scene, to detect a deadly moving target, and then drive down to destroy the target with 100 percent kill probability. Such futuristic combat techniques would not be possible without rapid advances in MEMS devices, nanoelectronics, and electro-optic devices [2].

Nanotechnology-based and MEMS-based sensors can play critical roles in monitoring the weapon health and battlefield environment parameters. This will require various embedded sensor suites capable of performing onboard diagnosis, maintaining a history of sensor data, and forecast of weapon health parameters. This requires the program managers or project directors to quantify the operational requirements for the remotely operated MEMS-based and NT-based sensors for a host of battlefield-related weapon systems such as missiles, mortar-locating radar, and remotely piloted vehicles (RPVs) involved in critical battlefield missions. After establishing the operational requirements, current NT-based temperature, humidity, shock, vibration, and chemical sensors for monitoring the outgassing of weapons propellant as well as hazardous gases on the battlefield and in urban environments must be improved to meet the actual battlefield requirements. Before the actual deployment of these sensors in real battlefield environments, performance parameters such as power consumption, reliability, maintainability, survivability, procurement cost, and size, along with technical challenges for these sensors operating in stringent military environments, must be carefully evaluated to satisfy the cost-effective design of the sensor. The final designs of the sensors must incorporate the laboratory test results obtained on a wireless sensor array using a thin film of functionalized carbon nanotube (CNT) materials. The detection of biohazardous agents or materials is possible using NT-based active and passive wireless sensors based on monitoring the reflected phase from the monitoring sensors deployed.

The sensors discussed above are best suited to identify the out-of-specification weapons, predict remaining useful shelf life, improve the reliability through minor design modification, and to assess the readiness of the weapon stockpile. Remote miniature sensors using MEMS technology are required to alert soldiers of harmful chemical gases and biological agents in battlefield or urban environments. It is desirable to mention that the concept of embedded prognostics and diagnostics sensors will be found most useful in performing routine maintenance checks to reduce soldier involvement, in obtaining readiness data from each missile or offensive weapon, and in improving missile reliability and kill probability [2]. Design requirements of an embedded sensor could vary from sensor to sensor depending on the specific parameters to be measured. The natural environmental parameters such as temperature, pressure, and humidity are very similar for various weapon systems, except for the salient environments inherent to ground, air, and underwater operations. This also holds true for bioterrorism and hostile environments involving chemical, biological, and suicide bomber threats. It is important to deploy a low-power battery-operated interrogation system involving a computer-controlled electronically steerable-phased array antenna using MEMS devices, onboard data acquisition module, and real-time data processing, and storage unit. The interrogator antenna focuses the electromagnetic energy and establishes robust and covert communications link with individual onboard sensors in an automatic sequential order. NT components that are vital in the design and development of battlefield monitoring sensors include single-wall carbon nanotubes (SWCNTs) and multiwall carbon nanotubes (MWCNTs), fiber optic cables with unique design parameters, and MEMS-based phase shifters and switches, acoustic monitoring devices, and electrochemical sensors.

1.4.1 Material Requirements for Fabrication of MEMS Devices

Material requirements for RF-MEMS or photonic-MEMS or optical-MEMS devices will be more stringent to meet stated performance requirements under extreme battlefield environments. High-performance substrates, improved semiconductors, and composite structural materials are vital to accomplish the battlefield mission requirements. Unique semiconductor material such as antimonide-based compound semiconductor material with narrow bandgap and carrier mobility ten times greater than silicon, if used in fabrication of MEMS devices, can allow operation under less than 1 Vdc, leading to minimum power consumption. Wideband semiconductors under development such as gallium nitride and silicon nitride (Si_3N_4) will be best suited for MEMS-based RF, photonic, and electro-optic sensors, electronically steerable-phased array antennas, radar actuators, and covert communications systems. It is necessary to develop new classes of semiconductor materials and metallic films for fabrication of MEMS devices to extend the military's dominance of the electromagnetic spectrum for development of state-of-the art radar, covert communications systems, electronically steerable-phased array antennas, wideband ELINT receivers, EW suites, and electronic sensors for critical military missions (Figure 1.7).

The Advanced Technology Office at DARPA is investigating new materials based on NT, which will enable small- and medium-caliber lightweight projectiles to be fired at very high velocities without the need of internal moving parts. Such a

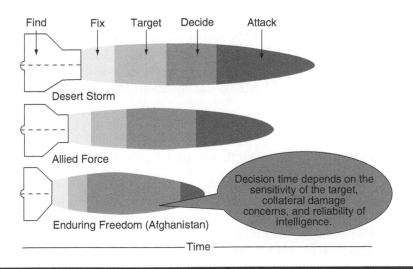

Figure 1.7 **Sensor-to-shooter cycle times for various missions deploying MEMS-based electronic sensors.**

Table 1.3 Structural Materials, Sacrificial Materials, and Etchant Techniques Widely Used in Fabrication of MEMS Devices

Structural Materials	Sacrificial Materials	Etchant Process
Aluminum (Al)	Single-crystal silicon	Electronic data processing (EDP), XeF_2
Aluminum	Photoresist (PR)	Oxygen plasma
Copper (Cu) or nickel (Ni)	Chrome	HF
Polyimide	Aluminum	Al etch (nitric acid)
Polysilicon	Silicon dioxide	HF
PR	Aluminum	Al etch (nitric acid)
Silicon dioxide (SiO_2)	Polysilicon	XeF_2
Silicon nitride (Si_3N_4)	Undoped polysilicon	Potassium hydroxide (KOH)

projectile will have a minimum velocity of 1600 m/s or about 5 M with a firing capability of 600 rounds or more per minute. Deployment of NT in the development of projectiles will eliminate field-reliability problems and will significantly improve the target kill probability.

Requirements for structural materials, sacrificial materials, and etchant techniques in the fabrication of MEMS devices are dependent on the operating frequency, wavelength, bandwidth, performance, and reliability specifications. Potential material requirements and etchant techniques widely used in the design and development of various RF-MEMS and optical-MEMS devices are summarized in Table 1.3.

1.4.2 Types of Nanostructures and Their Properties

Nanomaterial requirements for nanostructures will vary from application to application. However, material requirements will be most stringent for MEMS devices operating under harsh conditions and space environments. Important electrical, mechanical, and thermal properties of nanomaterials must be known before their deployment for specific MEMS-based system application. Electron mobility and thickness properties of materials such as porous silicon, high-resistive silicon, Si_3N_4, benzocyclobutene (BCB), platinum alloy, and polymide must be known for various applications. Most of these materials can be used in the fabrication of critical components of MEMS devices. Optical properties of nanomaterials must be known before their use in fabrication of mirror membranes, electrically pumped vertical-cavity surface-emitting laser (VCSEL) diodes and VCSEL optical amplifiers. Material requirements for NWs, nanocrystals, nanoribbons, and nanotubes will be specified to meet specific performance requirements.

Scientists at Portland State University, Oregon, have obtained near-IR emissions centered at 393 nm from the zinc oxide (ZnO) nanostructures when embedded in polymer. These ZnO-NW LEDs [3] could lead to rapid development of large-area lighting on flexible substrates. These ZnO NWs can be produced at temperatures as low as 85°C, which could yield significant savings in production costs. These temperatures and the present fabrication process are compatible with glass and flexible polymer substrates. Current research activities are focused to develop ZnO-NW LEDs using silicon substrates. Such emitters could be useful for on-chip optical interconnects, which are best suited for MEMS-based optical signal processors to meet higher data rates.

Scientists at Naval Research Laboratory, Washington, DC, have reported successful fabrication of electrically driven organic LEDs using standard lithographic techniques. These MEMS devices, according to research scientists, are best for chemical and biological sensing, direct-writing photolithography, and optical interconnects. Single 60 nm diameter organic LEDs could be used in the development of single-crystal light sources and uniform arrays of nano light emitters [3].

Japanese scientists have designed and developed a mode-lock fiber laser using CNTs operating at 1300 nm wavelength and they claim that this is the firstmock-lock fiber laser. CNTs act as a saturable absorber in the host laser resonator cavity and their performance is comparable to that of saturable absorbers including semiconductor saturable absorber mirrors and bleachable dyes. It is desirable to mention that CNTs offer ultrafast recovery times as fast as 500 fs, high optical damage threshold, excellent chemical stability, and minimum manufacturing cost.

Research activities focused on nanocrystals reveals that ZnSe nanocrystals doped with copper (Cu) and MnSe offer alternatives to CdSe nanocrystals. ZnSe nanocrystals have demonstrated lower toxicity and higher stability for the development of biomedical labels, light-emitting diodes, and minilasers. CdSe has been the material of choice in NT research and development activities. CdSe nanocrystals are relatively easy to produce. Their emission wavelength is continuously tunable over a visible range by controlling their physical dimensions. The toxicity of cadmium is a limitation to the application of these nanocrystals. The material also exhibits thermal quenching of its luminescence, which could affect its suitability for use in high-performance LEDs and lasers, where thermal quenching is not acceptable.

MEMS mirror membranes are formed on semiconductor chips and can be used to tune an IR electrically pumped MEMS-based VSCEL. The micromachined deformable membrane mirror made from silicon nitride is produced on a GaAs substrate and fixed to a silicon chip. The mirror membrane has a radius of curvature close to 5.5 mm, which eliminates the need for critical alignment between the two chips. The reflectivity of the mirror membrane is as high as 99.95 percent, but it does not offer optimum coupling. One chip contains the mirror membrane and its GaAs substrate, while the other chip contains the VCSEL module and amplifier. This laser offers lasing action in a single mode from 1553 to 1595 nm and is best suited for telecommunication, trace-gas sensing, metrology, process control, and medical diagnosis. It is important to mention that tuning speed is the most critical parameter in many applications. The tuning speed measured on a thermoelectrically tuned laser is slower than that for an electrostatically tuned MEMS laser. It is critical to point out that the micromachined membrane deformable mirrors have been used in ophthalmic applications, low-cost adaptive solar telescopes, and satellite tracking systems. It is desirable to point out that microdeformable mirrors are best suited for wave front control applications in adapting imaging or beam forming, for which aberrations are expected to be smooth and continuous. Such an application involves the correction of wave front of a high-power laser beam traveling through various regions of the atmosphere.

1.4.2.1 Surface Plasmon Resonance

Surface plasmon (SP) resonance occurs from the infusion colloidal metal particles in glass that produces wonderful brilliant colors for cathedral windows. The colors are due to excitation of plasma oscillation modes in the metal particles, which creates absorption mode bands and produces unique colors in the transmitted light beam. Plasmatic materials are nanostructures that support SP oscillation modes in the metals. The conduction electrons are essentially free to move with little interaction from their respective nuclei because of Coulomb shielding effects. Plasmonic nano-photonics technology essentially combines the capabilities of NT and photonics. Noble metals such as copper, gold, and silver have dominated the studies of plasmon effects because their resonances are the strongest and occur in the visible region of the electromagnetic spectrum. Many nanostructures support plasmon oscillations and these microstructures include small metallic spheres, thin films, continuous metallic films, nanorods, nanoshells, NWs, nanoribbons, and lithographically produced metal arrays.

SPs can be used to design efficient and sensitive MEMS-based biosensors. Biosensing, detection, and analysis of biomolecules such as proteins have become an area of greatest interest of scientific research. These sensors have potential applications for environmental testing, national security, airport passenger screening, optical bioimaging, nanophotolithography, and engineering process control. Plasmatic materials are characterized by the dielectric permittivity and magnetic permeability parameter. It is desirable to mention that negative-index materials are best suited for the development of biosensors. Plasmonic nanostructures could also be used as optical couplers across the nano/micro interface and bow tie-like nanoantennas.

Current research activities have been directed to develop intelligent materials, smart materials, and smart structures using NT. Passive and active damping functions are becoming increasingly important in terms of vibration control of the structures. Energy absorbing systems are receiving greatest attention because they seem to provide maximum protection to the shoulders from injury due to sudden impact in the battlefield. Micrometallic-closed cellular materials are considered the smartest materials because of their unique structures and properties. Cellular walls made from nickel (Ni)–phosphorous alloy have demonstrated highest energy absorbing capability of the metallic closed cellular materials, which are best suited for chest protection from bullets.

1.4.2.2 Ceramics for MEMS Sensors

Piezoelectric lead–zirconate–titanate (PZT) ceramics are widely used in the design of MEMS sensors and actuators capable of operating under harsh environments. These materials can be manufactured according to prescribed aspect ratios and are capable of satisfying multimode operations such as radial mode, thickness extension mode, length extension mode, and length/thickness extension mode. Important properties of PZT materials are elastic constants, dielectric constants, and electromechanical coupling factors as a function of temperature, sample geometry, frequency, and crystal structures of soft and hard PZT materials. MEMS sensors made from PZT materials are best suited for structural health monitoring of composite, repair patch in bridge rehabilitation, and damping circuits for high-speed turbomachine structures made from composites and smart materials.

1.4.3 Fabrication of Critical Elements of a MEMS Device

Fabrication requirements are very stringent for the elements of MEMS devices strictly used in military systems and satellite communications. RF-MEMS fabrication technique is well established on the current IC fabrication processes using sequential steps (a–d) as illustrated in Figure 1.8. Conventional 2-D IC fabrication process includes wafer fabrication, surface preparation, photomask involving photoresist (PR) polymeric light-sensitive material, image development, and transfer of the negative image on the photomask. IC fabrication steps need to be followed in the design and development of MEMS devices. It is desirable to mention that RF-MEMS is predicated on two mail techniques to create a third dimension, namely bulk micromachining. Material requirements must be defined for the film, sacrificial layers, mechanical support structures, cantilever beams, and activation mechanisms. Etchants such as KOH, XeF_2, and nitric acid must be identified for various MEMS devices. It is important to make sure that the spring forces of the mechanical structures overcome the wet etchant's surface tension forces to avoid the improper release and the sticking phenomenon also known as stiction, which is very common in an electrostatic actuation mechanism. In bulk micromachining, as shown in Figure 1.4, mechanical structures are created within the confines of the semiconductor wafer through selectively removing wafer material by wet and dry

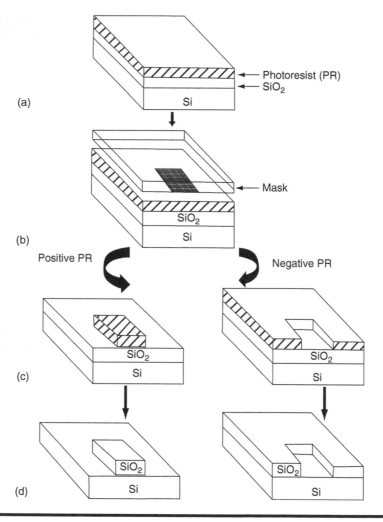

Figure 1.8 **Sequential fabrication steps (a–d) for MEMS devices employing IC technology.**

etching techniques. By integrating the 3-D freely movable mechanical structures in the same wafer with the electronics, it is possible to design highly functional MEMS with remarkable performance not feasible otherwise and without sacrificing in weight and size, power consumption, reliability, and production costs. The 3-D MEMS structures can be actuated via a variety of actuation mechanisms, which will be described in Chapter 2. In general, an actuation structure contains two pairs of electrodes: one pair for bias application to achieve the motion of the cantilever beam and the other pair for switching electrodes or contacts across which the RF signal is applied. When two pairs are insulated from one another, the device configuration is called a MEMS relay. If the direct current (dc) and RF signals share the same control lines, the configuration is known as a MEMS switch. From a circuit point of view, a relay is classified as a four-terminal device, whereas a switch is classified as a two-terminal device.

1.4.4 MEMS Technology for Electronic Circuits and Detectors for Military Applications

Use of MEMS devices incorporating NT in electronic circuits and detector elements will allow quick extraction of information from increasing amounts of data available, particularly from vertically integrated sensor arrays. Fast and accurate response from sensors using MEMS devices will make a firing decision after detection of a potential target. The MEMS technology will feature multispectral functionality and adaptability in response to a changing environment while performing a real-time data analysis. Adaptive focal planar arrays (FPAs) incorporating MEMS devices are best suited for frequency domain filtering and electronic circuits performing high-speed control functions. Photonics elements form the principal transducer, which consist of mercury–cadmium–telluride photodetectors with high sensitivity over a wide spectral range.

MEMS devices can play a critical role in detection of chemical and biological threats, thereby providing a sensing technology that offers multifunctionality. A sensor can detect chemical signals to identify explosives as well as individuals or group of people involved in terrorism. It is possible to develop a sensor capable of determining and identifying unique chemical signatures or explosives that emanate from specific individuals within groups of enemy troops or combatants. Israeli scientists have developed a means of identifying previously undetectable explosives, which are used by suicide bombers. The peroxide explosive tester (PET) developed by Israeli scientists can detect triacetone triperoxide (TATP), a chemical commonly used in suicide bombs.

The military authorities expect wide-ranging capabilities from technology advances involving every aspect of military life: arms, armor, armaments, communications, intelligence, and counter-intelligence, planes, submarines, UAVs, warships, robotics, and commercial-off-the-shelf (COTS) hardware and software products capable of meeting stringent reliability requirements. MEMS technology could allow design of hands-free weapons. Such weapons would consist of electronically fired caseless ammunition fired from the forearm. This sort of device could also act as a chemical and biological hazard detector that dispenses neutralizing agents.

1.4.4.1 Passive MEMS Devices for Commercial, Military, and Space Applications

A complex commercial or space or military system may contain both active and passive MEMS devices. Passive RF-MEMS devices include couplers, power dividers, matching networks, and CPW transmission-line sections. Passive components can use multilayer

CPW transmission lines fabricated on a polyimide as the insulating dielectric layer. CPW transmission lines are widely used in fabrication of RF-MEMS and optical-MEMS devices, because of low insertion loss, low frequency dispersion, and small overlapping requirements. Variation of line impedance as a function of frequency and overlapping can be seen from the data presented in Table 1.2.

Previously summarized data in Table 1.2 will accelerate the design and development of microwave millimeter (mm)-wave MEMS devices using CPW transmission lines, because these lines offer minimum insertion loss, lower dispersion at mm-wave frequencies, and minimum development time and costs.

1.4.5 Nanotechnology for Armors to Provide Protection to Soldiers

Studies performed by the author indicate that existing bulletproof vests made from composite materials provide needed protection to the soldiers to survive the rigors of combat, but the legs and arms remain exposed to bullets. NT-based CNTs can produce better and reliable armor materials for the protection of soldiers' arms and legs in the battlefield or in hostile combat areas. These extremities could receive serious and crippling, if not lethal, injuries. Battlefield statistics show that 75–80 percent of soldiers die from getting hit by shrapnel and ensuing excessive bleeding even in a short duration. To avoid this situation, NT is coming to the rescue of soldiers. Florida State University scientists are conducting experimental investigation to test the first-of-its-kind body armor to provide reliable protection for the men and women in the battlefield. Experimental data collected so far indicates that CNTs will improve the strength of the fabrics used to make bulletproof body armor. Battlefield commanders insist that time to attack the hostile target must be minimized [2] to reduce soldiers' injuries and collateral damage as shown in Figure 1.7.

Ballistic tests show that bound multiple layers of fabric made from CNTs and plastic thin sheets are better at stopping bullets than conventional bulletproof vests. In addition, they provide the necessary aesthetic and mechanical properties that make the armor more comfortable to wear even for long durations.

1.4.6 Nanotechnology-Based Biometric Structures to Monitor Soldier Health

Some research scientists think that biometric structures could be made for a new generation of ultrareliable armor materials by using NT. Development of biometric structures would lead to fabrics for soldier uniforms, which will not only demonstrate high strength properties, but also contain minisensors to monitor the soldier's health as well as to identify the presence of wide variety of chemical and biological weapons.

1.4.7 Nanomaterials for External Support Muscles and Artificial Muscles for Injured Soldiers on the Battlefield

Materials developed using nanotubes are called nanomaterials that have unique thermal, mechanical, and biomedical properties. MIT scientists have been working to develop nanomaterials that can act like exterior support muscles. These support muscles would boost the energy level of a soldier, which would help the soldier to run faster, to jump higher, to lift more weight, and to maintain physical stamina over extended durations. Even the demonstration of these unique capabilities may be several years away; nevertheless, many research scientists feel that the NT could one day find its way to demonstrate the above-mentioned unique capabilities.

Research scientists are currently looking at human enhancements that range from genetic engineering to molecular robotics based on NT. Research scientists at the NanoTech Institute of the University of Texas, Dallas, are conducting research studies to create chemically powdered artificial muscles for military applications. Electrically powered artificial muscles developed from conducting polymers and CNTs can exceed the performance of natural muscles by generating 100 times their force and elongate twice as fast. These unique capabilities of artificial muscles are essential to accomplish certain combat missions, which are not possible with natural muscles. Further research is necessary to achieve significant improvement, particularly in shorter life cycles and lower energy conversion efficiencies.

1.4.8 Robotic Arms for Battlefield Applications

The military authorities are aggressively looking at the integration of NT in unmanned military aircraft for ground, air, and sea combat missions. The military sees robotics as an essential ingredient of the mechanical army of the future. Research and development activities are funded by the Department of Defense involving more than 100 tractorized unmanned ground vehicles (UGVs) to perform a variety of military importance, such as surveillance, landmine detection and neutralization, and bomb disposal functions. Remote-controlled robots or UGVs can play an important role in searching for improvised electronic devices (IEDs) and dismantling bombs and mines laid by enemy soldiers. These lightweight robots weighing between 50 and 125 lb use robotic arm and vision systems (standard, IR, and night vision sensors) to perform their critical and a variety of assigned job functions. These robots can also use water cannons to blow apart the triggers of explosive charges. Military planners and field commanders envision a more expanded use of these UGVs, such as 24-hour guarding of base perimeters and conducting surveillance and reconnaissance missions behind enemy lines or in hostile territories. More sophisticated UGVs and more robotic-compatible sensors incorporating MEMS technology are needed to perform critical battlefield functions with high reliability. Robots are currently deployed in Iraq in disabling bombs, IEDs, and roadside mines. Battlefield soldiers control the robot functions via a suitcase-sized box located within the operating range of the robot. Sophisticated software can be created, which will enable the robots to communicate with overhead UAVs to accomplish critical mission objectives. A powerful robot called Special Weapons Observation Reconnaissance Detection System (SWORDS) is under development [2], which can carry different caliber weapons. This system can fire 5.56 mm rounds at the rate of 750 per minute or 7.62 mm rounds at the rates from 700 to 1000 per minute.

A compact robot weighing less than 40 lb is making news in Iraq. This robot can be augmented with IR cameras, batteries, and a robotic arm. It has played a key role in searching tunnels under Baghdad Airport, tracking for Iraqi soldiers hiding in agricultural buildings, and examining a booby-trapped airfield. This compact robot has demonstrated critical security functions with highest reliability.

Robots are playing critical roles in new UAVs and unmanned underwater vehicles (UUVs). UAVs include Global Hawk, the Desert Hawk, and the Predator. Currently, these aircraft are patrolling the skies in Iraq and Afghanistan at higher altitudes for strategic missions and at low altitudes for tactical applications. It is important to point out that unmanned aircraft equipped with MEMS-based sensors are best suited to carry a wide range of military systems, such as IR and electro-optical (EO) sensors for covert communications, weather/environmental monitoring, short- and long-range missile warning, surveillance of hostile areas, and target acquisition. A low-speed User Account Control (UAC) with a high-aspect-ratio configuration and equipped with MEMS-based sensors is most attractive for reconnaissance missions over hostile or unfamiliar territories. However, deployment of a wing morphing design to perform at high speeds and using MEMS-based smart structures converts the aircraft into a low-aspect-ratio configuration, which is best suited for tactical strike missions.

1.4.9 Portable Radar Using MEMS/Nanotechnology for Military Applications

Research scientists and radar engineers are currently working on an NT-based radar that can "see" through walls and can create an image of a person or persons hiding on the other side of the wall [4]. This will allow soldiers, police officials, or law-enforcement agents to see where the hostages or terrorists or criminals are congregated. This will also permit soldiers to know where the enemy is hiding or waiting to attack. This radar will significantly reduce the number of soldiers injured or killed, while conducting search and destroy missions in hostile territories or battlefield. It is important to mention that conventional radar can penetrate the building or room walls, but it cannot distinguish objects just ahead with high reliability. In addition, the conventional radar consumes high electrical power and emits too much RF energy. The operational equipment is of large size and weight, thereby making it useless for battlefield applications, where portability, minimum power consumption, and reliability are the most demanding requirements. Advances in MEMS-based digital signal processors and MEMS-based microwave ICs have made it possible to fit this complete wall-penetrating and imaging radar in a box size of two encyclopedia volumes. This is only possible through the use of MEMS devices incorporating NT circuits and smart printed circuit antennas. A change in software can significantly improve the signal-to-noise ratio (SNR) of the return signal and reduces considerably radar power consumption. In addition, this wall-penetrating radar makes difficult for the adversaries or bad guys to know they are being monitored, because signal detection devices cannot distinguish the low-power transmissions from the background noise, which is significantly higher than the signal.

Besides using the smart antennas, MEMS devices and NT passive devices, this wall-penetrating radar transmitter transmits several millions of 300–500 ps-long pulses every second using NT-based pulse compression circuitry, and each pulse transmits RF energy well below 100 mW that is sufficient to provide a detection range of 20–30 m. Furthermore, the radar transmitter pulses are spread across frequencies over a spectral region from 1 to 5 GHz using pseudorandom dithering in time. It is desirable to mention that dithering requires a time code, which is known only to the radar operator and which determines the position of the pulse within a time window. Dithering eliminates the spectrum interference and makes the radar even more immune to detection. This particular wall-penetrating radar is most suitable for battlefield applications and scanning disaster sites.

1.5 MEMS for Commercial, Industrial, Scientific, and Biomedical System Applications

Material presented in this section identifies the potential applications of MEMS devices in commercial, industrial, scientific, and biomedical sectors. Current research indicates that nanotubes, NWs, nanocrystals, nanoskins, and nanofilms will play key roles in the design and development of sensors best suited for above applications. "Nanoskin" polymer was created by research scientists at Rensselaer Polytechnic Institute (RPI) in New York. This new material can be used to build highly efficient electronic components for flexible electronic displays best suited for commercial and industrial applications. This material might have many other applications, all the way from adhesive structures to nanotube interconnects for electronics, most suitable for biomedical and scientific system applications.

1.5.1 Nanotubes and Nanotube Arrays for Various Applications

CNTs are made from rolled-up sheets of tightly bound carbon molecules and they exhibit high mechanical strength and excellent electrical conductivity. Nanotubes have diameters of few nanometers and are very flexible. Nanotube arrays normally do not maintain their shape because they are held together by weak forces. But the latest procedure developed at RPI allows them to grow them in an array on a separate platform and then fill the array with soft polymer. When the polymer hardens, a flexible skin with embedded-organized arrays of annotates is achieved. This skin can be bent, flexed, and rolled up like a scroll, all the while maintaining their ability to conduct electricity with minimum loss. These skins are most ideal materials for manufacturing electronic paper and other flexible electronics.

RPI scientists have developed a more malleable annotate-based skin by using larger multiwalled carbon annotates with diameters close to 10 nm. Then, annotates are deposited by chemical vapor deposition (CVD) process on to a silicon dioxide layer, which leaves the tubes pointing vertically up from the surface. Then a liquid polydimethylsiloxane (PDMS) is poured over the surface. This liquid covers and fills the CNTs, before hardening. Finally, the flexible polymer infused with CNTs is peeled away from the substrate. Development of composites consisting of nanotubes and polymers is possible, but it is difficult to engineer the interfaces between these two materials. The research results published in the March 2006 edition of the *Journal of Nano Letters* indicate that it is possible to develop arrays of annotates into a soft polymer matrix without disturbing the shape, size, or alignment of the annotates.

These arrays can be used in the design and development of artificial foot or arm with a sticking power capacity of 200 times. These arrays can also be used in the development of miniature pressure sensors and gas detectors for industrial applications.

When a voltage is applied to nanotubes of certain materials, electrons are pulled out from the surface, which can be used to produce high-resolution electronic displays. Nanotubes are very good field emitters because they have very low threshold for emission and they produce high currents. When nanotubes are placed very close to each other, each nanotube tends to shield its neighbor from the electric field. This effect has limited the development of field emission devices based on densely packed, aligned nanotubes. This effect can be eliminated when the nanotubes are embedded in an appropriate polymer medium. It is important to point out that threat detection or monitoring array sensor may deploy SWCNTs or MWCNTs or both, depending on the sensor monitoring capability.

1.5.2 MEMS-Based Video Projection System

MEMS-based optical switches are best suited for applications such as medical imaging, high-resolution tactical displays, and television screens, projection systems for conferences, and 2-D programmable array as video projection systems. A 1080-element linear grating light valve (GLV) array can project a standard 1080 by 1920 pixel high-definition television (HDTV) image at a 90 Hz refreshing rate. MEMS optical switches and other MEMS devices play the key role in the design and development of a television display with digital light processing (DLP) capability, The DLP-display technology accepts digital video and transmits as a burst of digital light pulses to the television screen. The heart of the DLP technology is an optical semiconductor chip (OSC) known as a digital micromirror device (DMD) containing close to 1.25 million tiny movable mirror elements that reflect light, electronic logic circuits, and digital memory. Each DMD mirror has 16 μm square sides and is made of aluminum (Al) for maximum reflection. DLP displays are widely used in HDTV sets because they offer improved resolution, brightness, contrast, color fidelity, and fast response due to silicon-based DMD chips.

1.5.3   Nanotechnology for Photovoltaic Solar Cells and 3-D Lithium Ion Microbatteries for MEMS Devices

NT can play a critical role in the design and development of photovoltaic (PV) cells, semiconductor solar cells [5], and 3-D lithium ion microbatteries (MBs) for integration in MEMS devices or sensors. Solar cells [6] using NT-based ZnO nanorods can provide an electrical power system for battery charging banks that support 12 V lighting and appliances, whereas a 3-D lithium ion battery power package can provide electrical power for automobiles, trucks, emergency lighting for homes in case of power failure or power blackouts cellular and mobile phones, laptops, computers, portable electrical appliances, and host of other electrical appliances. Large solar installation can be tied to an electrical grid system via a grid-tie inverter. A grid-tie system consists of mounting structure, safety disconnects, installation wiring, solar modules comprising of solar cells and a grid-tie inverter. Inverters integrate three functions: converting the dc from the solar modules to ac and synchronizing it with the electrical utility; tracking the maximum power point of the modules to operate the system at peak efficiency; and disconnecting the PV or solar cell system from the grid in case the power from the utility is interrupted. Integration of NT-based ZnO nanorods will significantly improve the conversion efficiency of the solar cells over the conventional solar cell's conversion efficiency. The conversion efficiency can improve from 8 to 12 percent for thin film cells and 14 to 20 percent for crystalline silicon cells, while keeping the manufacturing cost at less than 50¢ for both module types. The system price to the end user is expected to $3 per watt ac in 2010. Current research studies indicate that improved triple-junction (GaInP/GaAs/Ge) space solar cells could offer conversion efficiency as high as 37 percent [5]. Each of the cell's three layers captures and converts a different portion of the solar spectrum, thereby leading to higher conversion efficiency. Research scientists predict that a multijunction PV cell fabricated from InN and GaN films to form a InGaN ternary alloy would have a band gap of 0.7–3.4 eV. This multijunction PV cell would cover the entire solar spectrum and provide conversion efficiencies of 50–70 percent. Triple-junction amorphous PV cells (Figure 1.9) using amorphous silicon layers offer minimum manufacturing costs,

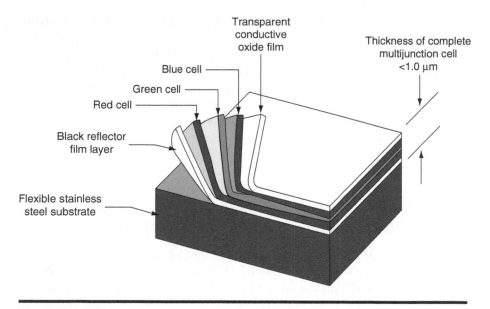

Figure 1.9   Triple-layer amorphous solar cell design using MEMS technology.

but suffer from lower conversion efficiencies. In summary, it is believed that NT-based multijunction solar cells using ZnO thin nanorods [6], which are currently under development, can provide conversion efficiency exceeding 50 percent with high yields, leading to minimum production costs.

1.6   MEMS Technology for Hard-Disk Drives

MEMS technology will significantly increase the recording density (gigabits per square inches) of the hard-disk drive (HDD). MEMS-based actuator on a silicon-on-insulator (SOI) wafer is best suited for HDD applications in high-speed computers [7]. The recording could be further increased by improving the bit-per-inch (BPI) recording rate and the track-per-inch (TPI) density. It is critical to mention that BPI can reach the upper limit soon, which is dependent on the data transfer rate of the electronics or thermal instability of the recording material used by the disk. This leaves TPI as the only available parameter for a moment for increasing the recording density by better designing the positioning mechanism. Currently, voice-coil motors (VCMs) are used for positioning the read/write head attached at the tip of the suspension-gimbal assembly. Thus, the servo bandwidth of the read-positioning mechanism is limited by the vibration modes of the suspension, which is usually in the kilohertz range. Therefore, the track density is limited to around 50 kTPI, which corresponds to a track width of 0.5 μm. Track density of more than

150 kTPI is needed for the future HDDs with capability exceeding 200 Gb/in.[2] The MEMS approach appears to overcome this problem.

The MEMS approach is referred as a piggyback system [7] or a dual-stage servo system, in which a microactuator carrying read/write head is used for positioning in collaboration with a VCM for course positioning. The simplest silicon fabrication process involves integration of MEMS piggyback actuators with the magnetic read/ write heads of HDDs. An electrostatic actuator of multiple parallel-plate (PP) mechanism when fabricated on an SOI wafer can provide a displacement of 0.5 μm at a driving voltage of 60 Vdc with a resonance frequency as high as 16 kHz. A displacement of ±0.5 μm is possible at a driving voltage of 30 Vdc with a resonance frequency in excess of 20 kHz. First-generation, second-generation, and third-generation piggyback actuators are discussed in Chapter 3. Piezoelectric (PZT) actuators, electrostatic linear actuators, and electrostatic rotational actuators, which are best suited for second-generation piggyback actuators, are described in great detail in Chapter 2. The third-generation actuator offers higher resonant frequency using a microactuator integrated with a read/write of a small mass. The most critical issue in the third-generation scheme is the total processing for integrating the microactuator with the read/write head. Monolithic integration is preferred, because the mutual position of the head and the actuator-slider need to be controlled within 10 nm. Furthermore, simple monolithic integration would be possible because the MEMS devices and magnetic head elements are formed on the identical surface of the silicon chip. A prototype electrostatic actuator of multiple PP mechanism has been developed on an SOI wafer to demonstrate a dc displacement of 0.5 μm on actuation or driving voltage of 60 Vdc and at a resonant frequency of 16 kHz. Trade-off studies must be performed in terms of displacement and resonant frequency to determine the lowest possible driving voltage to meet the minimum power consumption. One must use a smallest possible etching width for an actuator to generate a large electrostatic force and to meet small displacement requirements. The electrostatic actuator structural details are summarized under various actuation mechanisms.

1.6.1 MEMS Devices for Thermographic Nondestructive Testing

Thermographic nondestructive testing (TNDT) is possible using MEMS technology. The TNDT technique is widely used to detect and evaluate a range of subsurface defects, including voids, delaminations, inclusions, porosity, and corrosion. TNDT will be most suitable for quality control inspection of the semiconductor wafers and their yields. The TNDT technology using MEMS devices will be found most cost effective for the incoming inspection of the semiconductor materials and quality control of computer processing chips and other semiconductor chips used in various commercial, industrial, and military systems. The fundamental physical limitations that apply to all thermographic methods need to be identified. Various approaches of excitation, acquisition, instrumentation, and analysis are discussed. TNTD techniques for metals, alloys, composites, and ceramics must be provided with emphasis on reliability. Approaches including flash, step, modulated, lock-in, and sonic TNTD techniques are described in greater details. However, the major focus will be on the following critical aspects:

- Explain the underlined physical process that allows detection of structural defects on the surface of the objects under investigation
- Compare basic approaches to TNDT for a particular application
- Identify aberrations and artifacts that downgrade the results obtained during the testing
- Predict minimum detectable defective size and depth in a sample under inspection
- Estimate subsurface feature depth from an image sequence

1.7 MEMS Devices for Uncooled Thermal Imaging Arrays and Cooled Focal Planar Arrays for Various Applications

MEMS devices have potential applications in uncooled and cooled thermal imaging sensors and IR FPAs. MEMS technology could realize significant reduction in weight, size, and power consumption for the high-resolution thermal imaging arrays and systems, irrespective of cryogenic cooling. Uncooled FPAs could expand both the commercial and military markets for thermal imaging systems. Uncooled MEMS-based arrays including resistive and ferroelectric bolometric, pyroelectric, and thermoelectric arrays will be described. Performance capabilities and limitations will be identified with emphasis on dark current, detectivity, SNR, image quality, dynamic range, power consumption, weight, and size. It is critical to mention that IR FPAs and imaging sensors yield optimum performance when operating at cryogenic temperatures. However, uncooled imaging sensors and FPAs using MEMS technology with built-in CMOS readout circuits offer reasonably good performance with significant reduction in production costs and sensor complexity. A new generation of solid-state cameras integrated with MEMS technology and advanced FPA technology are best suited for commercial and military applications. Critical design and development issues of these imaging sensors and FPAs will be described with emphasis on the following aspects:

- Theoretical performance sensors and FPAs in terms of detectivity, responsivity, noise equivalent temperature difference (NETD), and response time
- Performance improvement at cryogenic temperatures
- Fundamental limits to their performance due to temperature fluctuations, noise variations, and background noise fluctuations

- Critical components using MEMS technology
- Performance degradation and cost benefits of uncooled IR FPAs using MEMS technology

RedShift Systems Corp., a U.S.-based company, has pioneered the latest technology to develop low-cost, high-performance thermal imaging sensors for mass-market applications. The Thermal Light Valve (TLV) technology developed by this company translates thermal energy into visible images with remarkable quality, thereby allowing the CMOS sensors using MEMS devices to "see heat" generated by the objects. This allows the manufacturers to include high-quality thermal imaging capability with minimum cost in a variety of sensors or products for the video surveillance and security of the premises, fire fighting, law enforcement, automobile safety, and other industrial security applications.

In summary, deployment of MEMS and NT-based devices in high-resolution imaging sensors and FPA-based low-cost thermal cameras will be found most beneficial for perimeter security, plant surveillance, and law enforcement applications with minimum cost and high reliability.

1.8 Applications of Nanotechnology in IR and Electro-Optical Sensors for Biometric and Security Applications

NT-based devices play a critical role in the design and development of EO imaging systems, security sensors, and tactical equipment. These sensors and systems have wide applications in commercial, industrial, and defense sectors, where security is of paramount importance. Integration of NT and MEMS technology in the imaging sensors has demonstrated significant improvement in resolution, linearity, responsivity, aperiodic transfer function (ATF), slit response function (SRF), random noise, jitter, uniformity, fixed pattern noise modulation transfer function (MTF), and minimum resolvable contrast (MRC). The performance improvement will provide improved biometric data and foolproof security checks of air travelers, cruise line passengers, and sport fans.

MEMS devices can play a critical role in biometric recognition capability. Biometric recognition is becoming increasingly important for passenger security. The biometric recognition capability offers a useful tool for developing important algorithms best suited for biometric recognition. In particular, common recognition algorithms for face, fingerprint, iris, and palmprint biometric are of critical importance, as far as screening of potential hijackers, suicide bombers, and terrorists is concerned. Critical issues and elements involved in biometric recognition include motivation for biometric recognition, pattern recognition basics, performance metrics for biometric recognition, biometric recognition algorithms, principal component analysis, linear discriminate analysis, correlation filters, artificial neural networks, reliable fingerprint recognition methods, iris code method for iris recognition, feature-based approaches for palmprint recognition, and multimode biometric recognition criterion. Biometric recognition technology will bar potential terrorists from entering a country or a movie theater or a sports stadium. Capabilities and limitations of different biometric modalities and associated algorithms will be summarized under appropriate heading.

1.8.1 Nanotechnology-Based Laser Scanning Systems

EO systems can be miniaturized through integration of optical-based MEMS devices, which will realize significant reduction in power consumption, weight, and size. Miniaturized EO imaging systems are best suited for space-based reconnaissance and surveillance functions, which are critical for tactical and strategic missions. Miniaturized EO microsensors could be used in UAVs and low-orbit space vehicles, where low-power consumption and minimum weight are the demanding requirements. Basic requirements for accurate prediction of the effects of thermal, structural, and servo system on the overall performance and quality of optical imaging systems must be defined. Multifunctional microsystems architectures (Figure 1.10) using MEMS technology will be found most attractive for spacecraft, robotics, and nanosatellites.

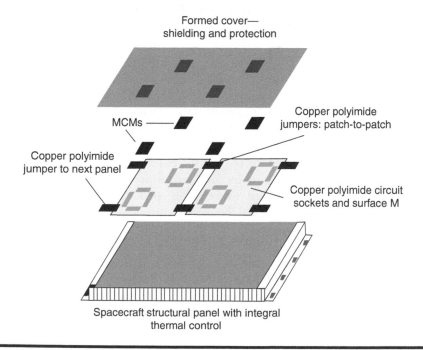

Figure 1.10 Multifunctional spacecraft structure containing thermal control and shielding cover using MEMS devices.

NT-based components and MEMS devices can be used in design and development of laser scanning image systems, which are best suited for commercial and military applications. Critical design issues involved in the design of laser scanning systems must be given serious consideration. These issues include cross-scan and along-scan optical errors, polygonal facet optical errors, stationary ghost images in the image format, and image distortion. Steps must be taken to minimize these errors to avoid adverse effects on the image quality. Various scanning mechanisms must be evaluated before selecting a particular scanning mechanism.

1.8.2 MEMS-Based Sensors for Detection of Chemical and Biological Threats

MEMS-based sensors are best suited for detection of chemical, biological, and toxic threats posed by Islamic terrorists and right-wing radical groups. These sensors can be used by civilian and military personnel to neutralize the threats posed by terrorists and to protect travelers or innocent citizens. Remote and sampled detection, discrimination, and identification techniques are introduced with the design parameters of these sensors to ensure reliability and to minimize false alarms. A sampling of specific technology applications for chemical point, chemical standoff, biological point, and biological standoff sensing are described to meet the sensor performance requirements. Brief definition for technical terms associated with these sensors such as ion mobility spectrometry, surface acoustic waves, Fourier transform IR spectrometry, differential absorption lidar, laser-induced fluorescence, and lidar backscattering are provided. In addition, a brief overview of chemical and biological agents and features must be presented to familiarize with the operating aspects of the sensors. Chemical and biological detection and discrimination techniques must be familiarized to ensure high operational reliability of these sensors. Comprehensive knowledge about the estimates of spatial, spectral, and temporal variations is essential for the sensor operators.

1.8.3 Potential Applications of Nanophotonic Sensors and Devices

Design and development of nanophotonic sensors and devices require thorough knowledge on the integration of performance parameters for both NT and photonic technology. Nanophotonics devices include light-emitting diodes (LEDs), solid-state laser diodes, and certain electro-optic devices. Incorporation of cadmium sulfide (CdS) into NWs in polymer structures leads to development of hybrid nanocomponents or devices. Doped ZnSe-based nanocrystals offer lower toxicity, but higher reliability for biomedical devices, LEDs, and solid-state laser diodes. LEDs using ZnO nanostructures embedded in polymer have demonstrated ultra-violet (UV) emissions. ZnO wires can be produced at temperature as low as 85°C with minimum complexity and cost. It is important to mention that RF-MEMS and optical-MEMS device integration and packaging at system-level architecture are possible using NT. This will accelerate the production development of RF and mm-wave smart antennas, resonators, switches, phase shifters, and CPW transmission-line sections. Nanophotonics and RF-MEMS technologies can be used in designing 3-D MEMS-based MMIC devices consisting of multilayered structures. These MMIC devices offer lower insertion loss, high yield, and minimum production cost.

1.8.4 MEMS Technology for Photonic Signal Processing and Optical Communications

The rapid development of broadband services and delivery technologies has been a response to business and consumer requirements for faster and easier access to relevant information and communications. Optical transmission and processing are the basic technology requirements for broadband operations, which provide the backbone of the Internet, network infrastructure for wireless communications, and high-capacity networks for government and business. Early optical fiber communications systems anticipated the need for higher data capacity and multiple channel capability. Optical fiber transmission can respond to future capacity expansion simply by adding more optical fibers [8].

Steady development and improved performance of electronic circuits and devices for optical transmission have been observed over the last two decades. Electronic circuits or devices such as solid-state lasers, photodetectors, laser drivers, optical modulators, transimpedance amplifiers, and signal processors still continue on a path of rapid development. The supporting electro-optic and opto-electronic components must keep pace with the rapid developments in the optical fiber, photonic signal processing, and device manufacturing methods capable of enhancing the performance of the optical transmission medium to meet future stringent system requirements.

Most switching, multiplexing, and modulation tasks have been handled electronically. This requires detection, signal processing, and re-modulation of the optical or lightwave signals. These functions require conversion of analog signal to optical signal as their first step and then converting back from optical to analog, which is not only a complex procedure but also very expensive. Photonic signal processing that handles the signals, although they remain light waves, is on the verge of mass production, particularly in switching, where MEMS technology is deployed to operate tiny mirrors that can direct light beams between input and output ports.

It is important to mention that true quantum processing of light waves is still in the laboratory, but it holds promise for additional optical signal processing functions, which are now being performed on the baseband electronic signal instead of light beams. Photonic signal processing using MEMS technology is expected to achieve terabit per second data rates, a significant step beyond the 10–20 Gbps rates currently under development. The optical switching technology that is posed for production now will allow new means of optical transmission. This technology has demonstrated how pulses or commutation can be used to achieve optical CDMA transmission in a manner that cannot be easily achieved by electronic signal processing and modulation techniques.

1.9 MEMS Technology for Medical Applications

Recent design and development of MEMS-based chips by medical research scientists have demonstrated their potential applications for detection and treatment of neurological disorders, epilepsy, seizures, and Parkinson's disease. Neurosurgeons have been successful in quelling epileptic seizures by electrically jolting the vagus verve, one of the twelve pairs of nerves that emerge from the brain instead of spinal cord [9].

Magnetic seizure therapy uses a powerful electromagnet instead of repetitive transcranial magnetic stimulations. Basically, it is a magnetic version of electro-convulsive therapy using NT-based biomedical chips. Magnetic seizure therapy, involving a magnetic force close to 2 T, induces a high-frequency current in a small portion of brain until it sparks a seizure. Medical researchers feel that a magnetically induced seizure will be more effective in treating depression. Neurosurgeons believe that about 25 percent of the patients with severe epilepsy also have chronic depression. In a recent NT symposium at San Diego, California (March 2, 2006), a neurosurgeon claimed that he has cured a man with Parkinson's disease using a biomedical MEMS-based chip.

Medical research scientists believe that deep brain stimulations are possible using MEMS-based implants, but it requires brain surgery. A MEMS-based stimulation implant in a patient's chest sends electrical pulses to tiny electrodes embedded deep within the brain. The stimulation switches off neurons within a few millimeters (mm) of the electrodes. It can cure severe depression by interrupting the malfunction of brain electronic circuits implanted. This method uses a dc source and the technique is called "transcranial direct current" (TDC) stimulation. Medical researchers believe that this technique is simpler than transcranial magnetic stimulation (TMS) to fight depression.

In the same NT symposium at San Diego (March 2006) a physician claimed that he used an endoscope with the shape of a tiny pill containing a MEMS-based IR camera to monitor the upper digestive track stomach lining with no discomfort to the patient. The pill is swallowed by a patient and the test is completed within 15 minutes in a doctor's office and the patient goes home within 30 minutes with no side effects. All these treatments are possible with chips or pills using MEMS or NT.

1.10 MEMS Technology for Satellite Communications and Space Systems Applications

Wireless satellite communications and space-based sensors are the major beneficiaries of the MEMS technology. Minimum size, power consumption, and high reliability are the critical requirements for both the applications, which can be satisfied by deploying MEMS-based devices and sensors. MEMS technology conserves energy and minimizes power consumption, weight, and size. RF-MEMS devices such as switches, phase shifters, switched antennas, filters, and amplifiers have become critical to increasing communications satellite capability while realizing significant reduction in launch weight and cost. Performance capabilities of widely used MEMS devices in space-based applications are summarized in Table 1.4. This is important not only in large satellites employing phased array antennas and switch matrices that are deployed in constellations to downlink data but also in the case of nanosatellites, which are very small, lightweight spacecraft containing microelectronic equipment, RF-MEMS components, and payloads.

Table 1.4   Performance Capabilities of Widely Used MEMS Devices in Space-Based Applications (Estimated Values)

Parameters	MEMS Devices			
	Reconfigurable Apertures	Phase Shifters	Switched Antennas	Filters
Insertion loss (dB),	0.75	0.70	0.35	0.35
Isolation (dB), min.	22	20	23	24
Return loss (dB), max.	16	15	15	17
Switching time (s)	0.001	1	1	0.8
Life cycles (billions)	0.5	0.5	20 (minutes)	400 (minutes)
Min. dwell time (s)	0.001	0.001	0.005	0.0001
Max. dwell time (s)	8	0.010	0.500	0.001
Frequency	(Ultrahigh frequency) UHF/K_a	K_a	X	X
RF power (mW)	>65	>60	>120	1100

These nanosatellites are attractive due to their lowest cost, lightweight construction, versatility of launching means, and applicability to national security missions [10], where covert communication is of critical importance. As a cluster of three or five or ten units, nanosatellites can function as smart distributed apertures for the geolocation of downed pilot in hostile territory. As a constellation of tens of units, they may allow Internet-in-the sky applications and continuous reconnaissance of a specific geographical area during a military operation needed to control hostilities [10].

It is important to mention that MEMS technology offers sub-miniaturization, thereby realizing significant reduction in weight, size, and power consumption. Parametric models indicate that MEMS device cost goes down with weight. Furthermore, MEMS devices are the next silicon revolution following the first IC technology introduced three decades ago. According to space engineers, MEMS devices are three-order cheaper and about 350 percent lighter in weight. Satellite cost is strictly dependent on payload requirements; launching means; batch fabrication; reliability of the devices and sensors deployed aboard; and tracking, monitoring, and orbital requirements. However, communications satellite designers feel that the batch fabrication process offers signification cost reduction for micro/nanosatellites based on the economy scale. This is opposed to conventional satellites, which have thousands of parts fabricated using various diversified manufacturing processes. In brief, conventional spacecraft assembly and manufacturing are labor-intensive processes. However, communications satellites incorporating MEMS and MMIC technologies will unquestionably offer not only cost-effective communications satellite designs but also ultrahigh reliability and longer life cycles.

It is important to mention that micro/nanosatellites are primarily designed for low earth orbiting (LEO) satellite systems, because of minimum weight and power consumption requirements. A LEO communications satellite allows 1600 times the bandwidth when compared to a GEO satellite system. Furthermore, a three-satellite constellation using NT offers 99.9 percent reliability, if each individual nano-satellite has 90 percent reliability. In addition, a nanosatellite-based constellation offers high redundancy, eliminates the risk of a $60-million global positioning system (GPS) satellite, improves refreshing rate, and provides multiple view angles and offers significant reduction in cost, weight, and electrical power consumption.

Unique advanced structures also called smart multifunctional structures (Figure 1.10) have been developed by NASA for potential applications in micro/nanosatellites. These multifunction structures involving 3-D electronic circuit layers are capable of integrating electronics, mini-sensors, power distribution, and thermal management involving cableless power bus, integrated power control circuits, and lightweight interconnects with minimum cost and complexity (Figure 1.11). Antenna interface circuits with quality factors (Q's) better than 20,000 can be most attractive for nanosatellite applications. MEMS-based tunable filters offer significant reduction in insertion loss and dc power dissipation and, therefore are best suited for nanosatellite applications. Micro-resonators using

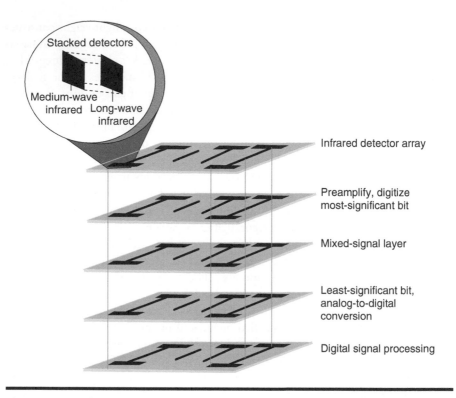

Figure 1.11 Three-dimensional electronic circuit layers using MEMS devices for various functions.

MEMS technology can be used in place of SAW filters and RF switches, which offer great reduction in weight, size, and power consumption.

1.11 MEMS Devices for Auto Industry Applications

MEMS devices are best suited for various sensors used in automobiles, where high reliability, minimum power consumption, and low production cost are of critical importance. MEMS technology can be used in the design and development of air bags, dashboard displays, security alarm systems, anticollusion radar, electronic ignition and timing control devices, miniaturized navigation systems, window drive mechanisms, and other sensors widely used by the automobile and trucking industries. MEMS devices offer significant reduction in power consumption and procurement costs for various control devices widely used in passenger automobiles and light trucks or pickups. Projected cost reductions from the second generation of MEMS devices and sensors to be used in automobiles are summarized in Table 1.5.

Table 1.5   Projected Cost Reduction from Second-Generation MEMS Sensors and Control Devices

Types of Sensors or Devices	Cost Reduction Factor
Infrared (IR) sensors	3.65
Hall-effect sensors	1.84
Surface acoustic wave (SAW) devices	3.10
Polymer-based gas sensors	3.15
Charged-coupled device (CCD)-based cameras	2.74
Anticollusion radar	2.90
Magneto-restive (MR) devices	1.89
Ultrasound devices	3.18
Devices using surface micromachining	4.12
Devices using bulk micromachining	3.18
Solid-state gas sensors	2.42
Chemical sensors	2.75

1.12   MEMS Technology for Aerospace System Applications

Studies performed by NASA scientists in 1994 indicate that MEMS-based devices or parts for passenger and military aircraft are six times cheaper, because it is easy to provide space qualification to aircraft parts with minimum material and labor costs, if modification is needed to meet revised performance specifications. Few 6-in. micro-artificial vehicles (MAVs) weighing less than 100 g have been fabricated to demonstrate technical feasibility utilizing DARPA grants. Such MAVs using MEMS technology are best suited for surveillance and reconnaissance missions over battlefield areas and hostile territories.

MEMS technology is currently used in the design and development of gyroscopes weighing less than 1.9 g, microradio control systems, microreceiver system comprising of command processor, actuators and UHF (ultrahigh frequency) printed circuit antennas, brushless motors, piezoelectric actuators, reduction gear boxes with 25:1 speed reduction capabilities, solid oxide fuel cells capable of providing two to three times the energy-density of lithium batteries, TV cameras on chips, flight control sensors, jet engine performance monitoring sensors, and charged-coupled device (CCD) cameras weighing less than 18 g. All these devices or components and sensors have displayed massive integration, electronic circuit miniaturization, multiple layer design techniques, and nanoelectronics to achieve significant reduction in weight, size, and electrical power consumption, which are the most ideal parameters for aerospace systems.

It is important to mention that computer-aided design (CAD) tools are of critical importance in prototype designing of aerospace systems and automobiles to minimize production costs. CAD tools are available such as an integral MEMS-CAD for integrity MEMS/MST. This particular MEMS-CAD modeling module can address all design issues such as coupled effects of package and device models. SPICE-like process-simulation tools are also readily available to achieve various simulation data before freezing the final design of the MEMS-based system or sensor. The foundry model uses intellectual proprietary technique to save simulation time and cost using reuse concept. Carnegie Mellon University in Pittsburgh, Pennsylvania, has developed a design tool in 1994 to generate a MEMS-based device or component layout with semiconductor process accurate up to 1 μm (10^{-4} cm).

1.13   Summary

The term MEMS stands for microelectromechanical systems that deploy NT in the design and development of MEMS-based devices and sensors for various applications. This term refers to the micro/nano techniques and application of 3-D devices having physical dimensions in the millimeter range. MEMS sensors or devices can be produced using micromachining technique. Manufacturing of MEMS devices or components can be achieved using VLSI process best suited for the development of submillimeter wave and mm-wave devices. MEMS fabrication process involves a modified ICs technique, which is a stress-controlled polysilicon process. MEMS can integrate microelectronics on a chip with minute electromechanical components such as transistors with nanometer dimensions or a MEMS-based accelerometer. Nano/micro spacecraft can use other advanced technologies including micro-multiple chip module (MCM) and micro-optics (MO). It is critical to mention that a nanospacecraft system can integrate several functions such as guidance, navigation, and control functions on a chip.

NT can be developed in two ways: (1) microminiaturization of existing technology involving printed circuit antennas or microelectronics circuits and (2) deployment of NT-based materials such as NWs, nanotubes, nanocrystals, nanoparticles, and nanostructures. Studies performed by the author indicate that MEMS resonant circuits or cavities offer unloaded Q's better than 20,000, which are best suited for MEMS devices operating at mm-wave frequencies.

As stated earlier, wireless cell phones, mobile phones, base stations, and communications satellites are the principal users and major beneficiaries of MEMS technology and NT. Studies performed by the author indicate that MEMS devices operating at RF and microwave frequencies are best suited for cellular phones, mobile phones, base stations, and satellites communications, while MOEMS are most attractive for optical communications and optical signal processing applications. Optical signal processing is essential, where coherent signal processing, large data-handling capacity, ultrahigh resolution, and fast response are the principal performance requirements.

Radio frequency identification (RFID) technology involving MEMS devices is currently receiving the greatest attention in the commercial sector. RFID technique offers an automatic way to collect product place, time, and transaction data instantly without intervention or error. In other words, RFID can play a critical role in inventory control of the products sold by the retail stores and to avoid merchandise theft in the shops. An RFID system comprises of two parts: a portable reader and a tiny transponder (costing about 5¢ apiece in production). The reader uses the embedded antenna to transmit RF energy to interrogate the transponder consisting of a radio tag and an RFID card. The transponder is embedded or attached to a product being tracked. The transponder does not have a battery, but receives its energy from the incoming RF signal. The radio energy is used to extract data stored in IC chips and send it to the reader from where it can be fed into a computer for processing. An RFID system uses the elements of MEMS technology or NT and offers the potential of "real-time supply chain visibility." RFID systems using MEMS technology have potential applications in food industry, retail merchandise shops, and grocery stores. With radio-tagged grocery items in a shopping cart, the whole cart can be read instantly without the cumbersome laser-based bar code scanning of each item. An RFID technology will significantly reduce the operating costs for retail shops, grocery stores, and part inventories and will eliminate merchandise theft from the store premises.

In summary, the MEMS technology involving smart materials and nanostructures such as NWs, CNTs, nanocrystals, and nanorods offer significant reduction in production cost, weight, size, and power consumption, while retaining the highest device reliability. MEMS devices have potential applications in commercial, industrial, military, and space systems. It is important to mention that MEMS technology is best suited for the design and development of devices and sensors for deployment in satellite communications, space surveillance, airborne reconnaissance, and battlefield drones, where minimum weight, size, power consumption, and ultrahigh reliability are the principal performance requirements. MEMS devices can play key roles in the development of medical diagnostic equipment, mini-robotics, lithium ion battery electrodes, ZnO-solar cells, MEMS-based nanobatteries, and a host of other applications.

References

1. H.J. De Los Santos, G. Fischer, H.A.C. Tilmans, J.T.M. van Beek, RF MEMS for ubiquitous wireless connectivity, *IEEE Microwave Magazine*, December 2004, 53–63.
2. Sr. Editor, MEMS based switched filter banks, *Electronic Design*, March 3, 2005, 44–46.

3. W. Jones, No place to hide, *IEEE Spectrum*, November 2005, 20.

4. D. Tuite and Technical Editor, Bright outlook for solar power generation, *Electronic Design*, December 16, 2004, 44–50.

5. S. Dhar et al., Synthesis of orientated ZnO nanostructure arrays for solar cell applications, *Proc. SPIE*, 6172, February 2006.

6. H.T. Yousi, A MEMS piggyback actuator for hard disk drives, *IEEE Journal of MEMS*, 11, November 6, 2002, 64–65.

7. Editor of Technical Report, Optical processing using electronics and photonics for high data rates, *High Frequency Electronic*, January 2006, 28.

8. Editor, Transcranial magnetic stimulation and therapy for seizers, *IEEE Spectrum*, March 2006, 28–30.

9. W.B. Scott, Nanosatellite technology ready for National Security Missions, *Aviation Week and Space Technology*, January 14, 2004, 59.

10. Editor, ZnO nanowire light emitting diodes (LEDs) having UV output, *Photonics Spectra*, January 2006, 135.

Chapter 2

Potential Actuation Mechanisms, Their Performance Capabilities, and Applications

2.1 Introduction

This chapter describes various actuation mechanisms, their performance capabilities, and potential applications. Various actuation mechanisms will be described with major emphasis on cantilever beam reliability, force-generating capacity, response time, design complexity, and drive voltage or input power requirements. Cantilever beam configuration and design requirements for each actuation mechanism are identified to meet specific output force level and actuation voltage requirements. A radio frequency microelectromechanical systems (RF-MEMS) switch is selected just to understand the critical roles played by various elements of the actuation mechanism deployed in the design of the switch. The microelectromechanical systems (MEMS)-based RF switch has two operating states similar to a conventional RF semiconductor switch. Switching between these two states is accomplished through the mechanical displacement of a freely movable, microstructural flexible membrane called as an armature or a single beam microactuator. One end of the beam is attached to a top electrode, while the other end to a bottom electrode as illustrated in Figure 2.1 [1]. In brief, the core of a MEMS switch is the flexible

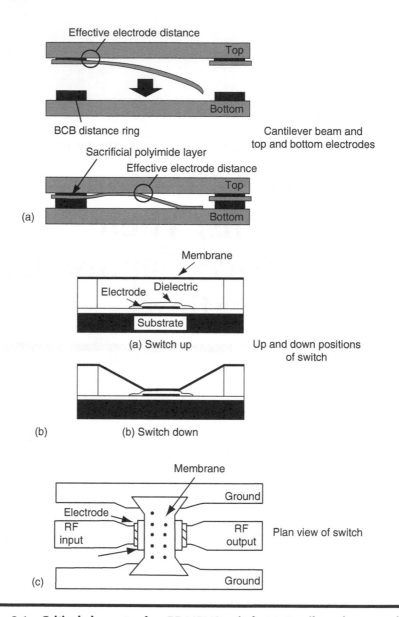

Figure 2.1 **Critical elements of an RF-MEMS switch. (a) Cantilever beam, and top and bottom electrodes; (b) up and down positions of switch; and (c) plan view of switch.**

membrane with metallic contacts moving between the top and bottom electrodes as shown in Figure 2.1. The planar view of an RF-MEMS switch shows the flexible membrane, top and bottom electrodes, and input and output terminals. The mechanical displacement is provided by the microactuator using the force provided by an appropriate actuation mechanism depending on the air gap, spring constant of the beam, and driving voltage applied.

2.2 Classification of Actuation Mechanisms

Six distinct actuation mechanisms are available for the design and development of MEMS devices and sensors. Some actuation mechanisms are most suitable for RF-MEMS switches; some are best suited for mechanical resonators; some are most attractive for micropumps (MPs); and some are ideal for MEMS devices capable of monitoring the health of structures and buildings, or detecting the chemical and biological threats. The following distinct actuation mechanisms are widely used in the design and development of various RF and optical-MEMS devices best suited for commercial, industrial, military, and space sensor applications:

- Electrostatic (ES) actuation
- Piezoelectric actuation
- Electrothermal actuation
- Electromagnetic actuation
- Electrodynamic actuation
- Electrochemical actuation

2.3 Structural Requirements and Performance Capabilities of Electrostatic Actuation Mechanism

Critical elements or components of each actuation mechanism are identified and their structural requirements are briefly described. Performance capabilities of each actuation mechanism are summarized with emphasis on reliability, response time, actuation voltage, input power, and output force generated by the armature or cantilever beam structure. Potential applications of each actuation mechanism are identified in commercial, industrial, military, space, and clinical MEMS-based systems.

2.3.1 Electrostatic Actuation Mechanism

Research studies performed by the author indicate that ES actuation mechanism is widely used in the design and development of RF- and optical-MEMS devices because of high design flexibility, moderate power consumption, and high reliability. ES actuation scheme has been proven very effective in the design of RF-MEMS switches, MEMS-phase shifters, and tunable capacitors and is widely used by cellular phones, mobile phones, satellite communications systems, base transmitters, and other communication equipment. These MEMS devices contain movable elements that are set into motion using microactuators. The mechanical displacement of a freely moving structure, called a cantilever beam, is achieved through the microactuator, which is operated by an ES actuation mechanism as shown in Figure 2.2. This figure shows an ES actuation mechanism for (a) a microrelay,

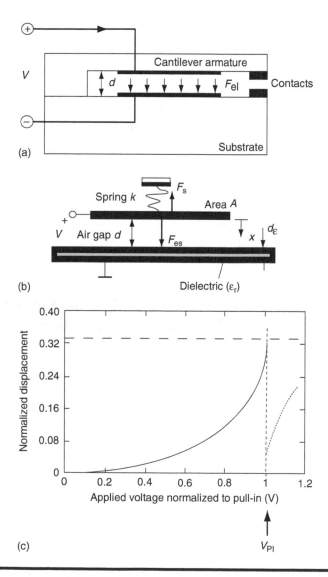

Figure 2.2 Typical elements of MEMS devices. (a) Microrelay, (b) force diagram of ES actuator, and (c) normalized beam displacement as a function of applied voltage.

(b) force diagram illustrating a lumped model for the actuator, and (c) displacement as a function of applied or actuation voltage [2].

2.3.1.1 Cantilever Beam Design Requirements

Optimum shape and size of a cantilever or beam must be selected to achieve best mechanical performance and stable RF performance under the desired operating environments. Preliminary studies conducted on cantilever shapes by the author indicate that the "arc-shaped" and "S-shaped" cantilevers must be thoroughly investigated with particular emphasis on forces of adhesion between the surfaces under variable environmental as well as surface conditions.

Cantilever beam design requires a comprehensive review of various dimensional parameters and potential shapes of the cantilever, which will lead to a design with high mechanical integrity and stable RF performance. It is important to mention that a cantilever beam is subjected to a combined actuation loading and interface adhesion force. Such actuation loading and interface adhesion force are strictly dependent on the cantilever beam shape, crack length, air gap, support post height, and overlap section of the beam. The beam deflection profiles are strictly dependent on crack length s $(L - d)$, applied force (F_a) or external force, contact zone length (d), interfacial energy of adhesion (joules per square meter), and region of interfacial adhesion forces ($0 \leq x \leq a$, where a represents the actuation pad length and the parameter x varies from 0 to crack length (s)).

The deflection profile (δ) for an S-shaped beam in the absence of applied or external force (i.e., $F_a = 0$) can be given as

$$\delta(x) = h\left[3\left(\frac{x}{s}\right)^2 - 2\left(\frac{x}{s}\right)^3\right]$$ (2.1)

where h represents the beam thickness and parameter x is a variable as defined above.

Assuming a beam thickness (h) of 2 μm and crack length (s) of 525 μm, computed values of the deflection profile are summarized in Table 2.1.

Power consumption for an ES mechanism is relatively low because power is used only during the motion of the cantilever beam or the armature. Other advantages of this actuation method include simple fabrication technology, inherent high degree of compatibility with a standard integrated circuit (IC) processing, rapid integration with planar and microstrip transmission lines, and fast response [1]. The major drawback of the ES actuation mechanism is the high actuation voltage requirement, which is in the range of 20–75 V in the case of actuators with large air gaps. However, there are certain performance advantages using low and high actuation voltages depending on the MEMS device configurations, biasing network requirements, and coplanar waveguide (CPW) transmission-line parameters, as illustrated in Figure 2.3 [3]. It is difficult to combine low actuation voltage with high isolation

Table 2.1   Computed Values of Deflection Profile of an S-Shaped Beam

Parameter x (μm)	Deflection Profile of the Beam (μm)
000	0000
100	189
200	648
300	1208
400	1710
500	1990
525	2

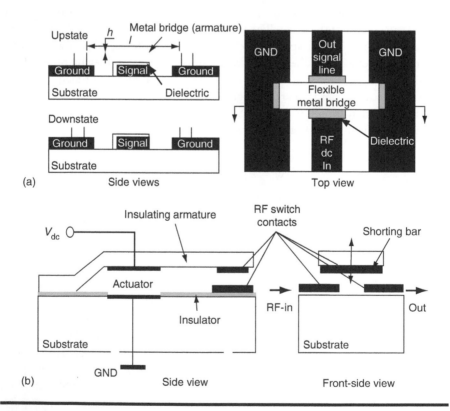

Figure 2.3   Configurations and biasing schemes for various RF-MEMS devices. (a) ES actuation for an RF shunt switch, (b) ES actuation for microrelay,

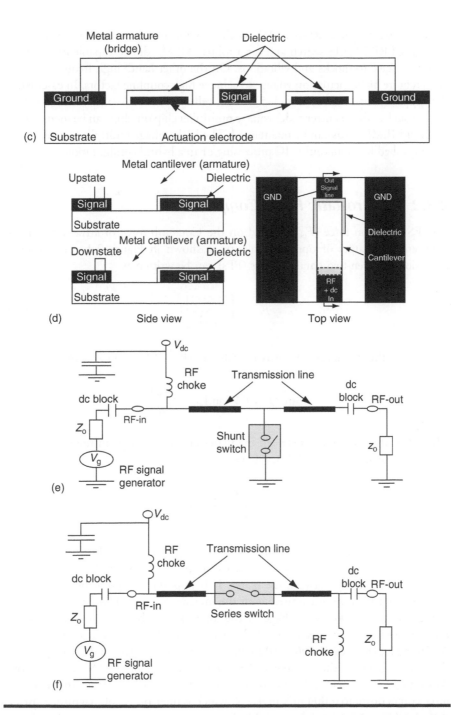

Figure 2.3 (continued) **(c) ES actuation of three-terminal switch, (d) RF switch using CPW line, and (e), (f) biasing network. (From De Los Santos, H.J. et al.,** *IEEE Microw. Mag.,* **40, 2004. With permission.)**

for a series-configured RF-MEMS switch or with low insertion loss for a shunt-configured RF-MEMS switch and for a robust MEMS device capable of providing high resistance to shocks and vibrations. On the other hand, high frequency resonators require higher actuation voltage to achieve high coupling factors. In cases where the supply voltage is limited to 3–5 V as in cellular or mobile phones, high-voltage devices such as direct current (dc) voltage multiplier chip on-chip can be incorporated and still MEMS status can be maintained. The high-voltage multiplier circuit can be accomplished in a monolithic IC processing or in a hybrid configuration.

2.3.2 Electrostatic Force Computation

The ES actuation force (F_{es}) is based on the Coulomb force of attraction existing between the charges of opposite polarity as shown in Figure 2.1. This force of attraction between two parallel plates (PPs) can be written as

$$F_{es} = \left[\frac{(0.5)Q}{\varepsilon_o A} \right] \tag{2.2}$$

where ε_o is the free-space permittivity (8.83×10^{-12} F/m) and A is the plate area.

$$\text{But } Q = CV \text{ and } C = \frac{\varepsilon_o A}{d} \tag{2.3}$$

where C is the capacitance between the plates and V is voltage in the air gap d.
From Equations 2.2 and 2.3, one gets

$$F_{es} = \left[\frac{(0.5)\varepsilon_o A V^2}{d^2} \right] \tag{2.4}$$

It is evident from Equation 2.4 that the ES force is proportional to the plate area, proportional to the square of voltage across the plate, and inversely proportional to the square of the gap between the plates. It is important to point out that the ES force will not increase indefinitely with the increase in applied voltage (V) but it is limited by the breakdown voltage of the air gap. The breakdown voltage in the micrometer-sized air gaps is 3×10^8 V/m, which leads to a maximum force. Assuming a plate area of 100×100 μm^2, an air gap of 1 μm, and a voltage across the gap of 300 V, the computed ES force comes to or 4 mN (millinewton). Assuming a more practical value of applied voltage equal to 10 V, the actuation force comes down to 0.004 mN or 4 μN, which is three orders of magnitude smaller than 4 mN at actuation voltage of 300 V.

Table 2.2   ES Forces as a Function of Various Parameters (μN)

Actuation Voltage (V)	Air Gap (μm)			Application
	1	*2.5*	*5.0*	
10	4.400	0.704	0.176	Tuning device
20	17.600	2.816	0.704	HF resonator
30	39.600	6.300	1.584	MEMS switch
40	70.400	11.262	2.816	MEMS switch
50	110.000	17.500	4.400	Phase shifter

Computed values of ES forces as a function of practical actuation voltage and air gap for a plate area of $100 \times 100 \ \mu m^2$ are summarized in Table 2.2.

It is important to mention that microrelays equipped with low quality contacts require contact forces at least in the order of 75–100 μN (micronewtons) to achieve a stable actuator performance. This means that ES actuation is not suitable for microrelay applications unless a large actuation area ($1000 \times 1000 \ \mu m^2$) or a higher actuation voltage (>100 V) is selected.

ES actuation force requirements are discussed for specific MEMS device applications. ES force plots shown in Figure 2.4a are best suited for MEMS devices such as HF resonators, which require actuation voltages from 10 to 20 V across the PPs with area of $100 \times 100 \ \mu m^2$. ES force plots shown in Figure 2.4b will be found most useful in the design of ES actuators operating at moderate actuation voltages from 30 to 40 V. ES force plots shown in Figure 2.4c will be most attractive in the design of RF-MEMS switches requiring high actuation voltages from 40 to 60 V. In each case, the plate area is assumed $100 \times 100 \ \mu m^2$ and the air gap is varied from 0 to 10 μm.

ES force versus actuation voltage curves shown in Figure 2.5a are best suited for MEMS devices requiring moderate actuation voltages with air gap from 2.5 to 5.0 μm. Curves shown in Figure 2.5b can be used in the design of MEMS device actuators having air gaps of 2.5 and 5.0 μm and requiring high actuation voltages. The air gap of 2.5 μm will yield actuation forces exceeding 100 μN at operating voltage greater than 120 V. In each case, a plate area of $100 \times 100 \ \mu m^2$ is assumed.

ES force versus plate area curves depicted in Figure 2.6 will provide instant ES force requirement as a function of plate area, air gap, and actuation voltage. The curves shown in Figure 2.6a and b will be found most useful in performing design trade-off studies for the RF-MEMS switches requiring moderate actuation voltages with a specified plate area and specified air gap.

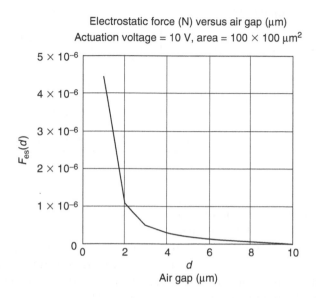

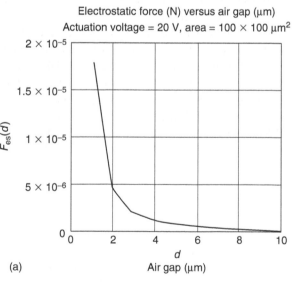

Figure 2.4 (a) ES forces as a function of voltage,

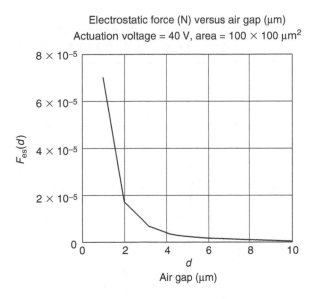

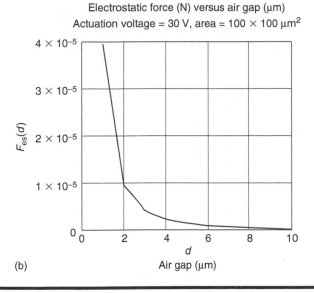

Figure 2.4 (continued) **(b) ES forces as a function of air gaps,**

(continued)

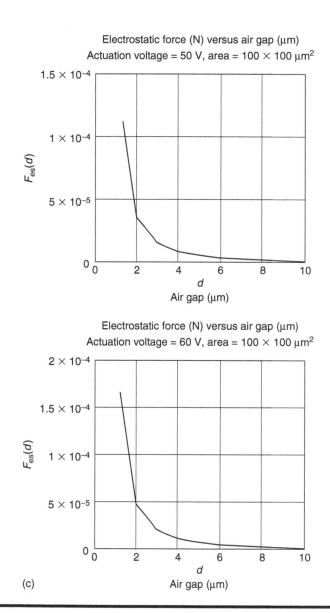

Figure 2.4 (continued) **(c) ES forces as a function of plate area.**

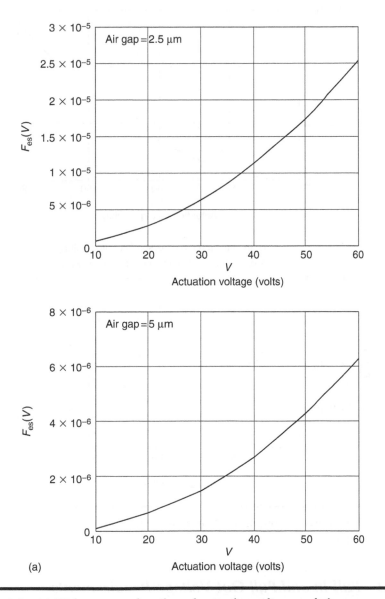

Figure 2.5 (a) ES forces as a function of actuation voltage and air gap,

(continued)

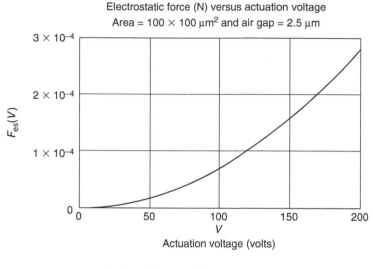

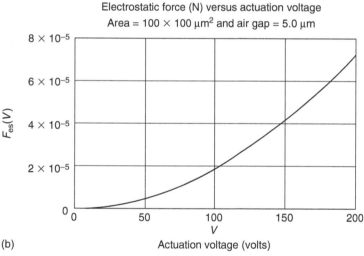

Figure 2.5 (continued) **(b) ES forces at much higher actuation voltages and for various air gap.**

2.3.3 *Pull-In and Pull-Out Voltage Requirements*

As mentioned earlier that the ES force will not increase indefinitely with the increase in applied or actuation voltage, but it is limited by the breakdown voltage in the air gap. The nonlinear relationship [2] between the normalized displacement and applied voltage normalized to pull-in voltage as illustrated in Figure 2.3 causes instability in the cantilever beam, when the actuation or applied voltage exceeds

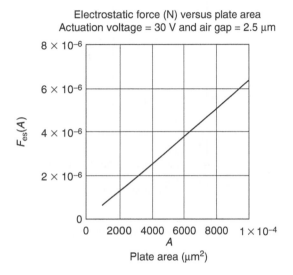

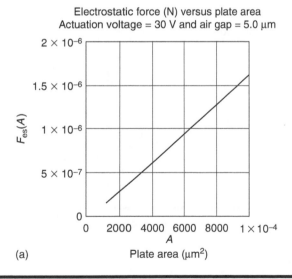

(a)

Figure 2.6 **(a) ES forces at actuation of 30 V as a function of plate area and air gap,**
(continued)

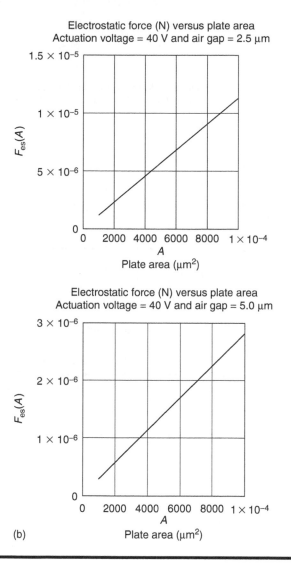

Figure 2.6 (continued) **(b) ES forces at 40 V as a function of plate area and air gap.**

the pull-in or snap-down voltage. In other words, the control of an electrostatically actuated cantilever armature or beam is lost owing to loss of equilibrium between the ES and spring forces. This phenomenon is quite evident from the curve shown in Figure 2.2c and also from Equation 2.4 shown for the pull-in voltage showing the relationship between the ES and spring forces and whose solution yields the displacement of the movable element or cantilever armature as shown in Figure 2.3a.

Note that the expressions provided for the pull-in and pull-out voltages are only exact for a lumped spring-mass system shown in Figure 2.2c.

2.3.3.1 Pull-In Voltage

It is important to mention that the expression for pull-in voltage or the actuation voltage is based on Hooke's law. This law is based on the assumption that the ES forces acting on the cantilever beam are the same as those in a PP capacitor. In other words, the expression for the pull-in voltage assumes that the ES force is equal to the restoring force of a mechanical spring given by Hooke's law. The expression for the pull-in voltage without a dielectric cover on the lower electrode can be written as

$$V_{\text{PI}} = 544 \left[\frac{kd^3}{e_r A} \right]^{0.5} \tag{2.5}$$

where
  k is spring constant (N/m)
  d is the air gap (μm)
  A is the plate area (μm^2)
  e_r is the relative dielectric constant for medium between the plates

The pull-in voltage with dielectric cover can be written as

$$V_{\text{PI}} = 544 \left[\frac{k(d - d_\varepsilon)^3}{e_r A} \right]^{0.5} \tag{2.5a}$$

where d_ε is the dielectric cover thickness, which varies between 8 and 10 percent of air gap.

It is evident from Equations 2.5 and 2.5a that the pull-in voltage is proportional to the square root of spring constant, inversely proportional to the square root of relative dielectric constant, inversely proportional to the square root of plate area, and directly proportional to the cube root of air gap.

Sample Calculations
Assuming a plate area of 100×100 μm^2, air gap of 2.5 μm, dielectric constant of 7, and spring constant of 10 N/m, the pull-in voltage magnitude without the dielectric cover can be written as

$$\begin{aligned} V_{\text{PI}} &= 500 \left[\frac{10 \times 2.5^3}{7 \times 100 \times 100} \right]^{0.5} = 544 \left[\frac{22.32}{10,000} \right]^{0.5} \\ &= \left[\frac{544 \times 4.72}{100} \right] \\ &= 25.7 \text{ V} \end{aligned}$$

Table 2.3   Computed Values of Pull-In Voltages without a Cover and as a Function of Plate Area, Spring Constant, and Air Gap

Spring Constant k (N/m)	Air Gap d(μm)	Dielectric Constant	Plate Area A (μm^2)	Pull-In Voltage V_{PI} (V)
10	2.5	7	10,000	25.7
10	2.5	9	12,000	20.7
10	5.0	7	10,000	72.7
5	5.0	9	18,000	33.8
5	5.0	9	8,000	50.7
5	5.5	7	10,000	51.3

Assuming various different parametric values, magnitudes of pull-in voltage without dielectric cover are computed, which are summarized in Table 2.3.

These computed values indicate that the devices using ES actuation mechanisms will require higher applied voltage, and consequently, higher power consumption. Higher input voltage requirement is the major drawback of the ES actuation.

In addition to sample calculations, pull-in voltage plots as a function of spring constant (k) and plate area (A) have been provided for a parametric trade-off analysis. Figure 2.7a and b shows the pull-in voltage curves for various air gaps (d) as a function of spring constant and plate area of 8,000 and 12,000 μm^2, respectively, assuming a dielectric constant or permittivity of 11.8. The pull-in voltage curves shown in Figure 2.8a and b are generated as a function of air gap for plate area of 16,000 and 20,000 μm^2, respectively, assuming the same dielectric constant. In addition, pull-in voltage curves shown in Figure 2.9a and b are generated for a plate area of 8,000 μm^2 and permittivity of 11.8 and plate area of 18,000 μm^2 and permittivity of 7, respectively, just to demonstrate the combined effects of plate area and permittivity on the pull-in voltage.

Pull-In Voltage with Cover
Equation 2.5a yields the magnitude of pull-in voltage with dielectric cover. Assuming the same parameters used in the sample calculations for the pull-in voltage and the cover thickness of 0.2 μm, the magnitude of the pull-in voltage with dielectric cover is

$$V_{PI} = 544 \left[\frac{10 \times (2.5 - 0.2)^3}{7 \times 10,000} \right]^{0.5} = 544 \left[\frac{17.38}{10,000} \right]^{0.5}$$

$$= \left[\frac{544 \times 4.169}{100} \right]$$

$$= 22.7 \text{ V}$$

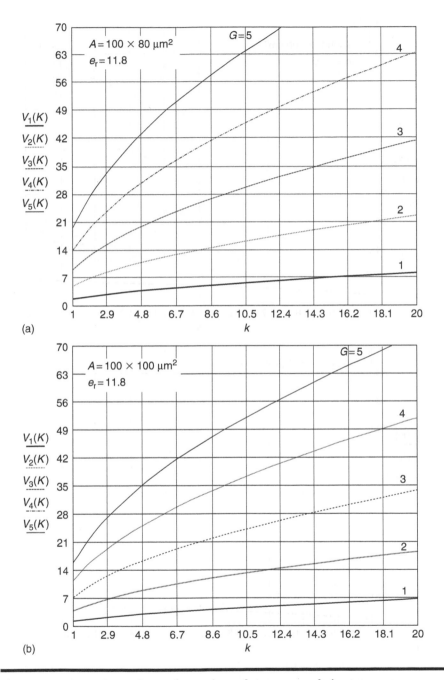

Figure 2.7 Actuation voltages for various plate areas and air gaps.

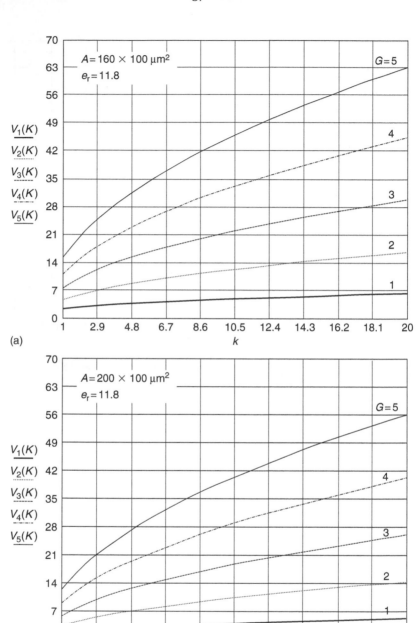

Figure 2.8 Pull-in voltage as a function of spring constant (*k*).

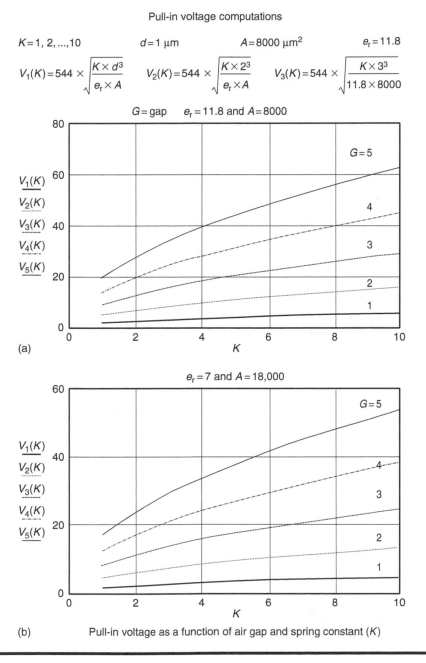

Pull-in voltage computations

$K = 1, 2, ..., 10$ $\quad\quad d = 1\ \mu m$ $\quad\quad A = 8000\ \mu m^2$ $\quad\quad e_r = 11.8$

$$V_1(K) = 544 \times \sqrt{\frac{K \times d^3}{e_r \times A}} \quad\quad V_2(K) = 544 \times \sqrt{\frac{K \times 2^3}{e_r \times A}} \quad\quad V_3(K) = 544 \times \sqrt{\frac{K \times 3^3}{11.8 \times 8000}}$$

$G =$ gap $\quad e_r = 11.8$ and $A = 8000$

(a)

$e_r = 7$ and $A = 18{,}000$

(b) $\quad\quad$ Pull-in voltage as a function of air gap and spring constant (K)

Figure 2.9   Pull-in voltage as a function of several parameters.

Now assuming a spring constant of 10 N/m, air gap of 2.5 μm, dielectric constant of 9, cover thickness of 0.2 μm, and plate area of 12,000 μm^2, one gets the magnitude of pull-in voltage with dielectric cover as

$$V_{PI} = 544 \left[\frac{10 \times (2.5 - 0.5)^3}{9 \times 12,000} \right]^{0.5} = 544 \left[\frac{13.519}{12,000} \right]^{0.5}$$

$$= \left[\frac{544 \times 3.677}{109.54} \right]$$

$$= 18.3 \text{ V}$$

It is evident from the computed values shown in Table 2.2 and above sample calculations that the pull-in voltage without dielectric cover in the first sample calculation is 25.7 V compared to 22.7 V with cover and in the second sample calculation, the pull-in voltage without cover is 20.7 V compared to 18.3 V with cover. In these calculations, it is evident that the pull-in voltage without the dielectric cover is slightly higher than that with the dielectric cover.

2.3.3.2 Pull-Out Voltage

It is important to mention that after the pull-in occurred, the displacement becomes imaginary and the slope of the applied voltage no longer controls the displacement. Under these circumstances, the cantilever beam releases at much lower voltage called pull-out voltage (V_{PO}) because of hysteresis experienced by the actuator (Figure 2.10). The pull-out voltage for an ES actuation is given as

$$V_{PO} = \left[\frac{2kdd_{\varepsilon}^2}{e_r^2 \varepsilon_o A} \right]^{0.5} \tag{2.6}$$

where
  k is the spring constant (N/m)
  d is the air gap (μm)
  e_r is the relative dielectric constant or permittivity of the dielectric cover on the electrode
  d_{ε} is the cover thickness (μm)
  ε_o is the free-space permittivity (8.85 × 10^{-12} F/m)
  A is the plate area (μm^2)

Sample Calculations
Assuming a spring constant of 10 N/m, air gap of 2.5 μm, dielectric cover thickness of 0.2 μm, dielectric constant of 7, and plate area of 10,000 μm^2, the pull-out voltage is computed as

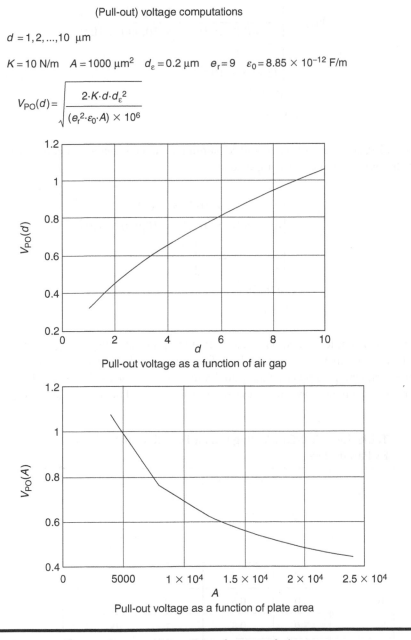

(Pull-out) voltage computations

$d = 1, 2, \ldots, 10 \; \mu m$

$K = 10 \; N/m \quad A = 1000 \; \mu m^2 \quad d_\varepsilon = 0.2 \; \mu m \quad e_r = 9 \quad \varepsilon_0 = 8.85 \times 10^{-12} \; F/m$

$$V_{PO}(d) = \sqrt{\frac{2 \cdot K \cdot d \cdot d_\varepsilon^2}{(e_r^2 \cdot \varepsilon_0 \cdot A) \times 10^6}}$$

Pull-out voltage as a function of air gap

Pull-out voltage as a function of plate area

Figure 2.10   Pull-out voltage as a function of area and air gap.

$$V_{PO} = \left[\frac{2 \times 10 \times 2.5 \times 0.2^2}{7^2 \times 8.85 \times 10^{-12} \times 10,000 \times 10^6} \right]^{0.5}$$

$$= \left[\frac{2}{433.65 \times 10^{-2}} \right]^{0.5}$$

$$= [0.46]^{0.5}$$

$$= 0.68 \text{ V}$$

Assuming an air gap of 5 μm and remaining parameters being same, the magnitude of the pull-out voltage is computed as

$$V_{PO} = \left[\frac{2 \times 10 \times 5 \times 0.2^2}{49 \times 8.85 \times 10^{-12} \times 10,000 \times 10^6} \right]^{0.5}$$

$$= \left[\frac{4}{433.65 \times 10^{-2}} \right]^{0.5}$$

$$= [0.922]^{0.5}$$

$$= 0.96 \text{ V}$$

These two sample calculations indicate that the pull-out voltage is significantly lower than the pull-in voltage at least by a factor ranging from 10 to 20 depending on other relevant parameters. Computed values of pull-out voltage as a function of various parameters are summarized in Table 2.4. The magnitudes of pull-out

Table 2.4 Pull-Out Voltages as a Function of Various ES Parameters

Spring Constant (k) (N/m)	Air Gap (d)	Dielectric Cover Thickness (d_e)	Relative Dielectric Constant (e_r)	Plate Area (A) (μm²)	Pull-Out Voltage (V_PO) (V)
10	2.5	0.2	7	100 × 100	0.68
10	5.0	0.2	7	100 × 100	0.96
10	5.0	0.5	9	100 × 100	1.87
5	5.0	0.5	7	100 × 100	1.69
10	5.0	0.1	9	100 × 100	0.39
5	7.5	0.2	7	100 × 80	0.93
10	5.0	0.2	7	100 × 120	0.88

voltages remain extremely low compared to those for pull-in voltages, regardless of values of other parameters. As stated previously that once the pull-in has occurred, the beam or the cantilever structure releases at much lower voltage because of the presence of hysteresis effect in the ES actuator.

Pull-out voltage computations using all possible permutations and combinations of parameters involved reveal that the pull-out voltage most likely will not exceed 2 V, although the pull-in voltages will never be less than 30 V for most MEMS devices requiring ES forces exceeding 50 μN. Note a force of 1 N is equal to 4.45 lb approximately.

2.3.3.3 Electrostatic Microactuator Configurations for Generating Higher Force and Energy Density Capabilities

ES microactuators requiring controlled actuation or motion will be required for many MEMS components or sensors such as micropositioners, precision mechanical mirrors, force-rebalanced and resonant sensors, micromotors, MPs, valves, adaptive optics systems, and read/write actuation mechanisms for magnetic disk drive heads. Because of the scaled-down dimensions of the MEMS sensors or components, ES drives will be found most attractive for their high energy generation capabilities and for minimum fabrication costs. Vertical comb array microactuators (VCAMs) [4] are capable of generating a large vertical force for a given geometrical area on a wafer surface. VCAM can generate a force that approaches the PP drive and is at least an order of magnitude larger than that with a lateral comb drive mechanism. In the case of VCAM beams, the force generated is approximately proportional to $2L$, where L is the beam length, which is several 100 μm long. In the case of lateral comb drive, the force generated is proportional to $2t$, where t is the thickness of the polysilicon layer, which is typically 2–3 μm. In summary, the force-generating capability from a VCAM mechanism could be few orders of magnitude larger than that available from a lateral comb drive mechanism.

ES rotary microactuators with optimum shape design offer significantly improved actuation force-generating capability. In conventional rotary ES microactuators, the force-generating capability is limited because of large clearance (i.e., the distance between the two facing electrodes) at the outer region from the center of rotation. To overcome the limitations of conventional rotary microactuators and to enhance the force-generating capability, a tilted configuration is developed, which can be integrated into the optimum shape. This will allow the gap between the two facing electrodes as small as possible with current fabrication technology. The optimum shape of the ES rotary microactuator involving integrated tilting configuration [5] will increase the force-generating capability dramatically over the conventional-shaped rotary microactuator. Structural details, critical elements dimensional requirements, fabrication aspects and performance capabilities, and limitations of above ES micro-actuators will be discussed in greater details in Chapter 3.

2.4 Piezoelectric Actuation Mechanism

It is important to mention that the piezoelectric actuation is based on inverse piezoelectric effect, although the piezoelectric detection is based on direct piezoelectric effect. Note a piezoelectric material provides its own internal biasing require-ment, either due to absence of a center of symmetry in the case of single-crystal materials such as aluminum nitride (AIN) or zinc oxide (ZnO) or due to a per-manent polarization present in ferroelectric materials such lead–zirconate–titanate (PZT). The inverse piezoelectric effect is of critical importance because it offers mechanical deformation in the piezoelectric material when subjected to an external or applied electric field. Note that pull-in or snap-down does not occur in a piezoelectric actuator as seen in an ES actuation, and as a result the piezoelectric actuator does not suffer from mechanical instability.

It is essential to know the critical parameters of potential piezoelectric materials for possible applications in cantilever beams for MEMS devices, which include AIN, PZT, and ZnO. Piezoelectric coefficients and constants are expressed generally in MKS system. The most important parameter is the transverse piezoelectric strain constant (d_{31}), which is expressed in picocoulomb per newton (pC/N) or picometer per volt (pm/V). The constants e_{ij} are called piezoelectric stress constants or coeffi-cients and are expressed in coulomb per newton (C/N) or meter per volt (m/V). The C_{ii} constants such as C_{11}, C_{22}, C_{33}, and C_{66} are known as electric stress constants or stiffness constants and are expressed in newton per square meter (N/m^2). The constant ε_{11} is known as the permittivity or the dielectric constant and has no unit. It is important to point out that the constant or coefficient d_{31} relates an electric field to external strain in any direction to the polar direction, where as the piezoelectric coefficient or constant d_{15} relates an electric field normal to the polar axis with a shear strain in the plane containing the polar axis and signal field. Piezoelectric properties of some selected materials are listed in Table 2.5.

The magnitude of transverse piezoelectric strain constant d_{31} is temperature sensitive. Its value for the PZT material ranges from 123 at $-100°$C to 125 at $0°$C to 118 pC/N at $100°$C. One can expect the variations in the same parameter for other piezoelectric materials as a function of operating temperature. The value of parameter d_{31} for the AIN material is rounded to 3, for PTZ material to 100, and for ZnO material to 5 in some numerical examples to simplify calculations and for rapid comprehension by the readers.

If some constants or coefficients are expressed in a mixture of "CGS" and "FPS" units, one can convert these constants into the "kgs" system using the equivalent values shown in Table 2.6. These equivalent conversion factors will help the readers in rapid comprehension of the parametric values given in different unit systems.

Table 2.5 Important Properties of Selected Piezoelectric Materials

Material	Density 1000 kg/m³	Dielectric Constant (ε_{11})	Piezoelectric Strain Constant (d_{31}) (pC/N)	Young's Modulus (E) GPa	Elastic Stiffness Constants (N/m²)			
					C_{11}	C_{22}	C_{33}	C_{66}
Aluminum nitride (AlN)	3.26	9.5	3.2	320	17.40	6.05	19.80	5.54
Lead–zinc–titanate (PZT)	7.50	1475	120	70	13.90	15.90	11.50	14.60
Zinc oxide (ZnO)	5.68	8.33	5.2	83	21.10	4.75	21.10	4.43

Table 2.6   Equivalent Conversion Factors

Picocoulomb per Newton (pC/N)	Equivalent Value in Picometer per Volt (pm/V)
Gigapascal (GPa)	103.51 kg/mm^2
Pounds per square inch (psi)	6898 Pa
Newton (N)	4.45 lb
Square meter (m^2)	1550 in.^{2}
Newton per square meter (N/m^2)	19.83 Pa
Meter (m)	10^6 μm

Note: Pico $= 10^{-12}$.

2.4.1   Structural Material Requirements for Cantilever Beams

Structural materials with high thermal conductivity, tensile strength, fatigue resistance, and hysteresis-free response are best suited for the cantilever beam and other elements of the actuator. Aluminum (Al) is widely used as a structural material because of lightweight and low cost. Gold (Au)-plated contacts are preferred, where high operating currents, low contact resistance, and high reliability are the principal design requirements for the actuators. Nickel (Ni)-plated contacts for MEMS actuators are recommended to avoid contact surface deterioration under chemical environments. Tungsten (W) is best suited for the deployment in cantilever membranes, contacts, and electrodes, where ultrahigh reliabilities over extended durations and under harsh thermal and mechanical environments are the principal requirements. Thermal and mechanical properties of potential structural materials are summarized in Table 2.7.

Table 2.7   Thermal and Mechanical Properties of Structural Materials for Beams

Structural Material	Density (g/cm^3)	Thermal Expansion Coefficient (10^{-6}/°C)	Thermal Conductivity (W/cm °C)	Young's Modulus (E) GPa
Aluminum (Al)	2.7	22.9	2.18	70
Gold (Au)	19.3	14.2	2.96	72
Nickel (Ni)	8.9	13.3	0.90	207
Tungsten (W)	19.3	4.3	1.99	345

2.4.2 *Threshold Voltage*

The threshold voltage (V_{th}) is required to close the air gap and this occurs when the tip deflection (δ_{tip}) is equal to gap spacing (d). The expression for the threshold voltage can be written as

$$V_{th} = \left[\left(\frac{0.33}{d_{31}} \right) \left(\frac{h_c}{L} \right) \left(\frac{E_c}{E_p} \right) (d \times 10^6) \right] \qquad (2.7)$$

where

h_c is the thickness of the carrier layer (μm)
L is the beam length (μm)
E is Young's modulus (GPa)
d is the gap spacing (μm)
d_{31} is the transverse piezoelectric coefficient of the material used (C/N) or (m/V)

The subscript c stands for carrier layer and p stands for piezoelectric material layer. The beam length is typically about 1000 μm. The carrier thickness is assumed not to exceed 10 percent of the beam length and must be much greater than the thickness of the piezoelectric material layer.

Computed values of threshold voltages for three distinct piezoelectric materials are depicted in Table 2.8.

In addition to sample calculations, plots of threshold voltages as a function of air gap or spacing (d) for three piezoelectric materials are provided in Figure 2.11a

Table 2.8 Computed Threshold Voltages for Piezoelectric Materials

Material	Piezoelectric Strain Constant (d_{31}) (pC/N)	Young's Modulus (E_c) (GPa)	Young's Modulus (E_p) (GPa)	Threshold Voltage (V_{th}) for	
				$d = 1.0$	$d = 2.5$
Aluminum nitride (AIN)	3	70	320	2.406	6.105
Lead–zinc–titanate (PZT)	100	70	69	0.371	0.842
Zinc oxide (ZnO)	5	70	83	5.562	18.322

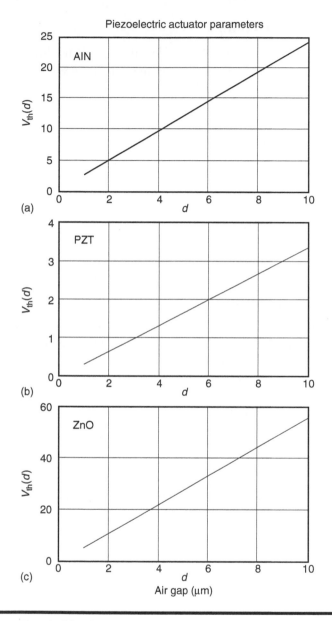

Figure 2.11 Threshold voltage (V_{th}) versus air gap for piezoelectric materials.

through 2.11c for materials AIN, PZT, and ZnO, respectively. These plots reveal that the threshold voltage for PZT material is the minimum, whereas it is the maximum for ZnO material, regardless of air gap values.

2.4.3   Tip Deflection of the Cantilever Beam

When an actuation voltage V is applied to a piezoelectric layer deposited on the top of the cantilever beam, a bending moment (M_{PE}) due to inverse piezoelectric effect is induced at the tip of the beam. The bending moment causes the cantilever beam to bend with a tip deflection (δ_{tip}). In fact, the tip deflection calculations cannot be completed till the bending moment values are available. The expression for the tip deflection of the cantilever beam can be written as

$$\delta_{tip} = \left[\frac{M_{PE}L^2}{2EI}\right] \tag{2.8}$$

where
  L is the beam length
  E is Young's modulus of the piezoelectric material
  I is the moment of inertia of the cross section of the beam ($I = bh^3/12$, where b is the beam width and h is the composite beam thickness involving the carrier layer and piezoelectric material layer)

Here EI is called the equivalent bending stiffness of the cantilever beam having a layer of piezoelectric material layer of given thickness. The tip deflection as a function of beam length for a piezoelectric cantilever beam of AlN material is shown in Figure 2.12 with actuation voltage of 15 V, whereas the tip deflection for a piezoelectric cantilever beam of PZT material is shown in Figure 2.13a and b, respectively. Because piezoelectric material ZnO is seldom used in the design cantilever beam actuator, no calculations or plots for bending moments or contact forces are provided.

2.4.4   Bending Moment of the Cantilever Beam

The bending moment expression can be written as

$$M_{BE} = [5d_{31}E_P bh_C V] \tag{2.9}$$

where
  d_{31} is the piezoelectric material coefficient (C/N) or (m/V)
  E_P is Young's modulus of the piezoelectric material
  b is the beam width
  h_C is the carrier material thickness
  V is the actuation or applied voltage

Note Equation 2.9 yields excellent results, when the piezoelectric layer thickness is much thinner than the carrier layer thickness. Computed values of tip deflection and bending moment as a function of various parameters for three distinct piezoelectric materials are summarized in Table 2.9.

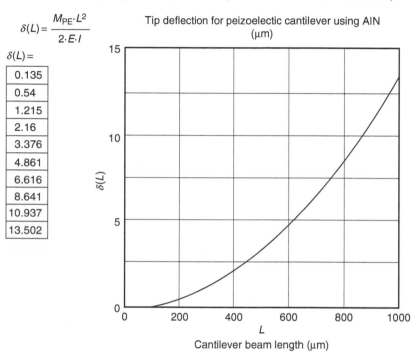

Cantilever bending with a tip deflection (δ)

Parameters assumed for AlN cantilever beam

$M_{PE} = 1.818 \times 10^{-3}$ for AlN material at actuation voltage of 15 V

$E = 1.616 \times 10^{-2}$ N/μm^2 (320 MPa) $I = 4166$ μm^4 $L = 100, 200, ..., 1000$ μm

$$\delta(L) = \frac{M_{PE} \cdot L^2}{2 \cdot E \cdot I}$$

Tip deflection for peizoelectic cantilever using AlN (μm)

$\delta(L) =$

0.135
0.54
1.215
2.16
3.376
4.861
6.616
8.641
10.937
13.502

Figure 2.12   Tip deflection for an AlN piezoelectric beam as a function of length and with an actuation voltage of 15 V.

These computed values of bending moments reveal critical mechanical performance requirements of the piezoelectric cantilever beams in terms of stiffness, tip deflection, and bending moment. Furthermore, these values will provide significant help in the selection and design of cantilever beams for MEMS devices using piezoelectric actuation mechanisms. Bending moment computations must be performed before tip deflections.

As stated earlier, the tip deflection is strictly dependent on the piezoelectric bending moment. Note the bending moment magnitude is dependent on piezoelectric material layer thickness, Young's modulus of the carrier layer, and the transverse piezoelectric coefficient. In case, when the piezoelectric material layer thickness (h_p) is much smaller than the carrier layer thickness (h_c), expression for the bending moment can be approximately written as defined by Equation 2.9.

Tip deflection in the PZT cantilever beam

Parameters assumed

$M_{PEPZT} = 4.419 \times 10^{-3}$ $E_{PZT} = 0.3535 \times 10^{-2}\,N/\mu m^2$ $I = 4166\,\mu m^4$,

$$\delta_{PZT}(L) = \frac{M_{PEPZT} \cdot L^2}{2 \cdot E_{PZT} \cdot I}$$ $\begin{array}{c} L = 100, 200, ..., 1000 \\ (\mu m) \end{array}$

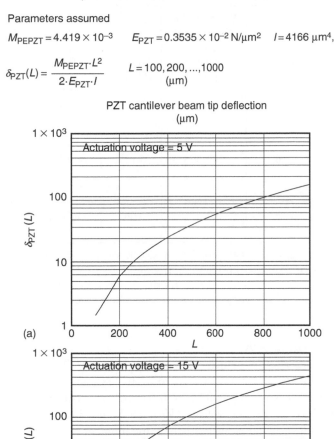

Figure 2.13   Tip deflection for a PZT cantilever beam versus beam length.

Table 2.9   Computation of Tip Deflection and Bending Moment

Piezoelectric Material	Young's Modulus (E) (GPa)	Piezoelectric Coefficient (d_{31}) Constant (m/V)	EI Product (N μm^2)	Tip Deflection (δ_{tip}) (nm)	M_{PE} Bending Moment (N/m) at 30 V
Aluminum nitride (AIN)	320	3.0–3.2	67.32	0.0270	0.3636×10^{-8}
Lead–zinc–titanate (PZT)	69	100–123	14.72	0.1234	2.64×10^{-8}
Zinc oxide (ZnO)	83	5.0–5.2	17.45	0.1043	0.156×10^{-8}

Note: Beam length of 1000 μm, beam width of 50 μm, piezoelectric layer thickness of 10 μm, and actuation voltage of 30 V are assumed for these calculations. Values of Young's modulus and piezoelectric coefficient are taken from various engineering materials handbooks.

Sample Calculations for AIN

Assuming the piezoelectric coefficient $d_{31} = 3$ pC/N (3 pm/V) for the AIN material, Young's modulus (E_p) = 320 GPa (or 1.616×10^{-2} N/μm^2), beam width (b) = 50 μm, carrier layer thickness (h_c) = 0.01L (10 μm) where the beam length, $L = 1000$ μm and actuation or driving voltage (V) = 15 V and inserting these values in Equation 2.9, one gets the piezoelectric bending moment as

$$[M_{PE}]_{AIN} = \left[5 \times 3 \times 10^{-12} \times 1.616 \times 10^{-2} \times 50 \times 10 \times 15\right]$$
$$= \left[24.24 \times 10^{-14} \times 0.75 \times 10^4\right]$$
$$= \left[18.18 \times 10^{-10}\right]$$
$$= 0.1818 \times 10^{-8} \text{ N m}$$

Sample Calculations for PZT

Assuming a piezoelectric coefficient of 100 pm/V for the PZT material, Young's modulus of 70 GPa (0.3535×10^{-2} N/μm^2), beam width (b) of 50 μm, carrier layer thickness of 10 μm, and actuation voltage of 15 V and inserting these values in Equation 2.9, one gets the piezoelectric bending moment as

$$[M_{PE}]_{PZT} = \left[5 \times 100 \times 10^{-12} \times 0.3535 \times 10^{-2} \times 0.75 \times 10^4\right]$$
$$= 1.325 \times 10^{-8} \text{ N m}$$

Sample Calculations for ZnO

Assuming a piezoelectric coefficient of 5 pm/V for the ZnO material, Young's modulus of 83 GPa (0.419×10^{-2} N/μm^2), and rest of the values remaining the same as above, and inserting these values in Equation 2.9, one gets the piezoelectric moment as

$$[M_{PE}]_{ZnO} = \left[5 \times 5 \times 10^{-12} \times 0.419 \times 10^{-2} \times 0.75 \times 10^4\right]$$
$$= \left[7.85 \times 10^{-10}\right]$$
$$= 0.0785 \times 10^{-8} \text{ N m}$$

These bending moment calculations indicate that ZnO provides the minimum bending moment compared to AlN and PZT piezoelectric materials.

Bending moment plots as a function of actuation voltage for piezoelectric AlN and PZT cantilever beams are displayed in Figures 2.14 and 2.15, respectively. It is evident from these plots that higher bending moments occur at higher actuation voltages, regardless of piezoelectric materials used by the beams.

2.4.5 Contact Force Requirements

Adequate contact force is required to make immediate and reliable contact as soon as the actuation voltage is applied. The expression for the contact force can be written as

$$F_c = \left[\frac{3M_{PE}}{2L}\right] + \left[\frac{3d_0 EI}{L^3}\right] \qquad (2.10)$$

where d_0 is the gap spacing (μm) and I is the moment of inertia (μm^4), which is given by Equation 2.11.

$$I = \left[\frac{bd^3}{12}\right] \qquad (2.11)$$

where b is the beam width (μm) and d is the beam thickness (μm), which is typically, equal to 10 percent of the beam length. Assuming a gap of 2.5 μm, a beam width of 50 μm, beam length of 1000 μm, beam thickness of 10 μm, and Young's modulus of 320 MPa (1.616×10^{-2} N/μm^2) for the AlN material, actuation voltage of 15 V and inserting these parameters into Equations 2.10 and 2.11, one gets the following value as obtained in the sample calculation:

Parameters assumed

$V = 5, 10, ..., 35$ $I = 4166 \ \mu m^4$ $d_0 = 2.5 \ \mu m$ $L = 1000 \ \mu m$

$E_p = 1.616 \times 10^{-2} \ N/\mu m^2 \ (320 \ MPa)$ $d_{31} = 3 \times 10^{-6}$ $b = 50 \ \mu m$ $h_s = 10 \ \mu m$

Bending moment in piezoelectric material AlN is given as

$$M_{PE}(V) = 5 \cdot d_{31} \cdot E_p \cdot b \cdot h_s \cdot V$$

$M_{PE}(V) =$

$6.06 \ \times 10^{-4}$
1.212×10^{-3}
1.818×10^{-3}
2.424×10^{-3}
$3.03 \ \times 10^{-3}$
3.636×10^{-3}
4.242×10^{-3}

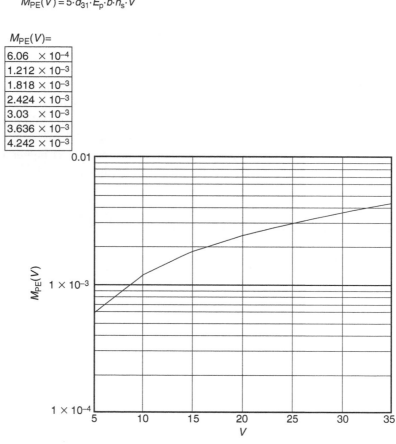

Figure 2.14   **Bending moment for an AlN beam as a function of actuation voltage.**

Sample Calculations for AlN

$$I = \left[\frac{50 \times 10^3}{12} \right]$$
$$= 4166 \ \mu m^4 \text{ (regardless of beam material)}$$

$V = 5, 10, \ldots, 35$ $I = 4166\ \mu m^4$ $d_0 = 2.5\ \mu m$ $L = 1000\ \mu m$

$E_{PP} = 0.3535 \times 10^{-2}\ N/\mu m^2$ (70 MPa) $d_{31} = 100 \times 10^{-6}$ $b = 50\ \mu m$ $h_s = 10\ \mu m$

$M_{PZT}(V) = 5 \cdot d_{31} \cdot E_{pp} \cdot b \cdot h_s \cdot V$

$M_{PZT}(V) =$

4.419×10^{-3}
8.838×10^{-3}
0.013
0.018
0.022
0.027
0.031

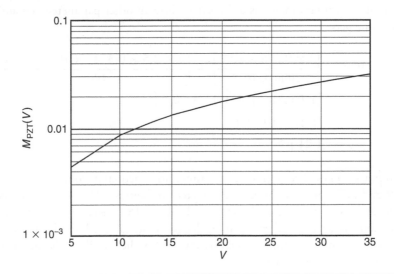

Figure 2.15   Bending moment for a PZT beam as a function of applied voltage.

$$[F_c]^{15\ V} = \left[\frac{3 \times 0.1818 \times 10^{-8}}{2 \times 1000 \times 10^{-6}}\right] + \left[\frac{3 \times 2.5 \times 1.616 \times 10^{-2} \times 4166}{1000^3}\right]$$

$$= [2.73 \times 10^{-6}] + [0.51 \times 10^{-6}]$$

$$= [3.24 \times 10^{-6}]\ N$$

$$= 3.24\ \mu N$$

Sample Calculations for PZT
Assuming a Young's modulus of 70 GPa while all other parameters remaining the same, one gets,

$$[F_c]^{15\,V} = \left[\frac{3 \times 1.32 \times 10^{-8}}{2000 \times 10^{-6}}\right] + \left[\frac{3 \times 2.5 \times 0.3535 \times 10^{-2} \times 4166}{1000^3}\right]$$

$$= [19.69 \times 10^{-6}] + [0.11 \times 10^{-6}]$$

$$= [19.80 \times 10^{-6}]N$$

$$= 19.80\ \mu N \text{ (for actuation voltage of 15 V)}$$

$$= 6.74\ \mu N \text{ (for actuation voltage of 5 V)}$$

Sample Calculations for ZnO
Assuming a Young's modulus of 70 GPa while all other parameters remaining the same, one gets

$$[F_c]^{15\,V} = \left[\frac{3 \times 0.078 \times 10^{-8}}{2000 \times 10^{-6}}\right] + \left[\frac{3 \times 2.5 \times 0.3535 \times 10^{-2} \times 4166}{1000^3}\right]$$

$$= \left[1.1 \times 10^{-6}\right] + \left[0.11 \times 10^{-6}\right]$$

$$= [1.21 \times 10^{-6}]N$$

$$= 1.21\ \mu N$$

Computed values of contact forces as a function of actuation voltage for piezoelectric AIN and PZT cantilever beams are depicted in Figures 2.16 and 2.17, respectively. Contact force in a piezoelectric actuator (Figure 2.18a) increases with an increase in actuation voltage, regardless of piezoelectric material used in the fabrication of cantilever beam. The contact force is lowest for the cantilever beam made from ZnO piezoelectric material.

2.5   Electrothermal Actuation Mechanism

This actuation mechanism is based on the expansion of the metal due to increase in temperature, which generates heat by an electric current (I) flowing through a resistor R as shown in Figure 2.18b. If this resistor is located in the upper region of the cantilever beam, a thermal wave occurs propagating in the thickness direction of the beam. This thermal wave creates a bending moment, which is used to actuate the cantilever beam. An approximate expression for the thermal bending moment for a cantilever beam can be written as

$$M_{TH} = \left[\frac{\alpha h^3 E_m I^2 R}{3KL}\right] \tag{2.12}$$

$$F_{CAIN}(V) = \frac{3 \cdot M_{PE}(V)}{2 \cdot L} - \frac{3 \cdot d_0 \cdot E_P \cdot I}{L^3}$$

$F_{CAIN}(V) =$

4.041×10^{-7}
1.313×10^{-6}
2.222×10^{-6}
3.131×10^{-6}
$4.04 \ \times 10^{-6}$
4.949×10^{-6}
5.858×10^{-6}

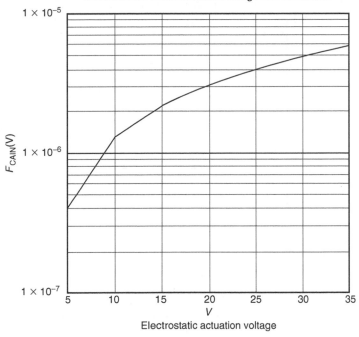

Contact force versus actuation voltage for AlN material

Figure 2.16   Contact force versus actuation voltage for an AlN beam.

where
  α is the coefficient of thermal expansion of the beam material
  h is the beam thickness (μm)
  E_m is Young's modulus of the beam material
  I is the current (ampere)
  R is the resistance (Ω)
  I^2R is the power dissipation in the resistor
  K is the thermal conductivity of the beam material (W/cm °C)
  L is the beam length (μm)

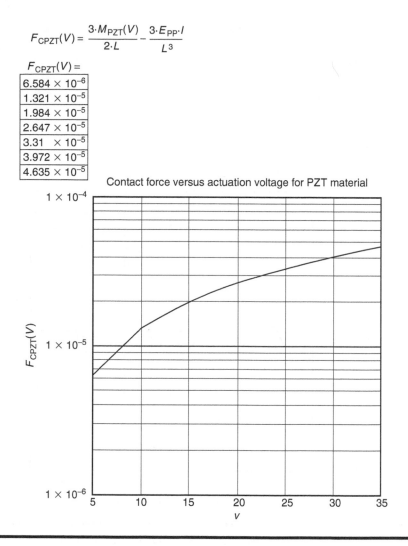

$$F_{CPZT}(V) = \frac{3 \cdot M_{PZT}(V)}{2 \cdot L} - \frac{3 \cdot E_{PP} \cdot l}{L^3}$$

$F_{CPZT}(V) =$

$F_{CPZT}(V) =$
6.584×10^{-6}
1.321×10^{-5}
1.984×10^{-5}
2.647×10^{-5}
$3.31 \ \times 10^{-5}$
3.972×10^{-5}
4.635×10^{-5}

Contact force versus actuation voltage for PZT material

Figure 2.17 Contact force (N) for a PZT beam as a function of applied voltage.

Pull-in or snap-down does not occur in electrothermal actuation mechanism as in the case of piezoelectric actuation method. Detection of motion of the beam can be achieved either through a capacitive detection method or piezoelectric detection technique, whichever is reliable and cost effective.

Major advantages of electrothermal actuation include low actuation voltage (typically from 5 to 10 V), lower construction cost of the device and simple process flow. Serious drawbacks are the slow response (typically 100–300 μs), high power

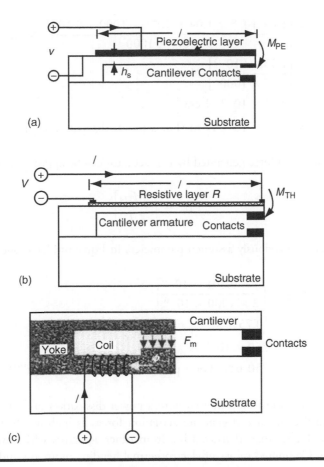

Figure 2.18 **Actuation mechanisms best suited for microrelays.**

dissipation, and continuous current drain in the actuated state. However, to avoid continuous current drain in a MEMS switch in the downstate, ES clamping technique can be integrated in the switch design, which will slightly increase device cost.

Because of extremely slow response, high power consumption, and continuous current drain, this particular actuation mechanism is used only in few thermally actuated MEMS switches, RF mechanical resonators, and tunable capacitors. Because of very limited and few applications of this actuation mechanism, electro-thermal actuation mechanism will not be discussed further, except for thermal bending moment calculations.

Assuming that nickel is the beam material and the coefficient of its thermal expansion of 13 (ppm/°C), beam thickness of 10 μm, Young's modulus of 206 GPa (or 30×10^6 psi), power dissipation (I^2R) in resistor 100 mW (0.1 W), thermal conductivity of 0.9 W/cm °C, and beam length (L) of 1000 μm and inserting these parameters in Equation 2.12, one gets

$$M_{TH} = \left[\frac{13 \times 10^{-6} \times 1.04 \times 10^{-2} \times 1000 \times 0.100}{3 \times 0.9 \times 1000} \right] \text{N cm}$$

$$= \left[\frac{1.352 \times 10^{-5}}{2.7 \times 1000} \right] \text{N cm}$$

$$= 0.501 \times 10^{-8} \text{ N cm}$$

$$= 0.501 \times 10^{-6} \text{ N m (For a power dissipation of 100 mW).}$$

The electrothermal force generated by this actuator can be approximately written as

$$F_{TH} = \left[\frac{3M_{TH}}{2 \, L} \right] \tag{2.13}$$

Inserting all the previously assumed parameters in Equation 2.13, one gets

$$F_{TH} = \left[\frac{3 \times 0.501 \times 10^{-6}}{2 \times 1000 \times 10^{-6}} \right]$$

$$= \left[0.75 \times 10^{-3} \right] = \left[75 \times 10^{-5} \right]$$

$$= 750 \times 10^{-6} \text{ N}$$

$$= 750 \, \mu\text{N (For a power dissipation of 100 mW).}$$

The above calculations indicate that with a power dissipation of even 10 mW, the electrothermal actuator can generate actuation forces as high as 75 μN, which is significantly higher than that available from other actuators with similar physical dimensions. Computed values of electrothermal bending moments and electrothermal force as a function of power dissipation (I^2R), beam thickness (h), and beam length (L) are summarized in Table 2.10.

Few MEMS devices using electrothermal actuation mechanism will be briefly discussed. It is critical to mention that this actuation method deploys a stress controlled membrane with pattern metallic contacts [6]. This actuation mechanism permits operation at lower voltages without sacrificing the spring constant (k). The membrane structure allows fabrication of resistive RF-MEMS switches operating at MM-wave frequencies. Lower operating voltages offer significant reduction in the spring constant requirements. Thermal actuation can be a good alternative, where lower spring constant and lower operating voltages are the principal design requirements. As stated earlier that an electrothermal actuation by definition is temperature sensitive operation, thermally triggered operations are insensitive to heating below a defined temperature and they usually snap-down above this temperature. MEMS switches using thermal actuators can operate at as low as 5 V. However, the turn-off time is about 50 μs, while the turn-on time is close to 300 μs. The "ON" state requires going through complete heating process below a defined temperature. In case

Table 2.10 Thermoelectric Bending Moment (μNm) and Actuation Force Generation (μN) as a Function of Beam Thickness (*h*), Beam Length (*L*), and Power Dissipation

I^2R (W)	$h = 10\ \mu m$, $L = 1000\ \mu m$	$h = 5\ \mu m$, $L = 1000\ \mu m$	$h = 2.5\ \mu m$, $L = 1000\ \mu m$	F_{TH} (μN)
1.0	5	0.625	0.0780	7500
0.8	4	0.500	0.0625	6000
0.6	3	0.375	0.0468	4500
0.4	2	0.250	0.0312	3000
0.2	1	0.125	0.0156	1500
0.1	0.5	0.062	0.0078	75

of thermally actuated RF-MEMS switch, typical isolation is better than 25 dB and the insertion loss is less than 2 dB up to 40 GHz. It is important to point out that standard thermal actuation offers linear displacement.

In case where it is required to generate large rectilinear displacements and substantial restoring forces, bent-beam electrothermal actuators are best suited to meet these requirements. The simplest manifestation [7] of bent-beam electrothermal actuator is a V-shaped structure. The displacement of the apex is dependent on the beam dimensions. The maximum displacement, which can be obtained in the absence of external force, is calculated from the beam bending equation. In a bent-beam actuator, beam length can vary from 500 to 2000 μm, both the beam width and thickness from 4 to 6 μm, and bending angle from 3° to 12°. Preliminary studies performed by the author reveal that bent-beam actuators could offer suitable compromises on displacements, actuation forces, power consumption, and actuation voltage and drive options best suited for several MEMS applications such as rotary and rectilinear microengines. Specific details on device structure, material requirements, peak-operating temperature, maximum loading force, maximum displacement, and fabrication issues will be discussed in Chapter 3.

2.6 Electromagnetic Actuation Mechanism

The electromagnetic actuator is also known as the variable reluctance actuator. The electromagnetic actuator shown in Figure 2.18c consists of a movable beam, a yoke, and a driving coil comprising of excitation winding of *N* turns. The magnetomotive force (MMF) generated in this coil by this current is equal to product of number of turns in the coil and current flowing in it (MMF = *NI*). When a current *I* flows in the coil, a magnetic flux is set up in the yoke and the air gap. The yoke and a part of the beam are made from high permeable material and they are separated by an air gap (d). The magnetic flux flows through the yoke and air gap as illustrated in Figure 2.18. The flux passing through magnetic circuit is approximately equal to the reluctance (R_{gap}), which can be written as

$$R_{\text{gap}} = \left[\frac{d}{A\mu_0}\right] \tag{2.14}$$

where
  d is the air gap
  A is the area of the yoke close to air gap
  μ_0 is the permeability of the free space, which is equal to 1.257×10^{-6} H/m

2.6.1 Pull-In and Pull-Out Magnetomotive Forces

Using the analogy with the ES force, the MMF or NI now can be replaced by the electromotive force (EMF) or V. But, due to nonlinearity dependence of EMF or F_{em} in the air gap region (d), instability can occur when the MMF exceeds the pull-in MMF or NI_{PI}. After the pull-in has occurred, the beam releases at much lower MMF or at much lower current (I). Because the electromagnetic actuator or the electromagnetic actuator displays a hysteresis phenomenon, the equations for the pull-in MMF and pull-out MMF can be written as

$$NI_{\text{PI}} = 0.544 \left[\frac{kd^3}{\mu_0 A}\right]^{0.5} \tag{2.15a}$$

$$NI_{\text{PO}} = \left[\frac{2\,k(d - d_{\text{R}})d_{\text{R}}^2}{\mu_0 A}\right]^{0.5} \tag{2.15b}$$

where
  d is called the zero-flux gap spacing
  k is the effective spring constant of the beam (N/m)
  A is the yoke area
  d_{R} is the rest-gap after closing of the contacts shown in Figure 2.18c

  The rest-gap typically varies from 8 to 10 percent of the air gap. It is interesting to note that these two equations are analogous to pull-in and pull-out voltages mentioned under ES actuation mechanism.

Sample Calculations
Assuming a spring constant of 10 N/m, air gap of 2.5 μm, rest-gap of 0.2 μm, number of turns in the winding of 100, and yoke area of 100×100 μm^2, one gets

$$NI_{PI} = 0.544 \left[\frac{10 \times 2.5^3 \times 10^{-6}}{1.257 \times 10^{-6} \times 100 \times 100} \right]^{0.5}$$

$$= 0.544 \left[\frac{12.43 \times 10}{10^4} \right]^{0.5}$$

$$= 0.544 \left[\frac{124.3}{10,000} \right]^{0.5}$$

$$= [0.544 \times 0.11144]$$

$$= [0.0606] NI_{PL} \text{ Turn-A (Turn-ampere)}$$

Therefore, the pull-in current $I_{PI} = \left[\dfrac{0.0606}{N} \right]$

$$I_{PI} = \left[\frac{0.0606}{100} \right]$$

$$I_{PI} = [0.000606 \text{ or } 0.061 \text{ mA}]$$

Inserting the assumed parameters in Equation 2.15b, one gets

$$NI_{PO} = \left[\frac{2 \times 10 \times (2.5 - 0.2) \times 10^{-6} \times 0.2^2}{1.257 \times 10^{-6} \times 100 \times 100} \right]^{0.5}$$

$$= \left[\frac{2 \times 2.3 \times 10 \times 0.04}{1.257 \times 10,000} \right]^{0.5}$$

$$= \left[1.4638 \times 10^{-4} \right]^{0.5}$$

$$= \left[\frac{1.2026}{100} \right]$$

$$NI_{PO} = [0.01203]$$

This gives $I_{PO} = \left[\dfrac{0.01203}{100} \right] = 0.00012 \text{ A}$

$$= 0.12 \text{ mA}$$

These calculations indicate that the pull-in current is much higher than the pull-out current, which is similar to that the pull-in voltage is much higher than pull-out voltage under ES actuation method.

2.6.2 Actuation Force due to Induced Magnetic Force

The actuation force generated by the magnetic induced force in the air gap can be expressed as

$$F_{m} = \left[\frac{0.5 \mu_{0} A(NI)^{2}}{d^{2}} \right] \qquad (2.16)$$

It is important to mention that the magnetic force F_{m} will not increase indefinitely with the increasing current flowing in the coil, but will be limited by the flux saturation in the yoke material, which can be written as

$$\Phi_{sat} = [B_{sat} \times A_{s}] \qquad (2.17)$$

where B_{sat} is the magnetic induction density of the yoke material and A_{s} is the smallest cross section of the yoke. Ferromagnetic materials with high saturation flux density or magnetic induction density will provide maximum actuation forces. Ferromagnetic material "permalloy" offers maximum saturation flux density ranging from 0.5 to 1.0 T (Tesla is the unit for magnetic flux density in the MKS system and is equal to 10,000 G in CGS system), leading to maximum magnetic forces in the range of 1–4 mN for a electromagnetic actuator with an area of $100 \times 100 \ \mu m^{2}$, pull-in current levels in the range of 4–8 mA, and air gap of 1 μm and number of turns of 100. Computed values of electromagnetic forces are as follows.

Sample Calculations
Assuming an air gap of 2.5 μm, yoke area of 10,000 μm^{2}, and the NI product of 0.0606 and free-space permeability of 1.257×10^{-6} H/m and inserting these parameters in Equation 2.16, one gets magnetic force as

$$F_{m} = \left[\frac{0.5 \times 1.257 \times 10^{-6} \times 10,000 \times (0.0606)^{2}}{2.5^{2}} \right]$$

$$= \left[\frac{0.5 \times 1.257 \times 10^{-2} \times 36.72 \times 10^{-4}}{6.25} \right]$$

$$= [3.67 \times 10^{-6}]$$

$$= 3.67 \ \mu N$$

Now assuming air gap of 1 μm, a coil of 100 turns, a current of 4 mA, and all other parameters remaining the same, one gets the magnetic force as

$$F_{m} = \left[\frac{0.5 \times 1.257 \times 10^{-6} \times 10,000 \times (100 \times 0.004)^{2}}{1^{2}} \right]$$

$$= \left[\frac{0.6285 \times 0.01 \times 0.16}{1} \right]$$

$$= [0.1 \times 0.01]$$

$$= \frac{1}{1000} \ \text{or 1 mN}$$

Now assuming a coil current of 8 mA and all other parameters remaining the same, one gets the magnetic force as

$$
\begin{aligned}
F_m &= \left[\frac{0.5 \times 1.257 \times 10^{-6} \times 10,000 \times (100 \times 0.008)^2}{1} \right] \\
&= [0.6285 \times 10^{-2} \times 0.64] \\
&= [0.4 \times 0.01] \\
&= \frac{4}{1000} \text{ or } 4 \text{ mN}
\end{aligned}
$$

2.6.2.1 Parametric Trade-Off Computations

Parametric trade-off computations have been performed to obtain pull-in and pull-out currents and magnetic force, assuming various variables, namely, spring constant (k), number of turns (N), and air gap (d). Computations are summarized in Tables 2.11 and 2.12.

These calculations indicate that when pull-in current level is about 0.06 mA and number of turns in the coil is 100, low pull-in magnetic force levels ranging from 3.68 to 0.73 µN are possible. When the pull-in current levels in the coil are in the range from 6 to 8 mA, maximum magnetic forces ranging from 1 to 4 mN can be achieved under the same values of yoke area, spring constant, and air gap. These calculations further indicate that maximum force levels are achievable for acceptable current levels in the milliampere range. These magnetic force levels are in three orders of magnitude higher than those possible under ES actuation under actuation voltage levels well below 15 V. Parametric plots of magnetic forces as a

Table 2.11   Pull-In and Pull-Out Currents (mA) and Magnetic Forces (µN) as a Function of Spring Constant k (N/m) and Coil Turns (N)

	Pull-In Current (N/m)		*Pull-Out Current (N/m)*	
Turns (N)	*k = 10*	*k = 5*	*k = 10*	*k = 5*
50	1.212	0.862	0.240	0.170
100	0.606	0.431	0.121	0.085
150	0.485	0.286	0.081	0.057
200	0.303	0.215	0.060	0.042
Magnetic force (µN)	3.68	1.84	0.148	0.074

Table 2.12 *NI*-Product and Magnetic Force (F_m) as a Function of Air Gap (d) and Spring Constant (k)

Current-Turn Product	$k = 10$ (N/m)		$k = 5$ (N/m)	
Air Gap, d (μm)	2.5	1.0	2.5	1.0
NI	0.0606	0.0153	0.0429	0.0108
NI^2	0.00368	0.00023	0.00018	0.00012
F_m (μN)	3.68	1.48	0.184	0.073

Note: Parameters assumed: $A = 100 \times 100\ \mu m^2$ and $\mu_0 = 1.257 \times 10^{-6}$ H/m.

function of current flowing in a 100-turn coil on a yoke area of 10,000 μm^2 are shown in Figure 2.19a and b for air gap of 2.5 and 1 μm, respectively. These plots provide instant estimation of electromagnetic actuation forces as function of air gap and current flowing in the 100-turn coil.

2.7 Electrodynamic Actuation Mechanism

Electrodynamic actuation involves the Lorentz force acting on a conducting wire carrying current, which is placed in the external magnetic field environment. The operating principle of electrodynamic actuation is illustrated in Figure 2.20 showing the configuration of a cantilever-type actuator. The expression for the Lorentz force (F_L) acting on the current (I) can be written as

$$F_L = [I \times L \times B] \tag{2.18}$$

where I is the current flowing in the wire of length L, which is placed in the external magnetic field B. In the above equation the magnetic field or flux density is expressed by Tesla in MKS system, where 1 T is equal to 10,000 G in CGS system or equal to weber per square meter in MKS system. Computed values of Lorentz forces as a function of wire length, current flowing in the wire, and magnetic flux density are summarized in Table 2.13.

These calculations indicate that the magnitude of the Lorentz force increases with the increase in wire length and in the current flowing for a given magnetic field or flux density. To achieve Lorentz forces or actuation forces in the order of 100 μN needed for Au–Au contacts, one can select longer wire and minimum magnetic flux density or higher current. But higher current is not a good option under minimum power requirement. Plots of Lorentz forces or electrodynamic actuation forces as a function of current level in a 100 μm long wire and as a function of magnetic

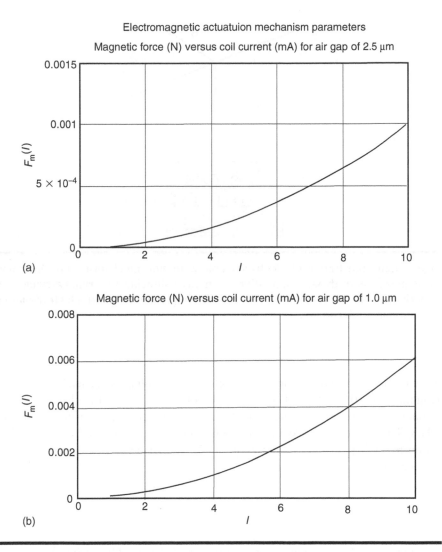

(a)

(b)

Figure 2.19 **Magnetic forces as a function current flowing in a coil having 100 turns.**

flux density are shown in Figure 2.21a and b, respectively. These plots will provide several options for an actuator designer to select the most cost-effective beam design.

It is important to mention that sensing of the cantilever beam motion is dependent on the change of magnetic flux linkage in the loop when the beam moves, which produces the induction voltage, thereby providing the detection signal. The actuation is strictly dependent on the current drive in the coil or wire, regardless whether an electromagnetic actuation or electrodynamic actuation

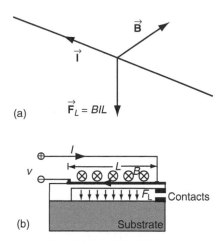

(a)

(b)

Figure 2.20 **Illustration of electrodynamic actuation mechanism. (a) Operating principle of electrodynamic actuation mechanism showing a current carrying wire experiencing the Lorentz force F_L and (b) implementation of electrodynamic actuation in a microrelay.**

mechanism is selected. In each case, the operating voltage is generally quite low ranging from 10 to 15 V. The low-voltage operation makes the electromagnetic actuation mechanism most attractive for many MEMS devices, including micro-relays. The major drawbacks include continuous current drain and complex fabri-cation technology involving integration of coil and ferromagnetic materials, which are normally temperature sensitive.

Table 2.13 **Lorentz Forces as a Function of Various Operational Parameters (μN)**

Magnetic Field (B) (T/G)	100 μm Long Wire				200 μm Long Wire			
Current (mA)	1	10	100	1000	1	10	100	1000
0.5/5,000	0.05	0.50	5	50	0.10	1	10	100
1.0/10,000	0.10	1	10	100	0.20	2	20	200
1.5/15,000	0.15	1.5	15	150	0.30	3	30	300
2.0/20,000	0.20	2	20	200	0.40	4	40	400

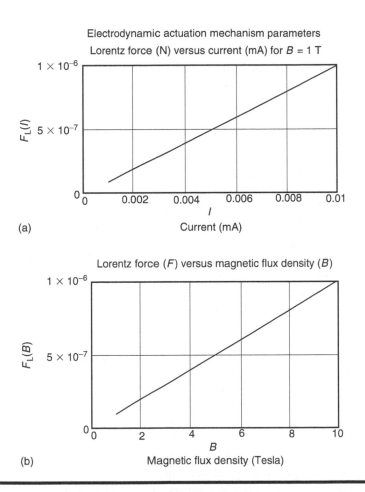

Figure 2.21 Magnitude of the Lorentz force (N) acting on a 100 μm long wire as a function of (a) current and (b) magnetic flux density.

2.8 Electrochemical Actuation Mechanism

Research studies performed by the nanotechnology (NT) scientists at the University of Cincinnati and University of North Carolina indicate that carbon nanotube (CNT) arrays can play a key role in the design of electrochemical actuators [8]. The studies further indicate that CNT arrays are considered "smart" materials because CNT have demonstrated high mechanical strength and electrical conductivity, in addition to unique piezoresistive and electrochemical sensing and actuation capabilities. Potential applications of nanotubes have spurred intensive research activities in the area of processing, device fabrication, and material characterization. Vertically aligned arrays of CNT have demonstrated well-defined properties with uniform length and diameter dimensional parameters. Thermally driven chemical vapor deposition (CVD) is best suited to grow high-density arrays on silicon substrates. Note the growth mechanism affects the length of the CNTs and is strictly dependent on the interaction between the nanoparticle and the supporting material.

2.8.1 Classification and Major Benefits of CNT

CNT can be classified into two distinct categories, namely, single-wall carbon nanotube (SWCNT) and multiwall carbon nanotube (MWCNT). Regardless of CNT type, the most effective method to grow CNTs requires the use of intermediate buffer layer of alumina (Al_2O_3) to prevent diffusion from the substrate. The efficient CVD synthesis of SWCNT must focus on the activity of the catalyst used to maintain quality control in the development of CNT. The first actuator was made from the nanotube sheets. This actuator produces strain due to the change in physical dimensions of the nanotube in the covalently bonded direction caused by the application of electric potential. The CNT actuation performance uses vertically aligned arrays of CNTs of well-defined properties with uniform length and diameter.

2.8.2 MWCNT Arrays and Electrochemical Actuator Performance

MWCNT-patterned arrays can be made on a silicon wafer. The most effective way to grow long high-density MWCNT arrays is to use an intermediate buffer layer of Al_2O_3, which prevents diffusion from the substrate. A chemically inert layer of silicon oxide permits longer MWCNT arrays with high purity. Keeping the posts straight poses a big problem regardless of their shapes.

The electrochemical actuator performance is dependent on the nanotube tower-shaped electrodes. The fabrication steps needed to construct these electrodes must be well defined to meet the response and stability requirements of the electrochemical actuator. Test and evaluation of the actuator requires a nanotube tower, which can be fixed on a glass substrate. The nanotube test tower must provide a Ag/AgCl reference electrode and a platinum-based counter or auxiliary electrode. The performance of the actuator can be evaluated in terms of test tower displacement and applied voltage. A vertically aligned CNT array has demonstrated promise for a low detection limit and fast response for chemical, environmental, and biosensors.

2.8.3 Fabrication and Material Requirements for the Actuator

Test tower requires the use of structural materials with high values of Young's modulus, which can offer wide displacements of the electrochemical actuators needed for fast actuator response. The overall performance of the actuator is contingent upon the strain uniformity along the length of the beam and the actuator outer surface of the nanotubes. If all the shells of the MWCNT array could be made to actuate at the same time, the force generated by the electrochemical actuator will be maximum. Furthermore, higher the applied voltage, higher will be the increase in strain. Higher actuation voltage increases the chemical charge accumulation on the electrical interface, which will lead to a faster response. Any generation of bubbles on the nanotube surface will reduce both the actuator lifetime and reliability. Improved mechanical properties of the structural materials and rapid strain generation of the nanotube actuator are key to the reliability and efficiency of the electrochemical actuator. Compared to other actuation mechanisms, the low driving voltage of the CNT tower actuator is a major advantage for various MEMS applications such as smart structures, active catheters for medical applications, artificial muscles, MPs, and molecular motors, power harvesting systems, strain sensors, and nanorobots [7]. Nanotube arrays made from high tensile strength materials can generate large forces during actuation. In summary, highly aligned MWCNT arrays offer excellent mechanical properties, high electrochemical sensitivity, high strain-generating capability, and improved electrically conductive probes best suited for many applications ranging from nanomedicine to space exploration. Performance comparison and unique capabilities of various actuation mechanisms are summarized in Table 2.14.

It is important to mention that highly aligned MWCNT arrays have been utilized in the development of smart materials, which play a key role in the design

Table 2.14 Performance Comparison Data of Various Actuation Mechanisms

Actuation	Actuation Drive Voltage (V)	Output Force (μN)	Response (μs)	Reliability	MEMS Application
Electrostatic (ES)	>30	1–4	10–50	Pull-in	RF switches
Piezoelectric	5–15	3–15	50–100	No pull-in	Resonators
Electrothermal	<12	100–10,000	100	No pull-in	Micropumps (MPs)
Electromagnetic	<10	2–5	100–200	Hysteresis problem	Microrelay
Electrodynamic	<10	>100,000	>1,000	Highly stable	Microrelay
Electrochemical	<12	0.1–2	<10	No data	Biosensors

and development of biosensors, electrochemical actuators, and NT-based probes for scientific research. CNT arrays are considered as smart materials because they have high mechanical strength, enhanced electrical conductivity, and improved piezo-resistive and electrochemical sensing and actuation properties. The small size as well as the unique properties of nanotubes are best suited for many applications such as environmental scanning electron microscopy, energy dispersive microscopy, mass spectroscopy, high resolution imaging sensors, smart sensors including chemical, environmental and biosensors, electrochemical actuators and NT-based sensors for weapon health, and harsh battlefield environmental monitoring applications. These NT-based sensors can identify the out-of-specification weapons including missiles and smart bombs, predict remaining useful shelf life, improve the reliability and readiness of the weaponry stockpile, and perform onboard readiness tests for critical elements of the offensive and defensive weapon systems. Note the NT-based sensors are best suited for harsh environmental conditions in military and commercial locations. These sensors can play a key role in sensing and monitoring of nuclear radiation and the outgassing of weapon propellant chemical/biological agents and toxic gases that can be potentially encountered in both military and civilian environments. Remotely located smart sensors can alert soldiers of harmful chemical and biological agents and explosives in the battlefield and its vicinity areas.

The biosensor can be formed by casting suitable epoxy into a nanotube array and polishing the ends of the CNTs. The nanotube electrodes can be used to design a label-free immunosensor based on an electrochemical impedance spectroscope, and the sensor has good sensitivity. Nanoprobes are widely used for biosensors and electrophysiology applications. The electrochemical actuators are very sensitive and require very low actuation or drive voltage.

2.9   Summary

Potential actuation mechanisms are summarized with emphasis on actuator response, force-generating capability, reliability, drive requirements, and suitability for specific applications. Performance capabilities and limitations of six distinct actuation mechanisms such as ES actuation, piezoelectric actuation, electrothermal actuation, electromagnetic actuation, electrodynamic actuation, and electrochemical actuation are briefly discussed with emphasis cost and complexity. NT-based electrochemical actuators require actuation voltage well below 5 V, whereas the other actuation mechanisms require actuation voltages close to 10 V or above. The mechanical displacement provided by the NT-based electrochemical actuator is dependent on the air gap, spring constant of the beam or membrane, and the drive or actuation voltage. Because the NT-based electrochemical actuator requires drive or actuation voltage between 2 and 4 V, its design is fully compatible with the monolithic microwave integrated circuit (MMIC) and microelectronic technology.

References

1. H.J. De Los Santos, et al., RF MEMS for ubiquitous wireless connectivity, *IEEE Microwave Magazine*, December 2004, 40.
2. J. Oberhammer and G. Stemme, Design and fabrication aspects of a S-shaped film actuator-based DC to RF MEMS switch, *IEEE Journal of MEMS*, 13(3), June 2004, 426.
3. H.J. De Los Santos, et al., RF MEMS for ubiquitous wireless connectivity, *IEEE Microwave Magazine*, December 2004, 39.
4. A. Selvakumar and K. Najafi, Vertical comb array microactuator, *IEEE Journal of MEMS*, 12(4), August 2003, 440–448.
5. S. Jung and J.U. Jeon, Optimal shape design of a rotary microactuator, *Journal of MEMS*, 10(3), September 2001, 460–468.
6. Z. Fent et al., Design and modeling of MEMS tunable capacitors using electro-thermal actuators, *IEEE MTT-S, Digest*, 2000, 1507–1510.
7. L. Que, Bent-beam Electro-thermal actuator—Part-I: single beam and cascaded devices, *Journal of MEMS*, 10(2), June 2001, 252.
8. Y.H. Yun, A. Bange et al., Carbon nanotube array smart materials, *Proceedings of SPIE*, 2006 (6172).

Chapter 3

Latest and Unique Methods for Actuation

3.1 Introduction

This chapter focuses on the latest versions of actuation mechanisms capable of providing higher actuating forces to provide displacements in rectangular dimensions. Performance capabilities of the unique actuation mechanisms described in this chapter are not readily available from the actuating mechanisms discussed in Chapter 2. Electrostatic (ES) rotary microactuator, bent-beam electrothermal (BBET) microactuator, vertical comb array microactuator (VCAM), and electrochemical actuator are described with major emphasis on unique performance capabilities, design configurations, operational benefits, design simplicity, fabrication aspects, and reliability. Optimum design configurations of electrodes capable of providing uniform and reliable actuation force for the actuators are identified with particular emphasis on improved force-generating capability and higher tracking accuracy over wide bandwidths.

Potential actuator configurations capable of generating higher actuation forces are discussed. Preliminary studies performed by the author on various actuators indicate that the widely used parallel-plate (PP) configuration may generate an actuation force by an order of magnitude greater than that of the interdigitated configuration, in addition to minimum cost and complexity. This leads to a conclusion that a PP actuator configuration is best suited for applications demanding a large output force over small displacements as in the case of hard-disk drives (HDDs).

It is important to distinguish between certain terms used in the design and fabrication of actuators. For example, the most common term "clearance" is used to denote the separation or physical distance between two facing electrodes, whereas the term "capacitance gap size" refers to a normal between two facing electrodes. Attempts will be made to provide precise definition or meaning of the technical terms or parameters for easy understanding or rapid comprehension by the readers. Potential benefits and performance improvements of the latest microactuators are described with emphasis on output force-generating capability, displacement, and reliability and operating bandwidth.

3.2 Electrostatic Rotary Microactuator with Improved Shaped Design

The electrostatic rotary microactuator (ESRM) is considered as one of the most promising candidates for the dual-stage servomechanisms as shown in Figure 3.1 [1] because of simplicity of design and lower fabrication cost. A rotary microactuator consists of two parts, namely, a rotor and a stator as illustrated in Figure 3.2, which generate the actuating force. In brief, the capacitive PPs integrated to the rotor and stator elements generate the actuation force to drive the rotor.

The beam springs impose the restoring force opposing to the actuating force, thereby limiting the displacement of the rotor. The direction of the motion is perpendicular to the longitudinal direction of the PP.

The ESRM is best suited for dual-stage servomechanisms (Figure 3.1) requiring wider operational bandwidth and high-tracking accuracy better than 25,000 track-per-inch (TPI). Several versions of microactuators for the dual-stage servomechanism [1] were recommended, but the ES design configuration of the microactuator was

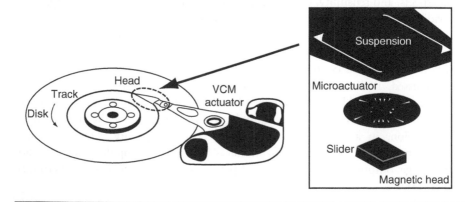

Figure 3.1 Critical components of a dual-stage servomechanism integrated with an ES actuation technique.

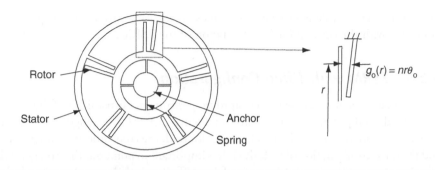

Figure 3.2 Schematic diagram of a conventional ESRM showing (a) elements of the rotary microactuator and (b) specific details on clearance variables.

preferred because of trouble-free and cost-effective fabrication procedures. In addition, several configurations of electrodes were evaluated to enhance the force-generating capability and tracking accuracy over wide bandwidths. Two electrode configurations, namely, PP and interdigitated configurations are available for implementation in the design and development of ESRMs. The principal difference between these two types of electrodes lies in the direction of motion of the moving electrode known as the rotor electrode. Critical physical parameters of a conventional ESRM and the clearance variation are shown in Figure 3.2 [1].

3.2.1 Performance Limitation of Conventional Parallel-Plate Electrodes

As far as the force-generating capability is concerned, the PP electrode configuration is capable of generating actuation force by an order of magnitude greater than the interdigitated electrode configuration. In addition, the PP electrode configuration is most desirable for applications demanding larger force over small displacement. However, in the case of a rotary microactuator using PP electrode configuration, the capacitive gap between the electrodes must increase proportionally to the radius from the center of the rotation, which requires large clearance at the far out region for the center of the microactuator assembly. Note the angular motion requires that the clearance increases with the radius. Therefore, the requirement of the gap size accommodating the angular motion puts limitation on the force-generating capability of a PP electrode configuration. Design techniques are available to overcome this limitation of the conventional PP rotary configuration (Figure 3.2). Most cost-effective techniques involve optimum shapes for the capacitive electrodes, which will allow the gap size to be evenly obtained at the minimum gap size compatible with the current fabrication technology and will significantly enhance the force-generating capability. In addition, a technique of integrating the optimum tilt angle into the rotary microactuator design will be found most attractive to overcome the problems associated with the conventional configuration of the ESRM.

3.2.2 ESRM with Tilted Configuration

This particular section deals with the improved performance capabilities of the ESRM with tilted configuration as illustrated in Figure 3.3 [1]. This figure shows the dimensions and location of the PP electrodes for both the conventional configuration and the tilted configuration of an ESRM. It is important to point out that rotary mode of microactuator is preferred because of its stiffness needed to reduce the device sensitivity to a lateral external shock, which is not possible with a PP configuration. As stated earlier, to integrate a conventional PP actuator concept into a rotary micro-actuator design, the capacitive gap between the facing electrodes must increase in proportion with radius from the center of the rotation, which demands the increase in clearance with increasing radius. A rotary microactuator with optimum shapes of electrodes will overcome the problems associated with a conventional ESRM, in addition to higher force generation capability over wide bandwidths.

3.2.3 Requirements for Optimum Shaped Electrodes

As mentioned previously, tilted configuration of conventional PP microactuator is capable of generating large output force compared to nontilted configuration of microactuator over wide bandwidths, which is a principal requirement for a dual-stage servomechanism. Optimum shaped electrodes for a tilted rotary microactuator

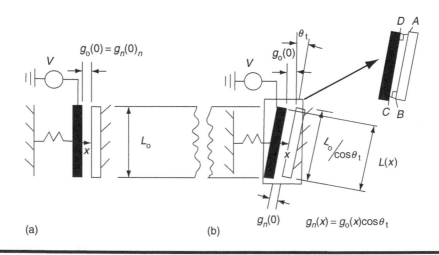

Figure 3.3 ESRM architecture showing (a) conventional actuator configuration and (b) tilted configuration, integrated with PP electrodes. (From Jung, S., Jeon, J.U., *J. MEMS*, 10, 460, 2001. With permission.)

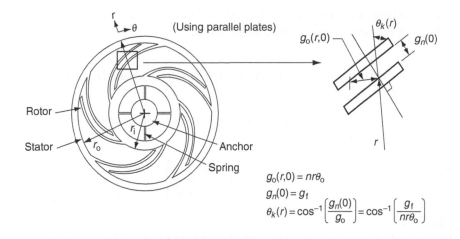

$$g_o(r,0) = nr\theta_o$$
$$g_n(0) = g_f$$
$$\theta_k(r) = \cos^{-1}\left[\frac{g_n(0)}{g_o}\right] = \cos^{-1}\left[\frac{g_f}{nr\theta_o}\right]$$

Figure 3.4 Exploded view of the tilted configuration of the microactuator integrated with optimum shaped electrodes and associated design parameters.

are illustrated in Figure 3.4 [1]. Critical elements of a conventional ESRM such as the rotor, stator, anchor, supporting structure, and spring are shown in Figure 3.2.

The beam springs shown in Figure 3.2 provide the opposing force to the actuation force generated by the applied voltage to the actuator mechanism. The direction of the rotor motion in the PP actuator configuration must be perpendicular to the longitudinal direction of the PPs, AB and CD, as illustrated in Figure 3.3. The gap is set to an appropriate value to ensure stability of the microactuator operation under various operating conditions. In general, the force-generating capability is dependent on the drive voltage, but in actual practice it can be limited by the gap size which is equal to the required clearance permissible by current fabrication technology. The gap size refers to the normal distance between the two electrodes as shown in Figure 3.3.

The optimum shape selected for the actuator electrodes must be free from kinematic constraints of the angular motion, which provides significantly increased actuation force over the electrodes used in the conventional rotary microactuator. It is important to mention that the ability of the gap size to accommodate an angular motion limits the output force of the rotary microactuator with PP configuration using unshaped electrodes. Capacitive electrodes with optimum shapes must be designed to allow uniform gap size, compatible with minimum gap size permissible with current fabrication technology.

3.2.4 Force Generation Computations of Rotary Actuator with Conventional and Tilted Configurations

Force-generating capability expressions for the conventional configuration and the tilted configuration of the ESRM will be derived involving various variables.

These mathematical expressions [1] clearly demonstrate the force-generating capabilities of the two microactuator configurations.

3.2.4.1 Actuation Force Computation for Conventional Configuration

The actuation force generated by the conventional configuration of the rotary microactuator can be expressed as

$$F_{\text{conv}}(X) = \left\{ \frac{e_o h L_o V^2}{2[g_o(x)]^2} \right\} \tag{3.1}$$

where

e_o is the free-space permittivity (8.85×10^{-12} F/m)
h is the "out-of-the plane" thickness of the electrodes
L_o is the overlapped length of the electrode
$g_o(x)$ is the clearance in the operational direction as shown in Figure 3.3
x is the rotor displacement in the operation direction

Note in the conventional configuration, the clearance $g_o(x)$ in the operational direction is equal to the capacitive gap or the normal distance between the two facing electrodes. Figure 3.5a shows the computed values of actuation force as a function of initial gap from a conventional rotary actuator with beam length of 100 μm electrode thickness of 10 and actuation voltage of 10 V, while Figure 3.5b shows the actuation force with beam length of 1000 μm and other dimensions remain the same. Figure 3.6a and b demonstrates the force-generating capability as a function of initial gap of the conventional configuration of the rotary actuator at an actuation voltage of 15 and 20 V, respectively, while other parameters remain the same.

3.2.4.2 Force Generation Computation for Tilted Configuration

The force generation expression for a tilted configuration of the rotary microactuator can be written as

$$F_{\text{tilted}}(x) = \frac{[e_o h L(x) V^2]}{\left\{ 2[g_o(x)\cos\theta_t]^2 \right\}} \tag{3.2}$$

where $L(x)$ is the overlapped length of the electrode, which varies with relative displacement of the rotor, and θ_t is the tilt angle as shown in Figure 3.3 and

$$g_o(x) = \left[\frac{g_n(x)}{\cos\theta_t} \right] \tag{3.2a}$$

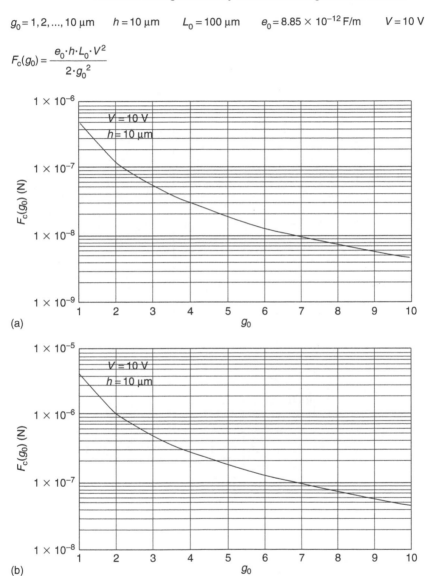

Actuation force generated by coventional configuration of actuator

$g_0 = 1, 2, ..., 10 \ \mu m$ $h = 10 \ \mu m$ $L_0 = 100 \ \mu m$ $e_0 = 8.85 \times 10^{-12} \ F/m$ $V = 10 \ V$

$$F_c(g_0) = \frac{e_0 \cdot h \cdot L_0 \cdot V^2}{2 \cdot g_0^2}$$

Figure 3.5 Actuation force generated by conventional rotary actuator (a) for a beam length of 100 μm and (b) for a beam length of 1000 μm.

$$g_n(x) = [g_0(0) \cos \theta_t - x \cos \theta_t] \qquad (3.2b)$$

$$L(x) = [L_0/\cos \theta_t - g_0(0) \sin \theta_t + X \sin \theta_t] \qquad (3.2c)$$

Actuation force generated by coventional configuration of actuator

$g_0 = 1, 2, ..., 10\ \mu m$ $h = 10\ \mu m$ $L_0 = 1000\ \mu m$ $e_0 = 8.85 \times 10^{-12}\ F/m$ $V = 15\ V$

$$F_c(g_0) = \frac{e_0 \cdot h \cdot L_0 \cdot V^2}{2 \cdot g_0^2}$$

Actuator generation force (N) versus air gap (μm)

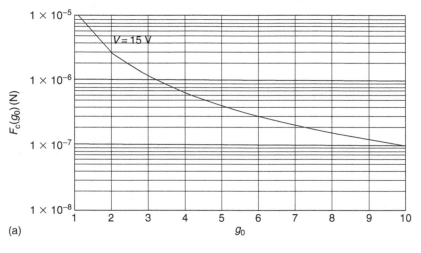

(a)

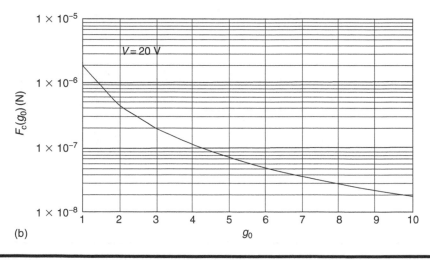

(b)

Figure 3.6   Force generated by the rotary actuator as a function of gap (g_o) for (a) beam length of 1000 μm and (b) beam length of 100 μm.

Computed values of actuation force by the tilted configuration of the rotary microactuator as a function of initial gap size are shown in Figure 3.7a through d at a tilt angle of 20°, 40°, 60°, 80°, respectively. The computational data is obtained at an actuation of 5 V. Actuation force computations at 20° and 80° tilt angles with actuation voltage of 10 V are shown in Figure 3.8a and b, respectively. Figure 3.9 illustrates the improvement in the actuation forces for the tilted rotary configuration as a function of tilt angle over the conventional configuration. These computations show significant improvement in the force generation capability

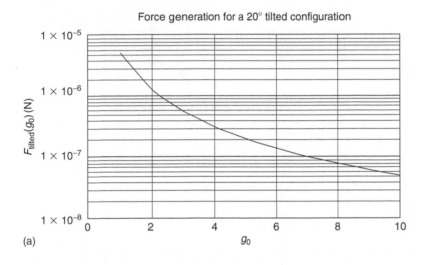

(a)

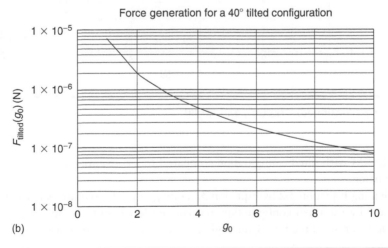

(b)

Figure 3.7 Force generated by the tilted rotary actuator as a function of air gap (μm) for a tilt angle of (a) 20°, (b) 40°,

(*continued*)

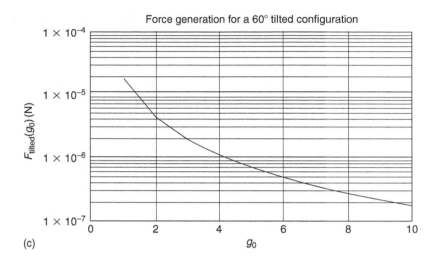

(c)

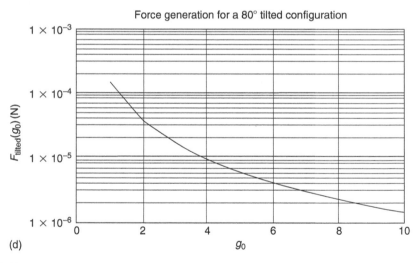

(d)

Figure 3.7 (continued)   **(c) 60°, and (d) 80°.**

of an ESRM with tilted configuration compared to a conventional configuration of the actuator.

Dividing Equation 3.2 by Equation 3.1, one gets the ratio of the actuation forces generated by the tilted configuration and conventional configuration of the rotary microactuator as a function of tilt angle (θ_t). This ratio can be written as

$$\frac{F_{\text{tilted}}(x)}{F_{\text{conv}}} = \left[\frac{L(x)/L_o}{\cos^2 \theta_t}\right] \tag{3.3}$$

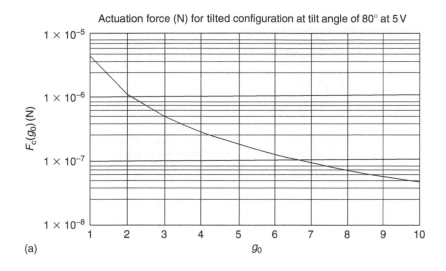

(a)

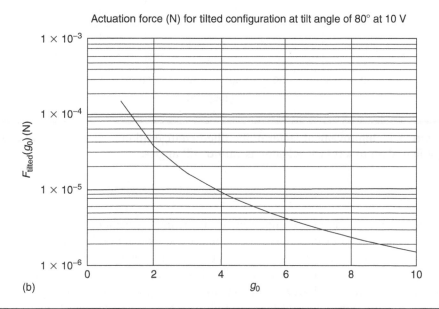

(b)

Figure 3.8 Actuation force generated by the tilted configuration as a function of clearance (μm) at actuation voltage (a) 5 V and (b) 10 V.

$$= \left[\frac{1}{\cos^2 \theta_t} \right] \tag{3.4}$$

In practice, the electrode lengths $L(x)$ and L_o are generally very close to each other, which will simplify the force ratio expressed by Equation 3.4. The ratio represented

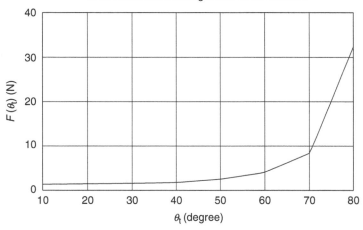

Improvement factor due to tilted configuration over the conventional configuration of the actuator as a function of tilt angle

$$\theta_t = 10, 20, ..., 80$$

$$F(\theta_t) = \cfrac{1}{\left[\cos\left(\cfrac{\theta_t}{57.3}\right)\right]^2}$$

Tilted configuration improvement factor versus tilt angle (°) over conventional configuration of the actuator

Figure 3.9 Improvement factor of the tilted configuration over the conventional configuration of a rotary actuator as various tilt angles.

by Equation 3.4 indicates an improvement in the force generation capability of the tilted configuration of the rotary microactuator and its magnitude is inversely proportional to the square of the cosine of tilt angle. It is evident from Equation 3.4 that theoretically the force generated by the tilted configuration of the microactuator could be unlimited. However, the actuation force-generating capability of the tilted configuration is limited by the minimum capacitive gap, which in turn is limited by the current fabrication technology. This means the optimum tilt angle $(\theta_t)_{opt}$ to maximize the actuation force can be expressed by

$$(\Theta_t)_{max} = \left\{ \cos^{-}\left[\frac{g_f}{g_o(o)}\right] \right\}, \quad \text{when } \frac{g_f}{g_o(o)} \text{ is less than unity} \qquad (3.5)$$

Here g_f is the minimum gap size permissible with the current fabrication technology and $g_o(o)$ is the initial clearance. Equation 3.5 indicates that tilt configuration is valid for the range where the minimum achievable gap size (g_f) is smaller than the clearance required for the rotary actuator operation.

The rotor is connected to the anchored column through the spring-loaded beam as illustrated in Figure 3.2. As stated earlier that in an ESRM than the initial clearance $g_o(o)$ in the θ-direction increases proportionately to the radius from the center of rotation. This is due to the fact that the angular motion leads to linearly increasing angular displacement with increasing radius r as illustrated in Figure 3.4. This means that the initial clearance for small rotation varies as

$$g_r(r, \theta_d) = [r(n\theta_o - \theta_d)]$$
$$= (nr\theta_o) \quad \text{when } \theta_d = 0 \tag{3.6}$$

where
   θ_d is the current angular displacement
   θ_o is the required angular displacement
   r is the radius
   n is the factor required for the static stabilization of the system in presence of external electric field

It is critical to mention that the electrode in an ES PP configuration has a stable range of ($g_o/3$), which implies that for an electrostatically driven PP actuator configuration the parameter has a value of three ($n = 3$). From the optimum curve shape for an electrode, one will notice that the initial capacitive gap (g_n) is required to be kept constant at g_f for any given curve radius and for the consistency of the capacitive gap size. It is the consistency of the gap size of the optimum electrode shape which enhances the force generation capability over the conventional configuration of the rotary microactuator, where the gap size increases with the radius of rotation.

3.2.5 Torque-Generating Capability of the Rotary Actuator with Tilted Configuration

Force-generating capability of an actuator can be predicted or estimated from the torque-generating capability of a microactuator simply by dividing the torque by the radius of rotation. The torque-generating capability is strictly dependent on the angular displacement and the capacitive gap size between the electrodes. The gap size changes with the angular displacement and can be written as

$$g_n(\theta_d) = g_f \left(\frac{1 - \theta_d}{n\theta_o} \right) \tag{3.7}$$

where

$g_n(\theta_d)$ is the capacitive gap size and is a function of angular displacement (θ_d)

g_f is the minimum gap size possible with the existing fabrication technology (see Figure 3.2 for its value)

n is a factor to maintain actuator stability

θ_o is the initial angular displacement

The ES-actuating torque, which is dependent on the variable g_n, drive voltage (V), and change in capacitance as a function of angular displacement can be written as

$$T(g_n) = \frac{1}{2}\left[\frac{dC}{d\theta_d}\right] V^2 = \frac{\left[e_o h\left(r_o^2 - r_i^2\right) V^2\right]}{4\, g_n^2} \qquad (3.8)$$

where

V is the drive or actuation voltage

h is the (out-of-plane) thickness of the electrode

r is the radius

the subscripts o and i stand for outer and inner radius, respectively

Because g_n is function of θ_d, the expression for the actuating torque with the help of Equation 3.7 can be written as

$$T(\theta_d) = \left[\frac{e_o h V^2}{4\, g_f^2 (1 - \theta_d/n\theta_o)^2}\right](r_o^2 - r_i^2) \qquad (3.9)$$

3.2.6 Optimum Curve Shape of the Electrodes

Studies performed on electrode shapes indicate that some optimum electrode shape offers improved force-generating capability as well as excessive radial force in the direction orthogonal to the operational direction (x). As a result of excessive radial force, the electrode shape may experience large deformation. To avoid deformation problems, an alternate electrode geometry with two optimum curve shapes, one setting the tilt in the clockwise direction and the other in the anticlockwise direction, must be deployed. In brief, an electrode shape involving a saw shape will not only provide enhanced force-generating capability but also higher electrode rigidity.

3.2.6.1 Potential Electrode Shapes

There are three distinct electrode shapes, which can be integrated in the design of a rotary microactuator to achieve higher actuation force. The three potential electrode shapes are (1) saw shape (type A), (2) unique shape with clearance increasingly proportional to inner radius (r_i) (type B), and (3) shape capable of providing constant clearance (type C).

Computer simulations data generated on the normalized actuating torque generated by these three shapes will illustrate that the type 1 electrode with "saw shape" generates the maximum actuating torque as a function of normalized angular displacement (θ_d/θ_o) as illustrated in Figure 3.10a through d.

3.2.6.2 Normalized Torque as a Function of Normalized Angular Displacement

Expressions for the normalized actuating torque, $T_N(\theta)$, as a function of normalized angular displacement (θ) for three distinct electrode shapes will be derived involving various actuator parameters. The normalized angular displacement θ is defined as θ_d/θ_o.

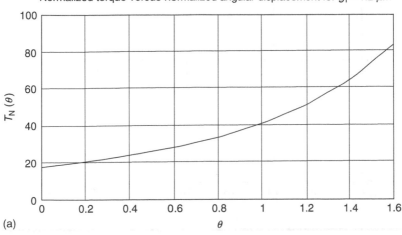

Normalized torque as a function of normalized angular displacement

$$\theta = 0, 0.2, ..., 1.6 \quad \theta_0 = 0.0019 \quad r_i = 800 \ \mu m \quad r_o = 1500 \quad n = 3 \quad g_f = 1.2 \quad \theta = \frac{\theta_d}{\theta_0}$$

$$T_N(\theta) = \frac{\left(r_0^2 - r_i^2\right) \cdot 0.000065}{4 \cdot g_f^2 \cdot \left(1 - \dfrac{\theta}{3}\right)^2}$$

Normalized torque versus normalized angular displacement for $g_f = 1.2 \ \mu m$

Figure 3.10 **Normalized torque as a function of normalized displacement for a capacitive gap of (a) 1.2 μm,**

(continued)

Normalized torque versus normalized displacement for $g_f = 2\ \mu m$

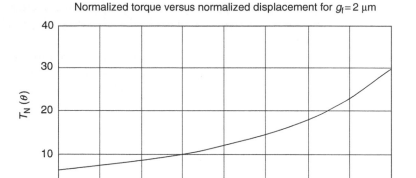

(b)

Normalized torque as a function of normalized angular displacement

$\theta = 0, 0.2, ..., 1.6 \quad \theta_0 = 0.0019 \quad \theta_d(\theta) = \theta \cdot \theta_0 \quad r_i = 800\ \mu m \quad r_o = 1500 \quad n = 3 \quad g_f = 4$

$$T_N(\theta) = \dfrac{\left(r_o^2 - r_i^2\right) \cdot 0.000065}{4 \cdot g_f^2 \cdot \left(1 - \dfrac{\theta}{3}\right)^2}$$

Normalized torque versus normalized displacement for $g_f = 3\ \mu m$

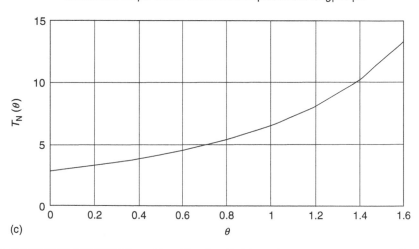

(c)

Figure 3.10 (continued)   (b) 2 μm (c) 3 μm,

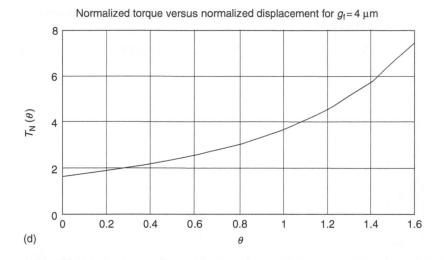

(d)

Figure 3.10 (continued)   (d) 4 μm (for type A shape).

Normalized actuating torque expressions for electrode shapes A, B and C are:

$$[T_{\mathrm{N}}(\theta)]_{\mathrm{A}} = \left[\frac{X}{4g_{\mathrm{f}}^2}\right]\left[\frac{r_{\mathrm{o}}^2 - r_{\mathrm{i}}^2}{(1 - Y)^2}\right] \qquad (3.10\mathrm{a})$$

$$[T_{\mathrm{N}}(\theta)]_{\mathrm{B}} = \left[\frac{X}{2(n\theta_{\mathrm{o}})^2(1 - Y)^2}\right]\left[\mathrm{Log}_{\mathrm{e}}\left(\frac{r_{\mathrm{o}}}{r_{\mathrm{i}}}\right)\right] \qquad (3.10\mathrm{b})$$

$$[T_{\mathrm{N}}(\theta)]_{\mathrm{C}} = \left[\frac{X}{4(nr_{\mathrm{i}}\theta_{\mathrm{o}})^2}\right](r_{\mathrm{o}}^2 - r_{\mathrm{i}}^2) \qquad (3.10\mathrm{c})$$

$$X = \left[e_{\mathrm{o}}hV^2\right] \qquad (3.11\mathrm{a})$$

$$Y = \left[\frac{\theta_{\mathrm{d}}}{n\theta_{\mathrm{o}}}\right] \qquad (3.11\mathrm{b})$$

Computed values of normalized actuating torque as a function of normalized angular displacement (θ) and initial capacitive gap size (g_{f}) for type A electrode shape are shown in Figure 3.10. Plots of normalized torque for capacitive gap size of 1.2, 2, 3, and 4 μm are displayed in Figure 3.10a through d, respectively. Plots of normalized torque for type B and type C electrode shapes are shown in Figures 3.11 and 3.12, respectively for g_{f} value of 2 μm. Normalized torque curves shown in

Normalized torque for type 2 rotary actuator shape with clearance increasing proportional to radius

$\theta = 0, 0.2, ..., 1.6$ $\qquad$ $n = 3$ $\qquad$ $r_0 = 1500$ $\qquad$ $r_i = 800$

$$T_N(\theta) = \frac{\ln\left(\dfrac{1500}{800}\right) \cdot 9}{(n - \theta)^2}$$

$T_N(\theta) =$

0.629
0.722
0.837
0.982
1.169
1.414
1.746
2.21
2.286

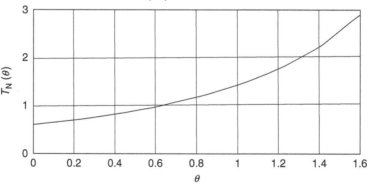

Normalized torque for type 2 electrode configuration of rotary actuator with clearance proportional to curve radius

Figure 3.11 **Normalized torque as a function of normalized angular displacement for type B electrode shape with a capacitive gap of 2 μm.**

Figures 3.10 through 3.12 indicate that type A electrode shape (saw shape) generates the highest normalized actuating torque as a function of normalized angular displacement. Normalized actuating torque when plotted against the normalized capacitive gap size $D(g_f)$, which is equal to $(g_f/nr_i\theta_0)$, indicates that type A electrode shapes offer much higher torque values (Figure 3.13) compared to type B electrode shape. As a matter of fact, type B electrode shape yields constant normalized actuating torque of 1.4143 as illustrated in Figure 3.14.

Normalized torque for type 3 electrode configuration with constant initial clearance

$$\theta = 0,\ 0.2,\ ...,\ 1.6 \qquad n = 3 \qquad r_o = 1500\ \mu m \qquad r_i = 800\ \mu m$$

$$T_N(\theta) = \frac{\left(r_o^2 - r_i^2\right) \cdot 4.5 \cdot \theta^2}{(n \cdot r_i \cdot \theta)^2}$$

$T_N(\theta) =$

0
1.258
1.258
1.258
1.258
1.258
1.258
1.258
1.258

Normalized torque for type 3 electrode configuration with constant initial clearance

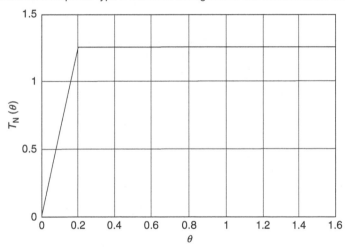

Figure 3.12   Normalized torque as a function of normalized angular displacement for a type C electrode shape with capacitive gap of 2 μm.

3.2.6.3   Parametric Requirements for Optimum Rotary Microactuator

Computer simulation data on three distinct electrode shapes is obtained as a function of critical actuator parameters, namely, initial capacitive gap size (g_i), normalized angular displacement (θ), normalized capacitive gap size ($g_i/nr_i\theta_o$),

Normalized torque when two angular displacements are equal and when normalized to g_f

(Tilted configuration of electrostatic rotary actuator with optimum shape of electrodes used)

$g_f = 1, 2, ..., 5$ $n = 3$ $r_o = 1500$ $r_i = 800$ $\theta_0 = 0.0019$ $\theta_d = 0.0019$

$$D(g_f) = \frac{g_f}{n \cdot r_i \cdot \theta_0} \qquad T_N(g_f) = \frac{\dfrac{r_o^2 - r_i^2}{r_i^2} \cdot 0.65}{16 \cdot 0.0361 \cdot [D(g_f)]^2}$$

$D(g_f) =$	$T_N(g_f) =$
0.219	58.866
0.439	14.716
0.658	6.541
0.877	3.679
1.096	2.355

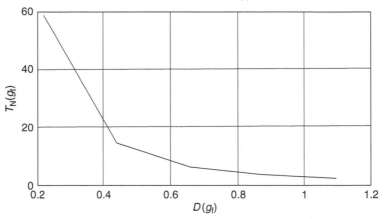

Normalized torque for type 1 (sawtooth configuration of rotary actuator)θ
Normalized to $D(g_f)$

Figure 3.13 Normalized torque versus normalized capacitive gap size $D(g_f)$ for type A electrode shape (sawtooth shape).

and the required angular displacement (θ_o). The rotary actuator engineer needs to know the following design guidelines, if maximum torque is the principal objective from a specific electrode shape:

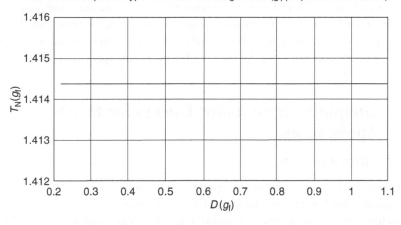

Normalized torque for type 2 configuration of electrostatic rotary actuator

$g_f = 1, 2, ..., 5$ $n = 3$ $r_o = 1500$ $r_i = 800$ $\theta_0 = \theta_d = 0.0019$

$$D(g_f) = \frac{g_f}{n \cdot r_i \cdot \theta_0}$$

$$T_N(g_f) = \frac{\left[\ln\left(\frac{r_o}{r_i}\right) \cdot r_i^2 \right] \cdot 9 \cdot \theta_0^2}{0.44444 \cdot g_f^2}$$

$$[D(g_f)]^2$$

$D(g_f) =$	$T_N(g_f) =$
0.219	1.414
0.439	1.414
0.658	1.414
0.877	1.414
1.096	1.414

Normalized torque for type 2 electrode configuration (g_f proportional to radius)

[Plot: $T_N(g_f)$ versus $D(g_f)$; vertical axis from 1.412 to 1.416, horizontal axis from 0.2 to 1.1]

Figure 3.14 Normalized torque versus normalized capacitive gap size $D(g_f)$ for type B electrode shape of the rotary actuator.

■ Type A electrode shape (or saw shape) offers the highest torque at initial capacitive gap size of 1.2 μm, which is best suited for a magnetic head requiring a track density exceeding 25 kTPI.

■ The factor n must have a value of 3 to ensure static system stability in presence of external electrical fields in the case of a conventional rotary microactuator using PP configuration. This factor must be set to 4.5 for type shape of the electrodes to achieve the same displacement for type A and type B shapes.

■ The geometrical dimensions of the actuator mechanism must be selected to achieve a ratio of overlapped length to gap size (ROLG) greater than 6, which satisfies the ROLG criterion. This criterion states that the saw shape (type A electrode shape) selected must retain the same force generating as the original optimum shape of both the electrodes.

■ The required angular displacement θ_o should be assumed around 0.0019 rad or 0.1106°, which will provide the linear displacement of 1.2 μm to achieve a tracking density of 25 kTPI.

■ The computer simulation data indicates that a tilted rotary actuator configuration with optimum electrode shape offers drastically increased force-generating capability over a conventional rotary microactuator. This type of rotary microactuator is best suited for HDDs requiring wideband bandwidth, improved tracking density, and high resonant frequency.

■ The required angular displacement for the HHD can be determined from the ratio of required linear displacement at the magnetic head (which is typically less than 1 μm) to the distance of the magnetic head from the center of rotation (which is close to 500 μm or equal to outer radius). This ratio is typically around 0.001 rad. This makes the angular clearance ($n\theta_o$) equal to or less than 0.001 at the factor n equal to 3. Under these assumptions, the initial capacitive gap size $g_n(0)$ of the designed electrode shape can be treated as the minimum gap size (g_f) achievable with the current fabrication technology.

3.3 Unique Microactuator Design for HHD Applications

3.3.1 Introduction

This particular section describes a unique microactuator design, which is considered most suitable for HHD applications and is widely known as a MEMS (microelectromechanical system) piggyback actuator. Currently, voice-coil motors (VCMs) are deployed to position the read/write head, which is attached to the suspension-gimbal assembly [2]. Upgraded versions of a MEMS piggyback actuator have demonstrated significant increase in both the recording rate in bit-per-inch (BPI) and tracking density in TPI. Current market surveys indicate that recording density for HDDs is increasing at the rate of 70–90 percent per year. However, the current recording rate requirements exceed 100 Gb/in.^{2} The recording rate is expected to reach the upper limit in near future, which can limit the data transfer rate of the electronics or the thermal instability of the recording material. Under these circumstances, TPI is the only parameter left to improve the recording density through an improved design of the positioning mechanism.

In addition, the servo bandwidth of the current actuators is limited due to the vibrations sustained in the suspension assembly. As a result of this, the track density is limited to 50 kTPI, which corresponds to a track width of 0.5 μm. Future HDDs may require a track density exceeding 150 kTPI to meet a recording density of 200 Gb/in.² An improved design of a MEMS piggyback, which is referred as a dual-stage servomechanism could effectively solve the above tracking density problems. In a MEMS approach, a microactuator comprising of a read/write head is used for fine positioning scheme in conjunction with a VCM feature for course positioning. A MEMS piggyback actuator has demonstrated a displacement better than 0.5 μm at an actuation voltage less than 60 V with a fundamental resonance frequency 16 kHz.

3.3.2 Benefits and Design Aspects of a Dual-Stage Servomechanism (or MEMS Piggyback Actuator)

The dual-stage servomechanism also known as a MEMS piggyback actuator will not only be able to eliminate the bandwidth problem associated with the head-positioning assembly but also will improve the track density of the recording system. There are several designs of piggyback actuators. The first generation of piggyback actuator, also known as a suspension-driven design, could deploy either an electromagnetic actuation mechanism or a pair of piezoelectric lead–zirconate–titanate (PZT) actuators mounted in between the VCM-driven arm and the suspension assembly. This actuation mechanism permits the swing of the suspension assembly by the induced distortion and adjustment of the spider mounted at the suspension tip, as illustrated in Figure 3.15. The second-generation actuator, known as the spider-drive actuator, deploys an actuation mechanism attached in between the spider chip and the suspension tip. The suspension arms and the spider chips as shown in Figure 3.15 are heavy for the microactuators and, therefore, the mechanical resonance can be expected to be in the lower kilohertz range.

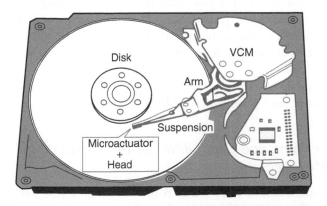

Figure 3.15 Third-generation MEMS piggyback microactuator design and its critical elements best suited for HDD applications. (From Toshiyoshi, H. and Mita, M., *J. MEMS*, 11, 648, 2002. With permission.)

3.3.2.1 Architecture of a Third-Generation Microactuator

A third-generation dual-servo system along with its critical elements is illustrated in Figure 3.15 and is considered most desirable for HDD applications. It is important to mention that a third-generation microactuator when integrated with a read/write head with low mass will offer higher mechanical resonance frequencies exceeding 30 kHz. Two types of MEMS piggyback microactuators are available, namely, the high-aspect ratio ES actuator using electroplated nickel and a microactuator fabricated with a dry-etch silicon. Monolithic integration is absolutely necessary because the mutual positioning of the head and the actuator spider must be controlled as small as the flight height of the head, which typically is about 10 nm or 0.01 μm. Read/write heads can be designed either using silicon (Si) substrate or aluminum (Al) titanium carbide (TiC) substrate. Electrical interconnections to the read/write head assembly on the chip is made from low-loss metal patterns on the actuator's suspension. Electrical bonding to the contact pads can be made from conventional wiring techniques. Read/write heads are integrated on a silicon substrate (Figure 3.16), which offers highest mechanical integrity under harsh thermal and mechanical conditions with no compromise in actuator performance. As far as physical dimensions of the actuator are concerned, the cross-sectional area and the height of the slider element are approximately 1000×1000 μm^2 and 500 μm, respectively. The dimensions of the slider are strictly dependent on the number of electrodes, each having a gap of 1–2 μm. The smaller the gap, the higher will be the actuation force. The larger the gap, the higher will be the slider dimensions. More powerful actuators use smaller electrode gaps, which will realize further reduction in overall size of the actuator. A microactuator structure with high-aspect ratio and narrow electrode gaps will yield optimum performance of the ES actuator. The silicon-on-insulator (SOI) wafer with 50 μm thickness will allow trench etching of 2–3 μm openings with photoresist mask, as shown in Figure 3.17. These openings must be set at least three times the maximum displacement to avoid spontaneous pull-in of the ES gap. Note the electromechanical performance of the microactuator is independent of the SOI thickness. The silicon substrate thickness varies from 500 to 550 μm. The sacrificial layer from the SOI wafer surface must be eliminated using appropriate chemical techniques.

These simple and inexpensive fabrication techniques will yield the most compact and cost-effective third-generation MEMS piggyback actuator design.

3.3.2.2 Performance Capabilities of the MEMS Piggyback Microactuator

Monolithic integration of these heads in the surface of a SOI wafer as shown in Figure 3.17 will make the MEMS piggyback actuator capable of providing displacements in 0.2–0.3 μm range with mechanical resonance frequencies exceeding 30 kHz and at drive voltages well below 30 V. The outstanding benefits of the MEMS piggyback microactuator cannot be matched by any other microactuator to this

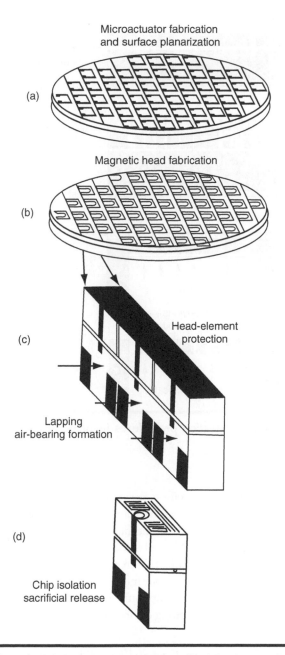

Microactuator fabrication
and surface planarization

(a)

Magnetic head fabrication

(b)

Head-element
protection

(c)

Lapping
air-bearing formation

(d)

Chip isolation
sacrificial release

Figure 3.16 Various fabrication processes involved for integration of magnetic read/write heads and MEMS actuators.

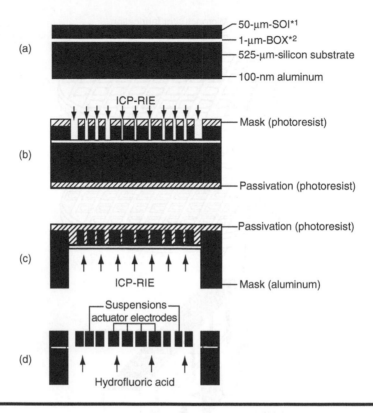

Figure 3.17 Various fabrication procedures involved in a MEMS piggyback ES actuator on an SOI wafer.

date. It is important to mention that for such small displacements less than 0.3 μm, one can use the smallest possible etching width of an actuator gap to generate large ES force at drive voltage close to 25 V or less, and mechanical resonance frequencies in excess of 30 kHz. Note this particular MEMS-based actuation system comprising of multiple PP microactuators can generate enormous ES force per given actuator size. Furthermore, the air bearing of the flying slider can be formed on the sidewall of the chip after dicing the SOI wafer, which will significantly not only improve the actuator performance but also the reliability under harsh operating environments.

3.3.3 Force Generation Capability, Displacement Limit, and Mechanical Resonance Frequency Range

Preliminary computation simulations indicate that it is not always possible to achieve both higher resonance frequency and larger displacement at a given actuation voltage. However, high force generation capability requires higher actuation or drive voltages and smaller gaps for a given electrode size.

3.3.3.1 Electrostatic Force Calculation

Expression for ES force (F_{es}) for a gap-closing microactuator can be written as

$$F_{es} = \left[\frac{0.5e_o V^2}{(g-x)^2} \right] (L_e H_e N) \tag{3.12}$$

where

  g is the initial gap (typically 3 or 2 μm)
  x is the linear displacement (maximum value 1 μm)
  L_e represents electrode length
  H_e represents the electrode height
  N indicates the number of electrodes (varies from 4 to 11) and all other parameters have been defined earlier

   Assuming the electrode length of 500 μm, electrode height of 50 μm, initial gap of 3 μm, actuation voltage of 30 V, displacement of 1 μm, and electrode number of 2 and inserting these parameters in Equation 3.12, one gets the following value for the ES force:

$$F_{es} = \left[\frac{0.5 \times 8.85 \times 10^{-12} \times 30 \times 30}{(3-1)^2} \right] [500 \times 50 \times 2]$$
$$= [0.995 \times 10^{-9}][5 \times 10^4] = [4.98 \times 10^{-5}]$$
$$= [50 \times 10^{-6}] = 50 \ \mu\text{N}$$

   Computed values of ES forces as a function of actuation voltage (V), initial gap size (g), number of electrodes (N) and displacement (x) are summarized in Table 3.1. The calculations above assumed maximum displacement of 1 μm.

   These calculations indicate that a MEMS piggyback actuator offers highest actuation forces even with two electrodes (or a pair of electrode) at drive voltage of 15 V compared to all other microactuators discussed so far. These forces will significantly increase with the increase in number of electrodes, with the decrease in initial gap size and with the decrease in displacement. Computer simulation data will reveal later that the maximum displacement is not more than 0.67 μm.

3.3.3.2 Mechanical Resonance Frequency Calculation

The mechanical resonance frequency (f_0) expression can be written as

$$f_0 = \left(\frac{1}{2\pi} \right)^{0.5} k/m \tag{3.13}$$

Table 3.1 ES Forces as a Function of Various Actuator Parameters (μN)

Actuation Voltage	30 V		15 V	
N	$g = 3$	$g = 2$	$g = 3$	$g = 2$
2	50	200	12.5	50
4	100	400	25	100
6	150	600	37.5	150
8	200	800	50	200
10	250	1000	62.5	250
11	275	1100	68.7	275

where

k is the spring constant (N/m)
m is the mass of the electrode (N)

The spring constant can be calculated using the following equation:

$$k = \left[\frac{12E\,I}{L_s^3}\right]N \tag{3.14}$$

where

E is the modulus of elasticity for the suspension structure material
L_s is the suspension length
I is the moment of inertia of the suspension structure and is defined as

$$I = \left[\frac{H_s W_s^3}{12}\right] \tag{3.15}$$

where H_s is the suspension height and W_s is the suspension width. Assuming the suspension length and width of 50 and 50 μm, respectively, one gets

$$I = \left[\frac{50 \times 50^3}{12}\right] = \left[\frac{625 \times 10,000}{12}\right]$$
$$= 52.08 \times 10^4 \ \mu\text{m}^4$$

Now assuming Young's modulus of 130 GPa (which is equal to 5.05×10^{-5} N/μm^2), suspension length of 500 μm and electrode number of 2, and inserting into Equation 3.14, one gets the value of spring constant, that is,

$$k = \left[\frac{12 \times 130 \times 5.05 \times 10^{-5} \times 52.08 \times 10^{4}}{500^{3}} \right] 2$$

$$= \left[\frac{60.60 \times 1354.06}{125 \times 10^{6}} \right]$$

$$= 656 \times 10^{6} \text{ N}/\mu\text{m}$$

$$= 656 \text{ N/m} \tag{3.14a}$$

3.3.3.3 Electrode Mass Computation

It is important to mention that the mass of the electrodes dominate the total mass (m) when the number of electrodes is large. The electrode mass can be computed from the following equation:

$$m = (\delta N)[W_e H_e (L_e + 3W_h)] \text{ kg} \tag{3.16}$$

where
δ is the density of the material used for the suspension structure (2330 kg/m^3 for silicon material)
L_e is the length of the electrode
W_e is the width of the electrode
H is the height of the electrode, subscript e stands for electrode
W_h is the width of the head

Assuming silicon for the structural material, an electrode length of 500 μm, electrode width of 50 μm, electrode height of 50 μm, and head width of 50 μm, and inserting these parameters into Equation 3.16, one gets

$$m = (2330 \times N)[50 \times 50(500 + 3 \times 50)]$$

$$= (2330 \times N)\left[1.625 \times 10^{6}\right]\left[2.202 \times 10^{-18}\right]\text{lb}$$

$$= (2.330 \times N)\left[0.804 \times 10^{-9}\right]\text{N}$$

$$= \left[1.873 \times 10^{-9} \times N\right]\text{N} \tag{3.16a}$$

Thus, the total mass for the 11 electrodes ($N = 11$) comes approximately to

$$m = \left[1.873 \times 11 \times 10^{-9}\right]\text{N}$$

$$= \left[20.6 \times 10^{-9}\right]\text{N} \tag{3.16b}$$

Inserting the values of parameters m and k in Equation 3.13, one gets the fundamental resonance frequency as

$$f_o = \left[(0.159) \left(\frac{656}{20.6 \times 10^{-9}} \right) \right]$$

$$= \left[0.159 \times 10^5 \times 1.7845 \right] \text{Hz}$$

$$f_o = 28.37 \text{ kHz at } N = 11$$

$$= 29.70 \text{ kHz at } N = 10$$

$$= 33.24 \text{ kHz at } N = 8$$

$$= 38.32 \text{ kHz at } N = 6$$

$$= 46.84 \text{ kHz at } N = 4$$

$$= 66.42 \text{ kHz at } N = 2$$

3.3.3.4 Displacement (x) as a Function of Gap Size (g) and Number of Electrodes (N)

Computation of actuator displacement (x) as a function of gap size (g) and number of electrode is very complex, because it involves structural dimensions, spring constant (k), and drive voltage (V). The actuator displacement can be computed using the following expression:

$$\left[x(g - x)^2 \right] = \left[\frac{0.5 e_o L_e H_e N V^2}{k} \right] \tag{3.17}$$

where x is the linear displacement, which is critically dependent on the gap size (g).

Assuming electrode length of 500 μm, electrode height of 50 μm, actuation voltage of 30 V, spring constant of 656 N/m (previously calculated value using Equation 3.14a), electrode number of 2, and gap size of 2 μm, and inserting these parameters in Equation 3.17, one gets

$$\left[x(g - x)^2 \right] = \left[\frac{0.5 \times 8.58 \times 10^{-12} \times 500 \times 50 \times 2 \times 900}{656} \right]$$

$$= \left[\frac{2 \times 10^{-4} \times 10^6}{656} \right] = \left[\frac{200}{656} \right]$$

$$= [0.305]$$

This means the third-order linear equation can be written as

$$\left[x^3 - 4x^2 + 4x - C \right] = 0 \quad \text{(for gap size of 2 μm)} \tag{3.17a}$$

$$\left[x^3 - 6x^2 + 9x - C \right] = 0 \quad \text{(for gap size of 3 μm)} \tag{3.17b}$$

Table 3.2 Computed Values of Resonance Frequency (f_0) and Actuator Displacement (x) as a Function of Electrode Number (N) and Gap Size (g)

Number of Electrodes (N)	Constant (C)	Resonance Frequency (f_0) (kHz)	Actuator Displacement × (μm)	
			$g = 3$ μm	$f = 2$ μm
2	0.306	66.42	0.034	0.082
4	0.612	46.84	0.075	0.185
6	0.918	38.32	0.115	0.334
8	1.224	33.20	0.150	0.675 (clipping)
10	1.530	29.74	0.200	0.675 (clipping)
11	1.683	28.38	0.244	0.675 (clipping)

where C is a constant and its value is dependent on the actuation voltage (V), electrode dimensions, number of electrodes (N), and spring constant (k). Computed values of this constant (C) are shown in Table 3.2.

Calculated data on resonance frequency and actuator displacement as a function of number of electrodes deployed by the actuator and gap size is summarized in Table 3.2.

It is critical to point out that clipping starts slightly before when N approaches to 8 and it occurs when gap is equal to 2 μm. However, the actuator displacement (x) is 0.450 μm for $N = 7$. It is important to point out that displacement values are higher for smaller gap size. However, the resonance frequency, which is independent of gap size, decreases as the number of electrode increases. These computations indicate that to achieve higher resonance frequencies and higher displacements, one needs either higher actuation voltages or reduced structure mass. This clearly indicates that silicon material must be selected over nickel-plated alloy for the actuator structure. Plot of these parameters as a function of electrodes are displayed in Figure 3.18.

3.4 Capabilities of Vertical Comb Array Microactuator

VCAMs are best suited for applications where reliable vertical actuation is the principal requirement. VCAM permits motion into and out of the wafer surface

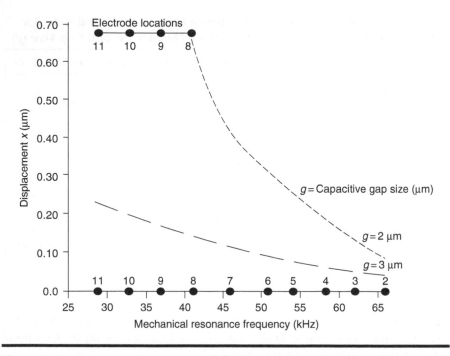

Figure 3.18   Resonance frequency and displacement as a function of electrodes in the actuator.

with minimum force [3]. Although large actuation forces are possible with PP actuators, these actuators suffer from drawbacks in their actuation mechanisms such as reduced range of motion, pull-in instability, nonlinear relationship between the actuation force, and displacement in the presence of electrical field and reduced frequency response due to damping effects. The vertical levitation or actuation force generated by a technique involving nonsymmetric fields is limited by a displacement range of 1 to 2 μm. VCAM provides a linear drive to deflection characteristics and a large throw capability, which are the basic requirements in many MEMS-based sensors, actuators, and micromechanisms. Note the actuation mechanism deployed in a lateral comb drive actuator (LCDA) design uses a variable electrode overlap area motion, which offers linear force to displacement, large actuation force, reduced damping effects, and higher pull-in stability. However, this particular design concept is not suitable for deflection in a vertical direction.

A new polysilicon technology combining both the bulk and surface micromachining techniques offers a unique design of a VCAM. As stated before, the VCAM provides large vertical actuation forces, while maintaining linear relationship between the actuation force and vertical displacement and high efficiency.

3.4.1 Structural Requirements and Critical Design Aspects of VCA Actuator

The architecture for the VCAMs shown in Figure 3.19 identifies the critical elements used and structural details of the actuation mechanism. The VCAM structure as illustrated in Figure 3.19 involves a set of moving mechanical polysilicon (MP) electrodes and a set of rigid p^{++} electrodes or chassis suspended over a pit. The actuating mechanism is fabricated using a trench-refilled-with-polysilicon process technology. When a driving voltage is applied between the chassis and MP, an ES force is generated between the offset electrodes causing the MP bridge to deflect upward and away from the wafer surface. Downward force into the wafer between the control region and other parallel surfaces is prevented by using a controllable poly-silicon (CP) layer. It is important to mention that the ES actuation forces are unidirectional, while the downward or negative actuation force is automatically achieved by the restoring force provided by the spring in the absence of ES pull-up forces. The MP structure can be moved in and out of the wafer surface by applying appropriate actuation voltage between the MP and p^{++} vertical electrode sets.

MP beam physical parameters and drive voltage play a key role in the design and development of VCAM with optimum performance. Large travel range requires fabrication of deep trenches with equal thickness and width. Maximum range can be achieved by setting the height of the etch stop electrode layer rather than other dimensions. Deep trenches with high-aspect ratios can be easily achieved using current fabrication and etching technologies. The actuation force generated is contingent on the electrode and beam parameters and the drive voltage levels. The dissimilar gaps on either side of the individual MP beams can result in imbalance in the lateral forces, leading to instability in the actuator structure. To avoid such instability, the MP beams should be made stiffer by increasing the beam width or

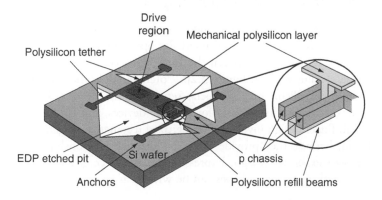

Figure 3.19 Isometric view of a VCAM showing the architecture of the micro-actuator and its critical elements.

widening the refill width of the trench or by decreasing the beam length. It is interesting to mention that higher drive voltages can be tolerated with stiffer beams, which can lead to maximum displacement. Typical drive voltage varies from 10 to 30 V; however, lateral pull-in instability is possible for drive voltage exceeding 25 V.

3.4.2 VCAM Performance Comparison with Other Actuators

Preliminary studies undertaken on LCDA, vertical comb drive actuator (VCDA), and PP drive actuator (PPDA) by the author indicate that the VCDA is capable of generating large vertical forces for a given array geometry on SOI wafer with minimum cost and complexity. Note the force generated by the VCDA or VCAM, which approaches the force generated by the PPDA, is at least one order of magnitude greater than that possible with LCDA. This is due to the fact that VCAM possesses larger perimeter per comb than the LCDA. In addition, the force generated by the VCDA beam is proportional to $2L$, where L is the length of the beam which can vary from 200 to 800 μm. In the case of LCDA, the force generated is proportional to $2t$, where t is the thickness of the polysilicon layer, which is typically 2–3 μm. In brief, the force generated by a VCDA can be a few orders of magnitude greater than the force generated by the LCDA depending on the length and thickness dimensions. The above comb drive microactuator requiring controlled actuation or motion are best suited for various micromechanical (MM) devices such as resonant sensors, micromotors, micropositioning systems, MEMS mirrors, micropumps (MPs), microvalve switches, adaptive optical systems, and read/write precision actuators for magnetic disk drive heads.

3.4.3 Potential Comb Finger Shapes

Comb drive actuation is the major building block for MEMS devices requiring vertical, lateral, or rectangular displacement. Its operating principle is based on the ES force, which is generated between the interdigital conducting comb fingers of specific shapes as illustrated in Figure 3.20. Potential applications of shaped comb fingers include polysilicon microgrippers, sensing probe devices, forced-balanced accelerometers, actuation mechanism for rotating devices, and RF filters. The most popular comb finger design involves interdigitized rectangular fingers with constant gap of 1 μm, because they provide [4] constant force relationship and the most simple and least expensive design. Expression for the force generated between the comb fingers with rectangular shapes can be written as

$$F_x = \left[\frac{e_o t V^2}{h(x_1)} \right] \tag{3.18}$$

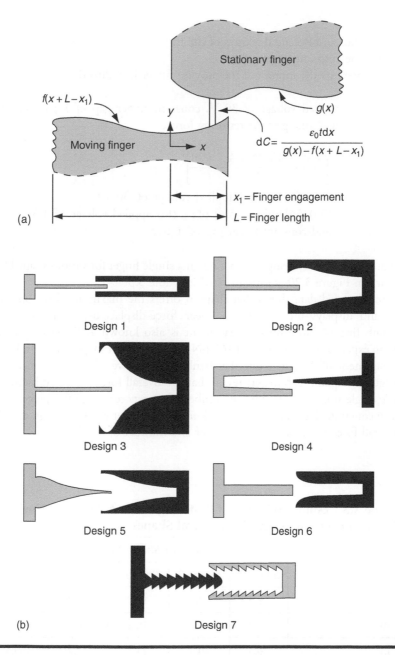

Figure 3.20   Comb finger design shapes. (a) Arbitrary shape and (b) potential finger shapes.

where

t is the out-of-plane thickness of the finger

$h(x_1)$ is the constant finger gap

x_1 represents the motion of the moving finger in x direction

Assuming applied voltage of 30 V, constant finger gap of 1 μm, and finger thickness of 4 μm, one gets the restoring force,

$$F_x = \left[\frac{8.85 \times 10^{-12} \times 4 \times 900}{1} \right]$$

$$= [31.86] \, \text{nN} \quad \text{(for applied voltage of 30 V)}$$

$$= [0.0354] \, \text{nN} = [35.4] \, \text{pN} \quad \text{(for applied voltage of 1 V}$$

and constant finger gap of 1 μm)

Estimated values of restoring force acting on a single finger for various shaped fingers as shown in Figure 3.20 are summarized in Table 3.3.

The tabulated data shows that shape 6 offers the maximum force on a single finger and displays more or less a linear force-displacement behavior between the comb fingers. Note finger engagement is also known as finger displacement. One can expect a restoring force of 160 pN per finger with shape 6 at a displacement or engagement of 25 μm. Note that banks of one, two, three, four, six, and ten rectangular shaped comb fingers reveal that the overall force predicted by multiplying the single-finger result with the number of fingers remained 0.2 percent. Thus, one can conclude that computer simulation for one finger will be enough to estimate the overall force for a bank comprising of any number of comb fingers.

Table 3.3   Estimated Values of Restoring Force Acting on a Single Finger for Various Finger Geometrical Shapes

Finger Engagement (μm)	Finger Shape Number				
	Shape 2	Shape 3	Shape 4	Shape 5	Shape 6
5	5.9	2.9	15.8	5.8	40.2
10	6.8	5.2	18.3	10.1	68.5
15	8.5	10.1	23.5	17.2	100.3
20	10.8	16.3	28.8	25.3	130.2
25	12.3	24.4	37.3	48.2	157.5

It is desirable to mention that the linear response of fingers of shape 6 makes this particular shape best suited for tuning the mechanical resonators. Furthermore, the shaped comb finger allows frequency shift either downward (a weakening case) or upward (a stiffening case) over a wide range of frequencies. The amplitude of the resonator is not limited to small variations. Frequency tuning of mechanical resonators involving shaped comb fingers is most desirable for the actuator requiring linear voltage displacement behavior, which is essential for a simple and least expensive design of a control system. In addition, comb fingers with linear response can be used to decrease the effective stiffness of an actuator carrying a load, thereby realizing significant reduction in power dissipation. Shaped comb fingers also yield most efficient design of accelerometers capable of operating over wide stiffness range.

3.5 Capabilities of Bent-Beam Electrothermal Actuators

MEMS actuators described in Chapter 2 require the use of magnetic or piezoelectric materials, or shaped memory alloys. Use of these materials in some cases is prevented by processing constraints, design complexities, and cost factors. Furthermore, the force generated by the ES actuators operating at lower voltage rarely exceed 10 μN in surface micromachined devices, and with other ES mechanisms the displacements are limited to 100 μm in resonance operations and 10 μm in nonresonance operations, unless gear trains or other steeping mechanisms are deployed to achieve higher displacement. Although ES actuators have several advantages such as low temperature coefficients and essentially zero direct current (dc) power consumption, but require higher actuation voltages (>30 V), which make them incompatible not only with microelectronic power supplies but also with silicon substrates and integrated circuits. BBET actuators [5] offer performance capabilities, which can alleviate some of these problems.

3.5.1 Performance Capabilities and Design Configuration of Bent-Beam Electrothermal Actuators

Brief studies performed on various actuators by the author indicate that the BBET actuator offers several advantages such as improved actuator performance, which is feasible by changing the beam geometry, peak beam displacement at a given temperature that can be increased using longer beams, or by reducing the bent angle and higher peak force using higher bent angle. It is important to mention that earlier versions of BBET actuators relied on the bending moment developed by heating two adjacent materials with different thermal expansion coefficients, which made the device more complex and expensive. The structural configuration and

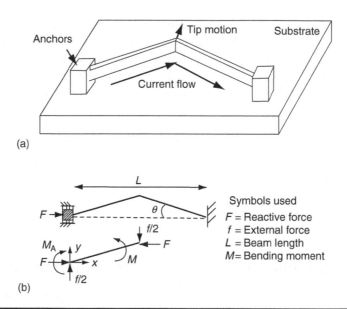

Figure 3.21   BBET actuator showing (a) structural details and (b) various forces acting on the beam in presence of external loading.

operating principle of the BBET actuator are shown in Figure 3.21. The design configuration of this actuator shows two adjacent arms of different widths, but of the same material. The actuator provides planar motion with displacements exceeding 20 μm for polysilicon devices. Polysilicon, electroplated metals, and shaped memory alloys are best suited as structural materials in the fabrication of this actuator.

In general, electrothermal actuators consume dc power exceeding 180 mW for a beam thickness of 4 μm and displacement of 5 μm compared to ES discussed previously, but they yield higher forces at lower drive voltages not exceeding 15 V, and thus are compatible with low-voltage microelectronic circuits. It is important to mention that BBET actuators provide a unique compromise of force, displacement, and scalability beside simplicity in the design. Important properties of the structural materials best suited for BBET actuators are summarized in Table 3.4.

3.5.2   Brief Description of the BBET Structure

BBET actuator incorporating a V-shaped beam (Figure 3.21) offers the simplest and cost-effective design. Its displacement of the apex as shown in Figure 3.21a is strictly dependent on the beam dimensions and the slope. The apex displacement is linearly proportional to the external loading force (f) as illustrated in Figure 3.21b. The maximum displacement in the actuator beam in the absence of external force

Table 3.4 Properties of Structural Materials Best Suited for BBET Actuators

Properties	Polysilicon	Electroplated Nickel	Platinum
Density (kg/m^3)	2.42	8.90	21.25
Thermal expansion coefficient (ppm/°C)	2.82×10^{-6}	13.30×10^{-6}	8.91×10^{-6}
Thermal conductivity (W/cm°C)	0.84	0.92	0.69
Young's Modulus (in various units)			
(psi)	15.7	30	21.4
(kg/mm^3)	11,000	21,000	15,500
(MPa)	130	185	234
($\times 10^{-2}$ N/μm^2)	0.547	1.045	0.745

can be calculated from the beam bending moment expression including the beam compression effect. The maximum displacement of the beam can be written as

$$d_{max} = \left\{ [2\tan(\theta)/k] \left[\tan\left(kL \times 10^{-6}\right) \right] \right\} - \left[\frac{L\tan(\theta)}{2} \right] \qquad (3.19)$$

where
  θ is the bending angle
  L is the effective beam length
  k is the spring constant

Note the effective beam length includes the impact of thermal coefficient of beam material. Typically, the bending angle varies from 2.86° to 11.46°. Computed values of maximum displacement as a function of bending angle for 250 and 500 μm beam length are shown in Figure 3.22a and b, respectively. Maximum displacement as a function of beam length at bending angle of 2.86° and 11.46° is shown in Figure 3.23.

The maximum loading force that can be sustained at the apex is defined as that at which the displacement of the apex is forced back to zero due to spring constant effect of the beam. The maximum loading force is directly proportional to sine square of the bending angle, cross-sectional area of the beam (product of beam width and beam thickness), and Young's modulus of beam structural material, but inversely proportional to the effective length of the beam. Peak output forces

Maximum displacement for a bent-beam actuator in absence of loading force (f)

$\theta = 0.01, 0.02, ..., 0.20$ $\qquad\qquad$ $k = 10$ N/m $\qquad\qquad$ $L = 250$ μm

$$d_{max}(\theta) = \frac{2 \cdot \tan(\theta)}{k} \cdot \tan\left(\frac{k}{4} \cdot L \cdot 10^{-6}\right) - \frac{L}{2} \cdot \tan(\theta)$$

Maximum displacement for a 250 μm long beam

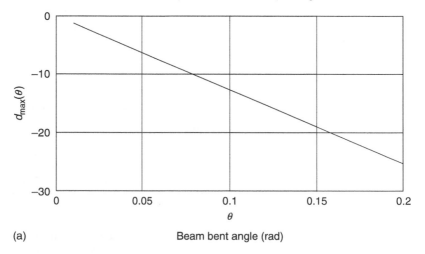

(a) $\qquad\qquad$ Beam bent angle (rad)

Maximum displacement for a 500 μm long beam

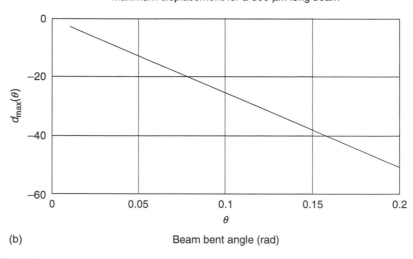

(b) $\qquad\qquad$ Beam bent angle (rad)

Figure 3.22 **Maximum displacement for a bent-beam actuator with beam length of (a) 250 μm, (b) 500 μm,**

Maximum displacement for a bent-beam actuator in absence of loading force (f)

$\theta = 0.01, 0.02, \ldots, 0.20$ $k = 10$ N/m $L = 1000$ μm

$$d_{max}(\theta) = \frac{2 \cdot \tan(\theta)}{k} \cdot \tan\left(\frac{k}{4} \cdot L \cdot 10^{-6}\right) - \frac{L}{2} \cdot \tan(\theta)$$

Maximum displacement for a 1000 μm long beam

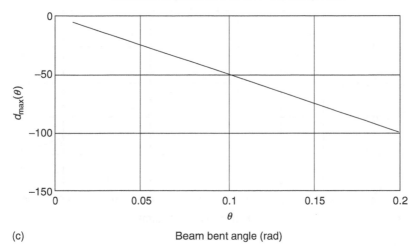

(c)   Beam bent angle (rad)

Maximum displacement for a 2000 μm long beam

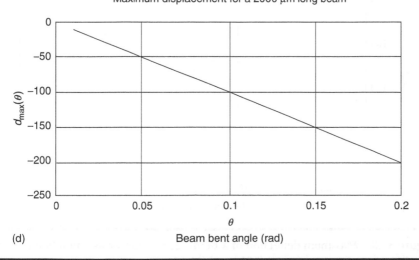

(d)   Beam bent angle (rad)

Figure 3.22 (continued)   (c) **1000 μm, and (d) 2000 μm as a function of bending angle.**

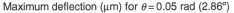

Deflection calculations for a bent-beam electrothermal actuator

$L = 500, 600, ..., 2000\ \mu m$ $k = 10\ N/m$ $\theta = 0.05$

$$d_{max}(L) = \frac{2 \cdot \tan(\theta)}{k} \cdot \tan\left(\frac{k}{4} \cdot L \cdot 10^{-6}\right) - \frac{L}{2} \cdot \tan(\theta)$$

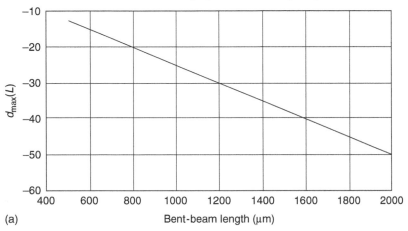

(a)

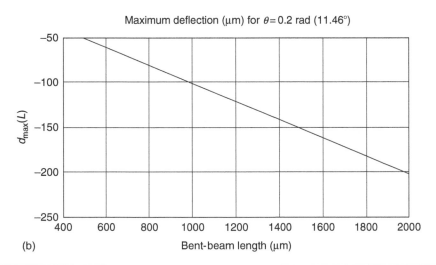

(b)

Figure 3.23 Maximum deflection of a bent-beam actuator as a function of beam length for bending angle of (a) 2.86° and (b) 11.46°.

available from a single beam actuator are in the range of few millinewton (mN), which can be increased by optimization of beam geometry. Further increase in force can be accomplished by placing BBET actuators in parallel. Larger displacements can be generated by cascading such actuators [5]. From the data presented in Table 3.4, it is evident that the thermal expansion coefficient of the electroplated nickel is about four times higher than that of silicon. This can result in larger displacements for nickel devices, but the position of the apex can shift due to change in ambient temperature and thermal mismatch with the substrate material. Note the thermal actuation using nickel on a silicon substrate has an operating limit of 350°C, which can be elevated to 450°C by electroplating the nickel structure with 500 Å gold film. Elevating the operating temperature limit will increase the peak displacements with nickel devices.

3.5.3 Input Power Requirements for BBET Actuators

It is critical to point out that these devices can be operated with driving voltage of 12 V, thereby permitting the use of standard electronic interfaces, which is not possible with other ES actuators. Input power requirements as a function of displacement for silicon and nickel devices are summarized in Table 3.5.

It is interesting to note that nickel devices with 400 μm long, 6 μm wide, and 3 μm thick beams yield static displacements exceeding 10 μms with 79 mW power consumption. Whereas, silicon devices with 800 μm long, 14 μm wide, and 3.8 μm thick beams offer static displacement of 5 μm only at 180 mW input power consumption. However, cascaded silicon devices offer identical displacements with 60 percent reduction in power consumption. Furthermore, the force generated are in the range of 1,000–10,000 μN (or 1–10 mN) over a bandwidth exceeding 700 Hz. The drive voltage for bent-beam actuators is less than 12 V, which offers excellent compatibility with standard electronics interface requirements. Regarding current and drive voltage requirements, nickel actuation devices require input current less

Table 3.5 Input Power Requirements as a Function of Displacement (in milliwatts, mW)

Device Material and Parameters	Displacement (μm)					
	1	*2*	*3*	*4*	*5*	*10*
Silicon Device	15	28	52	61	75	105
$W = 14$ μm, $H = 3$ μm						
$L = 800$ μm, $\theta = 0.2$ rad						
Nickel Device	18	33	45	53	65	85
$W = 6$ μm, $H = 3$ μm						
$L = 400$ μm, $\theta = 0.05$ rad						

than 100 mA and drive voltage close to 2 V, whereas silicon actuation devices require current levels below 10 mA and drive voltages around 12 V. The major drawback of the bent-beam actuator is its long response time, because its thermal time constant is much higher than the electrical and mechanical time constants.

3.6   Summary

This chapter has summarized the performance capabilities and limitations of ESRMs with conventional as well as tilted configurations, microactuators best suited for dual-stage servos, microactuators with optimum shaped electrodes, third-generation MEMS piggyback microactuators with integration of magnetic read/write heads, VCAMs and lateral comb array microactuators, comb drive actuators with shaped comb fingers capable or providing stable and constant output force, and BBET actuators to provide rectilinear displacements and force generated by leveraging deformations caused by localized thermal stresses developed in the actuator structure. Computerized data on ES force generated by the conventional and tilted configurations of rotary microactuators as a function of clearance in the operational direction and tilt angle is presented. Improvement factors in the actuation force generated are specified. Normalized torque plots as a function of normalized angular displacement and normalized capacitive gap generated by a rotary microactuator with optimum shaped electrodes are presented. Calculations and plots of displacement and mechanical resonance frequency calculation as a function of piggyback actuators dimensions, number of electrodes, and structural material's properties are provided. Performance comparison data on VCAMs and lateral comb array microactuators is summarized. Maximum displacement calculations and plots as a function of bent-beam angle and beam length in the absence of external loading force are provided. Most of the actuating mechanisms described here operate at drive voltages close to 12 V, which is not only compatible with low-voltage microelectronic interface but also can be easily supplied by the standard integrated power supplies readily available in the market at minimum cost.

References

1. S. Jung and J.U. Jeon, Optimal shape design of a rotary microactuator, *Journal of MEMS*, 10(3), September 2001, 460–468.
2. H. Toshiyoshi and M. Mita, A MEMS piggyback actuator for hard-disk drives, *Journal of MEMS*, 11(6), December 2002, 648–652.
3. A.S. Kumar and K. Najafi, Vertical comb array microactuators, *Journal of MEMS*, 12(4), August 2000, 440–441.
4. B. Jenser, S. Mutlu et al., Shaped comb fingers for tailored for electromechanical restoring force, *Journal of MEMS*, 12(3), June 2003, 373–382.
5. L. Que, Y.B. Gianchandani et al., Bent-beam electrothermal actuators—Part I: Single beam and cascaded devices, *Journal of MEMS*, 10(2), June 2001, 247–252.

Chapter 4

Packaging, Processing, and Material Requirements for MEMS Devices

---

4.1 Introduction

This chapter deals with the packaging, and processing requirements essential for the design and development of MEMS (microelectromechanical systems) devices incorporating nanotechnology (NT) or smart materials. Structural materials most desirable for fabrication of actuator beams are identified. Important properties of materials best suited for MEMS devices for aerospace and space applications are summarized. Critical properties for passivation layers widely used in MEMS switches are described. Thermal, mechanical, radio frequency (RF)/microwave, and optical properties of the materials needed in the fabrication of MEMS devices for critical military and aerospace applications are summarized. Important characteristics of soft/hard substrates, piezoelectric, and ferromagnetic materials for fabrication of MEMS devices are discussed.

Properties of important etchants best suited for specific MEMS applications are summarized. Properties of thin-film materials needed in the fabrication of MEMS devices such as MEMS transducers, resonators, and accelerometers are identified. Packaging and sealing material requirements are summarized with emphasis on coefficient of thermal expansion (CTE) match and mechanical integrity of the device. It is important to mention that rapid progress in the design, development, fabrication, and commercialization of MEMS and NT devices is contingent on the availability of advanced materials to the MEMS designers. In addition to traditional materials used in silicon microelectronics involving single crystals and polysilicon, silicon oxide (SiO), silicon nitride (Si_3N_4), and aluminum (Al), it is now possible to introduce and integrate a wide variety of metals, alloys, semiconductors, ceramics and glasses, polymers, and elastomers (polyimides) in the design of MEMS or NT devices. In summary, important properties of structural materials, high-quality factor (Q) dielectric ceramics, and low-loss electroplating materials best suited for fabrication of MEMS devices are summarized.

MEMS designers must have a choice of materials or set of materials to optimize performance capabilities of MEMS and NT devices/sensors with minimum cost and complexity. Studies performed by material scientists have identified three critical material requirements, namely, compatibility with silicon technology, desirable electromechanical properties, and minimum residual stresses. In brief, a MEMS designer must be able to select a material or materials best suited for MEMS pressure transducers, micropumping, shock resistance microbeams, micromechanical filters, force-detection sensors, and optical-MEMS and RF-MEMS switches.

Critical issues associated with integration of various metals and alloys into microsystems are discussed with emphasis on cost-effective design and reliability. Materials for mm-wave MEMS switches, phase shifters, and RF resonators are identified for the design of new generation of MEMS devices best suited for fighter/bomber radar, missile tracking radar, communication equipment, and electronic warfare (EW) systems using electronic beam steering capabilities vital for fast and precision response.

4.2   Packaging and Fabrication Materials

This section focuses on various types of materials required in packaging and fabrication of MEMS devices. Various materials may be required for specific MEMS elements such as cantilever beam, contact pad, substrate, wafer, cavity, feedthrough, packaging housing, seal cap, and electroplating film. The fabrication of a MEMS device may involve several materials and various processes, namely, surface machining, deposition of sacrificial film, wet etching process to dissolve the sacrificial layer, beam pattern on the polysilicon layer, installation of contact pads, photomask glass plate, and transfer of negative image of the mask to the resistive layer.

In surface machining process, thin metallic films are selectively added to and removed from the wafer surface at the designated temperatures. The film materials that are eventually removed are called sacrificial materials, whereas the materials that still remain are known as structural materials, which provide high mechanical integrity under severe thermal and mechanical environments. It is critical to mention that a silicon nitride film when deposited onto a polyimide sacrificial layer yields a reliable and flexible film membrane, which is widely used in the design of MEMS switches with high isolation and phase shifters with low phase error, while operating at mm-wave frequencies. Electroplated nickel structures are best suited for MEMS switches as structural materials for actuation pads. In the case of an RF-MEMS shunt switch, different structural materials are used for the top and bottom sections of the switch to achieve high design flexibility with minimum fabrication cost. In general, the bottom section can use glass substrates for cavities, membrane electrodes, separating posts, and clamping electrodes, although the top section of the switch uses silicon substrate or polyimide material known a benzocyclobutene (BCB). Note BCB can also be used as a switch cavity material. Polyimide can be used as a sacrificial layer in certain MEMS applications.

Several high-performance substrates are available for MEMS applications. Studies performed by the author indicate that glass substrates are most ideal for electrostatic (ES) microactuators to investigate the behavior of high-speed jet flows. Quartz substrates are best suited for V-band (50–75 GHz) phase shifters and tuning filters. Gold (Au)-plated contact pads yield optimum high-power performance over long durations with no compromise in reliability. Materials best suited for contact pads are shown in Table 4.1. Materials for metal-to-metal and metal-to-ceramic seals required to meet stringent hermetic performance requirements are described wherever possible.

Table 4.1   Properties of Metals Best Suited for Contact Pads

Metal	Melting Temperature (°C)	Coefficient of Thermal Expansion (CTE) (ppm)	Thermal Conductivity K (W/cm °C)	Tensile Strength at 25°C (kpsi)
Iridium (Ir)	2410	3.8	0.57	80
Rhodium (Rh)	1996	4.6	0.84	73
Platinum (Pt)	1769	4.9	0.71	21
Tungsten (W)	3410	2.2	1.67	300

Note: Tungsten offers the highest value of thermal shock resistance, highest value of thermal conductivity, and the lowest value of CTE and provides maximum tensile strength at room temperature.

4.2.1 Packaging Material Requirements and Packaging Processes

Packaging material requirements are dependent on MEMS performance and environmental parameters such as pressure, temperature, and humidity. These two factors can have a significant impact on the functionality and reliability of MEMS devices, namely, capacitive RF-MEMS switches, requiring high isolation and minimum insertion loss over a wide frequency range [1]. It is important to mention that packaging temperature profiles with higher temperature steps could have detrimental effects on a MEMS device performance, because of fabrication materials involved with different CTE values. Outgassing of the packaging materials can be the most serious problem, depending on the level of packaging. Packaging can involve soft or hard substrates, high-strength structural materials, low-loss semiconductors and fused glass, and silicon for wafers. MEMS device packaging can be classified into two categories, namely, zero-level packaging and first-level packaging, illustrated in Figure 4.1. Regardless of the packaging classification, the MEMS package must protect device functionality and required operation against external environmental factors such as pressure, temperature, humidity, dust, and toxic gases. The package design can be tailored to control pressure and temperature in the cavity, if necessary to achieve optimum performance. In precision or sensitive

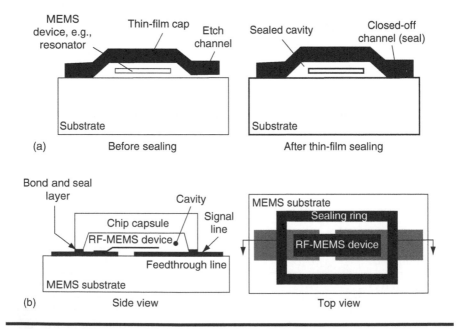

Figure 4.1 Critical elements of zero-level packaging using (a) thin-film technology, (b) bonded chip capsule,

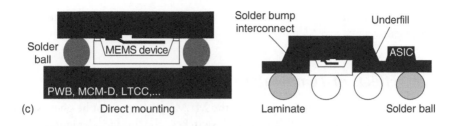

Figure 4.1 (continued)   (c) mounting techniques.

MEMS devices, encapsulation must be accomplished at the wafer level to prevent exposure of the free structures to particles and debris produced during the dicing and handling process.

4.2.1.1   Sealing Methods

Two sealing methods are available, namely, metal/soldering sealing and polymer (BCB) sealing, which are best suited for MEMS devices such as RF-MEMS and optical-MEMS switches. The metal sealing has the distinct advantage to be an excellent hermetic in some applications, but it requires processing on the MEMS substrate. The BCB sealing method requires processing on the top cavity, which provides electrical insulation but does not provide a good hermetic seal. In the case of RF-MEMS switches, packaging on-wafer level requires both the glass cap and the BCN seal to provide consistent RF performance over extended periods. In various MEMS applications, materials shown in Table 4.2 are capable of providing reliable

Table 4.2   Materials Best Suited for Ceramic-to-Metal and Glass-to-Metal Seals

Metal	Melting Point (°C)	Resistivity ($\times 10^{-6}$ Ω cm)	
		at 25°C	at 500°C
Platinum (Pt)	1769	10.03	29.10
Tungsten (W)	3410	5.44	18.50
Molybdenum (Mo)	2610	5.70	15.80
Palladium (Pd)	1556	10.79	31.30
Titanium (Ti)	1675	47.52	76.12

Note: Tungsten has high pitting resistance. Molybdenum oxidizes at elevated temperatures. Titanium has high strength and stiffness at elevated temperatures and provides most reliable ceramic-to-metal and glass-to-metal seals.

ceramic-to-metal and glass-to-metal seals under severe thermal and mechanical environments.

4.2.2 Effects of Temperature on Packaging

Impact of temperature on the integrity of the MEMS package must be given serious consideration. In the case of RF-MEMS switches, the bridge element is made of metals such as gold (Au), platinum (Pt), or aluminum (Al) alloy. In the case of MEMS switch packaging, the switch assembly can be subjected to different curing temperatures and different temperature step-cycles. In the case of the BCB process, a curing temperature of 250°C is required. So the switch designer must make sure that the metal or the alloy selected for the bridge can withstand this curing temperature and the bridge will not be subjected to excessive thermal stresses. It is important to mention that an aluminum–copper–magnesium–manganese alloy film when heated to 200°C and cooled back to room temperature will not be affected adversely. This shows that the MEMS bridge made of this particular alloy would not be deformed at this temperature. However, when the curing temperature is increased above 200°C, the thermal stress curve starts deviating from its linear elastic behavior at a certain temperature and the stress changes due to plastic deformations in the material. Upon cooling, significant increase in tensile stress can be observed. However, if the packaging or the curing temperature profile exceeds the critical temperature in a certain material or alloy, the stress developed in the bridge might develop sudden deformation leading to a catastrophic failure of the switch. In a bridge structure with zero-level packaging using BCB sealing material and glass cap, very small deformation can be expected.

4.2.3 Effect of Pressure on Packaging and Device Function

Adverse effects of pressure within the cavity on packaging integrity and MEMS device function must be carefully investigated. Complete investigation can be accomplished using two bridge designs, namely, a mechanical stiff with small holes and a less stiff with large holes. If the gas pressure within a cavity or housing is reduced below 0.0002 mbar N, the switching speed is increased drastically and at the same time the pull-in and pull-out times are much shorter. Under these conditions, the bridge goes down and touches the dielectric medium. A vibration effect or critical damping can be observed under high pressure conditions, which must be avoided to preserve the mechanical integrity of the bridge. At critical damping, the bridge structure moves very fast leading to high instability and device failure. In summary, the excessive gas pressure within the cavity or device enclosure can impact both the device operation and reliability. It is important to mention that the failure mode in a capacitive RF-MEMS switch can occur either due to excessive gas pressure within the cavity or due to an electrical charge that builds up in the insulator material involving a silicon oxide, titanium oxide, or silicon nitride, when the bridge is touching the dielectric as mentioned previously.

Regarding device longevity, the lifetime is approximately 100 times longer in the nitrogen gas environment. This effect is attributed to the humidity of the air and its influence on charge trapping in the insulator material. It is critical to mention that humidity enhances the charge trapping leading to a faster charge buildup, which will shorten the lifetime of a MEMS capacitive switch, and the charge effects occur at the ends of the insulator. Furthermore, the difference in the breakdown voltage between that of the bridge and the insulator can occur under different operating environments within the cavity or package. Studies performed by the author indicate that by selecting a low-pressure gas and the right stiffness material can lead to optimum design of the RF switch. In summary, to achieve optimum performance from a MEMS switch with zero-level packaging, optimum internal gas pressure is necessary. In addition, a constant, well-controlled gas pressure in the packaging is required for superior RF performance and high reliability. The studies further indicate that an RF-MEMS switch will exhibit a higher reliability when operated in nitrogen rather than in air. Furthermore, a fully hermetic package will maintain constant hermeticity during the lifetime of an RF-MEMS switch or any MEMS device.

4.2.4 Fabrication Aspects for MEMS Devices Incorporating Nanotechnology

Packaging and fabrication issues (Figure 4.2) must be addressed to achieve optimum design of MEMS and NT devices with minimum cost and complexity. MEMS and NT devices involve fragile, movable parts, which require fabrication and packaging in a clean and stable environment. Encapsulation of MEMS devices is possible using hermetic one-level ceramic or metal-can packages, but the cost is very high and the technology is very complex. Standard wafer sawing is possible because it will not destroy the delicate movable elements of the MEMS devices or sensors. This means that the packaging must be carried out during wafer processing before die singulation [2].

This particular packaging is known as zero-level packaging, which creates an on-wafer enclosure around the MEMS device, acting as a first protective interface. Two distinct approaches, namely, "thin-film capping" and "chip-capping" are involved in the zero-level packaging concept. Regardless of the approach used, the MEMS package must satisfy the following requirements:

■ Protection of various MEMS device elements under harsh environments during assembly and part screening processes
■ High tensile and shear strength capabilities
■ Controlled cavity environments

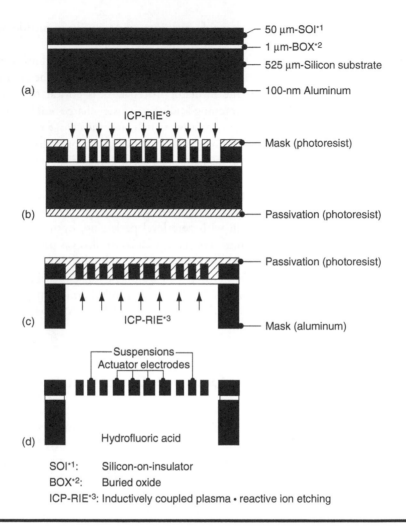

Figure 4.2 Fabrication processing steps for a MEMS actuator on SOI wafer.

- Hermetic cavity sealing capability to meet stringent humidity operating conditions
- Electrical feedthrough provisions such as direct current (dc), RF, and microwave with minimum loss
- Minimum impact on a MEMS device performance
- Temperature process not to exceed 350°C
- Rapid and efficient heat dissipation capability
- Avoidance of thermal stresses due to CTE mismatch between various materials
- Cost-effective integration with RF front-end process

4.2.4.1   Thin-Film Capping Requirements for MEMS Devices

Thin-film capping involves surface micromachining process and is widely used in the design of various MEMS devices such as MEMS resonators, pressure transducers, and accelerometers. Thin-film encapsulation requires a sacrificial layer. After removing of the sacrificial layer, the etched channel is sealed off using either a conformal coating of low-pressure chemical vapor deposition (LPCVD) nitride layer, plasma-enhanced chemical vapor deposition (PECVD) oxide layer, or simply a deposited metallic film. When aluminum or gold is used to fabricate a MEMS device, the process temperature exceeding 400°C cannot be used for deposition and sealing of the thin-film cap. Sealing at reduced pressure is not advisable, because the RF switches and the tuning capacitors must operate at standard pressures in an inert ambient condition, if reliable RF performance is the critical requirement. Metal thin-film encapsulation is most desirable for tight temperature budgets.

4.2.4.2   Chip Capping and Bonding Requirements

The studies performed by the author indicate that chip capping offers maximum flexibility in processing temperatures. In chip capping, it is necessary to bond a recessed copper onto a MEMS device wafer to retain most of the optical properties. Capping can be achieved using two distinct bonding methods. One method realizes on solder and the other on BCB polymer as the bonding and sealing layer. In the case of MEMS devices using aluminum or gold thin-film capping, processing temperatures not exceeding 400°C are recommended for the deposition and sealing of thin-film capping.

However, the bonding process must be performed at lower temperatures not exceeding 325°C to avoid adverse effects on the metallization and other materials used in MEMS devices. Standard bonding techniques include solder bonding, thermo-compression bonding, and other bonding concepts using low temperature glass seals, epoxy seals, and polymer adhesive materials. Note the bonding technique must provide adequate shear and pull strength, besides hermeticity of the seal. Strong bonds require shear strengths better than 15 MPa or 2175 psi. A zero-level package with good hermetic seals prevents moisture and other organic contaminants from migrating into the active regions of the MEMS device. Note true hermeticity is not possible with BCB polymers. Glass, silicon nitride, ceramics, and certain metals are known to achieve most effective seals.

4.2.4.2.1   Potential Sealing Processes and Associated Materials

Glass-to-metal seals are best suited for applications where excellent RF, optical, and insulating qualities, and high metallic conductivity are the principal design requirements. The glass must be wet and adhere to the metal, and the thermal expansion coefficients of the glass and metal should closely match over the entire temperature range. Differential expansion between the glass and metal must be kept well below 100 ppm to avoid compressive stresses in short seals.

Kovar seals use iron–nickel–copper alloy with low thermal expansion rate, which matches that of several high or low-expansion glasses. An advantage of Kovar seals is that they can be fired in air because the oxidation rate of Kovar is very low compared to the softening rate of hard glass. The problem of power losses at higher microwave frequencies and the resultant heating of the glass–metal joint can be eliminated by plating the Kovar with silver, gold, or chromium to lower its effective resistance. Kovar seals are widely used in vacuum technology and in packaging of solid-state MEMS devices.

Fused silica-to-metal seals are most suitable for glass-to-metal seals needed for high-temperature applications. Because fused silica has a very low thermal expansion rate, it produces minimum strain between the metal and glass, thereby ensuring high joint integrity of the seal even at elevated temperatures. These seals are best suited to seal large diameter (3 mm) tungsten rods to fused silica.

4.2.4.2.2   Development of Various Stresses in Seals

Development of stresses must be avoided to maintain integrity and reliability of the seal. The stresses that develop in a glass-to-metal seal are proportional to the differential contraction between the two materials. When molten glass is first joined to a metal, no stresses are developed until the glass becomes rigid. Glass is considered rigid at its setting point, which is about 20°C below its annealing temperature. Below the setting point, the differential contraction rate of the metal and the glass will determine the stresses developed in the seal. Tangential stress can be computed using the following expression:

$$\text{Tangential stress} = \left[\frac{E_{\text{ave}}\Delta L}{2L}\right] \tag{4.1}$$

where E_{ave} is the average elastic modulus for the two sealing materials used and $\Delta L/2L$ is the differential contraction between the two materials. Stresses between 715 and 2145 psi can be tolerated in tubular butt seals, depending on the size and the quality of the seal.

4.2.4.3   Transition and Feedthrough Requirements for MEMS Devices

Materials used in packaging must not have adverse effects on the device performance after capping of the package. Low-loss transitions are required for minimum overall insertion loss and detuning of transition lines due to proximity coupling to the cap.

Four distinct implementation of RF transitions are available, namely, coplanar horizontal feedthrough, buried horizontal feedthroughs, vertical vias through the substrate, and vertical vias through the capping chip made of high resistivity silicon substrate. Buried feedthroughs are not suitable for operations beyond 30 GHz due to increased insertion losses. Vertical vias technique offers the most compact packaging, but at higher complexity. Coplanar horizontal seethroughs are simple and most cost effective. A coplanar feedthrough using BCB as the dielectric substrate, acting both as bonding and seal layers, can provide a cost-effective package design. At mm-wave frequencies, the sealing and packaging materials and techniques play a key role in the reliability of MEMS devices. Plastic packaging designs are best suited for operation below 5 GHz or so. Ceramic packaging is most ideal for RF-MEMS and optical-MEMS devices operating at mm-wave frequencies and under harsh environments.

4.2.4.4 Material Requirements for Piezoelectric Actuators

It is important to mention that the actuator is the most critical element of a MEMS device. Even its intermittent operation and instability are not acceptable for MEMS devices. Materials used in the design of actuators must meet stringent performance requirements to preserve its mechanical integrity and performance stability under harsh operating environments. As stated previously, ES actuating mechanisms using conventional materials could suffer from pull-in, pull-out, and mechanical instability problems, thereby creating serious reliability problems for the MEMS devices. Studies performed by the author on various actuators indicate that the piezoelectric actuators do not suffer from these problems, if designated piezo-electric materials are used in the design and fabrication of actuator beams.

Therefore, it is desirable to examine potential piezoelectric materials for the possible use of actuator beams in MEMS devices, which include aluminum nitride (AIN), lead–zirconate–titanate (PZT), and zinc oxide (ZnO). Critical parameters of piezoelectric materials include piezoelectric stress constants or coefficients (e_{ye}), elastic stiffness constants (C_{IF}), which are also known as electric stress constants and transverse piezoelectric strain constants (d_{31}). These constants or coefficients have units expressed in MKS units. The piezoelectric stress constants are expressed in coulomb per newton (C/N) or in meter per volt (m/V), the elastic stiffness constants are expressed in newton per square meter (N/m^2) and the transverse piezoelectric strain constant is expressed in coulomb per newton (C/N). Important properties of potential piezoelectric materials best suited for actuator beams are summarized in Table 4.3. The dielectric constant or permittivity (ε_{11}) has no unit. Modulus of elasticity (E) is expressed in megapascals.

The magnitude of the transverse strain constant d_{31} for these materials is temperature sensitive. For example, its value for the PZT material ranges from 123 pC/N at −100°C to 125 pC/N at 0°C to 118 pC/N at 100°C. The alphabet

Table 4.3 Properties of Piezoelectric Materials for Actuator Beams

Material	Density (kg/m³)	Dielectric Constant or Permittivity (ε_{11})	Transverse Piezoelectric Strain Constant d_{31} (pC/N)	Elastic Stiffness Constants				Young's Modulus E (GPa)
				C_{11}	C_{22}	C_{33}	C_{66}	
Aluminum nitride (AlN)	3.26	9.5	3.2	17.40	6.05	19.80	5.40	320
Lead–zirconate–titanate (PZT)	7.50	1475	123	13.90	15.90	11.50	14.60	70
Zinc oxide (ZnO)	5.68	8.33	5.2	21.10	4.75	21.85	4.43	83

"p" stands for pico (10^{-12}). One can expect similar variations in this parameter for other piezoelectric materials as a function of operating temperatures. For simple calculations and for rapid comprehension by the readers, the value of this parameter has been rounded to 3 for AlN, to 120 for PZT, and to 5 for ZnO in numerical examples with errors not exceeding 1 percent.

4.2.4.5 Material Requirements for Structural Support, Electrodes, and Contact Pads

Material requirements for structural support, housing, electrodes, and contact pads are quite different from that needed for a actuator's cantilever beams. Materials for mechanical support structures and housing must have high mechanical strength and must be capable of maintaining structural integrity of the structure under severe thermal and mechanical environments. Aluminum and nickel are best suited for structural materials. Gold-plated contact pads made from copper-plated nickel and tungsten will able to withstand high compressive loads with no surface damage under severe thermal and mechanical stresses. Important thermal and mechanical properties of materials best suited for cantilever beams, structural supports, electrodes, and contact pads are summarized in Table 4.4.

4.2.4.6 Requirements for Electrodeposition and Electroplating Materials

Requirements for electroplating or electrodeposition materials must meet the stringent reliability and electrical performance of the critical parts of the MEMS devices

Table 4.4 Thermal and Mechanical Properties of Structural Materials Best Suited for MEMS Devices

Material	Density (g/cc)	Thermal Conductivity K (W/cm°C)	Tensile Stress (psi)	Coefficient of Thermal Expansion (CTE) (ppm/°C)
Aluminum (Al)	2.71	2.18	9,000	23.1
Copper (Cu)	8.96	3.94	32,625	16.5
Gold (Ag)	19.32	2.96	16,445	14.2
Nickel (Ni)	8.90	0.91	46,189	13.3
Platinum (Pt)	21.45	0.69	22,882	8.9
Tungsten (W)	19.32	1.99	130,000	4.3

Table 4.5 Thermal and Mechanical Properties of Electroplating Materials

Material	Melting Point (°C)	Coefficient of Thermal Expansion (CTE) (10⁻⁶/°C)	Thermal Conductivity K (W/cm °C)	Elastic Modulus (10⁶ psi)
Copper (Cu)	1083	16.6	3.74	17
Nickel (Ni)	1452	13.5	0.58	30
Platinum (Pt)	1769	8.9	0.68	22
Silver (Ag)	961	19.7	4.08	11

such as electrodes and contact pads. Thermal and mechanical properties of these materials [3] best suited for MEMS applications are summarized in Table 4.5.

4.3 Impact of Environments on MEMS Performance

This section focuses on adverse effects due to environmental factors on MEMS device performance during the curing and encapsulation processes. Most of the properties of metals, alloys, substrates, and semiconductors are specified at normal temperature and pressure (NTP). Mechanical, electrical, dielectric, and thermal properties of the materials used in MEMS and NT devices are most likely to be affected by temperature, pressure, chemicals, toxicity, and electromagnetic environments. It is important to point out that the one-level package provides mechanical and environmental protection to MEMS devices it holds, with no degradation in performance under NTP conditions. A zero-level package for a MEMS device using small cavity volume with stable gas pressure offers optimum performance and reliability.

Physical cavity parameters, cantilever beam dimensions, and mechanical stresses such as tensile stress, compressive stress, and torsional stress are affected by the increase or decrease in operating temperature inside the cavity or package. In the case of semiconductors, the electron velocity, transconductance, gain, and noise figure undergo changes as a function of temperature. In the case of structural materials, tensile and compressive yield strengths are reduced at elevated temperatures. Furthermore, when the temperature exceeds the recrystallization point, the effects of prior heat treatment are altered. Under certain thermal conditions, heat can cause plated coatings to diffuse into grain boundaries because of the "stress alloying" phenomenon, thereby changing the physical characteristics of the base metal. Most metals will experience permanent deformation in the form of creep, which is the time-dependent part of the strain resulting from the stress, excluding the elastic, thermal, and instantaneous strain components. Change in physical properties of the structural metals and alloys are possible because of cryogenic temperatures and extreme elevated temperatures. However, a well-designed MEMS package will not see such radical changes in the operating temperature, which is required for good electrical performance and excellent mechanical stability.

Variations in tensile and compressive strengths for various structural materials as a function of temperature can be seen in standard military handbooks. For example, the ultimate tensile strength for aluminum alloys drops by 143 percent of the room temperature strength at −425°F to 4 percent at +800°F. The effects at elevated temperatures on the grain boundaries in certain thin-film coatings are very alarming. The effects of metallic coatings such as nickel plating commonly used for MEMS actuator electrodes on the grain boundaries of certain metals at elevated temperatures have been observed. For specific details about grain boundary, refer to any mechanical engineering handbook or to the *Military Handbook (MIL-HDBK-5)*.

In the case of integrated circuits (ICs) or semiconductor devices used in MEMS sensors, reliable and proven techniques must be used to detect and remove harmful contaminations that can occur during the fabrication of the plastic thin films or thin sacrificial layers. Failure to remove such contaminations will affect the MEMS device reliability and overall performance. The presence of certain gases in MEMS cavity during the encapsulation phase can inject undesirable gas vapors. For example, the presence of ammonia gas (NH_3) will inject unwanted transmission of NH_3 into the semiconductor or plastic materials during the encapsulation process, which can significantly degrade the device performance. This is due to the fact that this gas permeability contributes to the microporosity in certain plastic or semiconductor thin films. The permeability characteristic is responsible for the retention of unwanted vapors, which is a direct result of gassing phenomenon during the curing process. This effect causes the IC package to maintain reverse leakage currents, which generate reverse bias stress.

4.3.1 Impact of Temperature Variations on Coefficient of Thermal Expansion

The CTE parameter (α) for a material provides [5] change in length per degree Celsius from the length at 0°C. The coefficient of volumetric expansion (β) is approximately equal to three times the linear expansion ($\beta = 3\alpha$) in solids. The coefficient of volumetric expansion for the liquid is the ratio of the change in volume per degree Celsius to the volume at 0°C. Mathematical expressions for the linear and volumetric CTEs involving various parameters can be written. The expression for the linear CTE is given as

$$\alpha = \left[\frac{L_t - L_0}{L_0 t}\right] \tag{4.2}$$

where
L_t is the length at temperature t
L_0 is the length at 0°C

Table 4.6 Thermal Properties of Metals Best Suited for MEMS Devices for Possible Applications in Aerospace and Satellite Systems

Metal	Service Temperature (°C)	Thermal Conductivity K (W/cm °C)	Coefficient of Thermal Expansion (CTE) × 10^{-6} per	
			°C	°F
Aluminum (Al)	660	2.18	23.1	13.3
Magnesium (Mg)	651	1.55	25.2	14.1
Nickel (Ni)	1535	0.91	13.3	7.4
Tungsten (W)	3410	1.99	4.3	2.3

Note: Aluminum and magnesium are the lightest materials and are best suited for MEMS devices for aerospace and satellite system applications.

A more complex and accurate expression for length at temperature *t* with linear expansion can be written as

$$L_t = L_0\left[1 + at + bt^2 + ct^3\right] \qquad (4.3)$$

where *a*, *b*, and *c* are the coefficients, which can be determined empirically. The length at any operating temperature can be computed using Equation 4.3.

Thermal properties of various metals best suited for MEMS support and housing applications in aerospace and satellite systems are shown in Table 4.6. Major emphasis on the metals used for aerospace and satellite applications is placed on lightweight, long operating life, and high mechanical strength. It is important to mention that residual stresses in the packaging seals must be minimized, if long operational reliability of a MEMS device is the principal requirement. To ensure long MEMS device life, the differential thermal expansion mismatch between substrate and components, interconnections and seals should be as small as possible, if minimum residual stresses in the seals are the principal design requirements.

4.3.2 Effects of Temperature on Thermal Conductivity of Materials Used in MEMS

Thermal conductivity plays a key role in designing a reliable package for a MEMS device, because it promotes enhanced heat dissipation capability and improved thermal shock resistance. Substrates with high thermal conductivities permit greater component packaging density, which is the most desirable design requirement for

Table 4.7   Impact of Alumina Contents on the Thermal Conductivity of Alumina-Based Substrates

Alumina Content (Percent)	Thermal Conductivity K (W/cm °C)	Change in Thermal Conductivity (Percent)
99	0.292	None
98	0.256	−13
97	0.218	−26
96	0.181	−39
85	0.147	−50

MEMS and NT devices. With greater component packaging, surface temperatures can be lowered and the danger of hot spots is minimized through the use of miniaturized heat sinks. Reduction in thermal conductivity of a substrate at elevated temperatures can degrade the device thermal performance. In the case of alumina substrate with 100 percent silicon oxide content, the thermal conductivity is approximately 0.315 at 20°C, 0.275 at 100°C, and 0.118 at 400°C (W/cm °C). However, at cryogenic temperatures, one can expect higher values of thermal conductivities.

Alumina-based substrates are widely used in the design of mm-wave MEMS switches and phase shifters. Studies performed by the author on alumina-based substrates indicate that higher alumina content is required to achieve a high thermal conductivity. Any reduction in alumina content will lower the thermal conductivity of the substrates. The studies further indicate that the thermal conductivity decreases roughly by 13 percent, when the alumina content is lowered from 99 to 98 percent, and by 50 percent when the alumina content is lowered to 85 percent, as illustrated in Table 4.7.

Note ceramic substrates have higher thermal conductivities than glass substrates. But glazing ceramic substrates to improve surface finish will decrease their thermal conductivities. Furthermore, beryllium (Be) has a higher thermal conductivity than alumina (Al_2O_3), but it is very expensive compared to alumina.

Thermal shock resistance of substrates is very important because it affects the ability of a substrate to withstand soldering and joining operations in the MEMS package. Thermal conductivity, thermal expansion, and mechanical strength largely determine the ability of a substrate to withstand thermal shock. The high thermal shock resistance of beryllium is due to its high thermal conductivity, whereas that of fused silica or alumina is due to its low rate of thermal expansion. This means that a substrate with high thermal conductivity and low thermal expansion rate is best suited for MEMS devices for possible applications in aerospace and satellite systems.

Silicon is another substrate which is widely used in the fabrication of MEMS devices because it is relatively cheap and offers reasonably good thermal performance at elevated temperatures. Impact of temperature on thermal conductivity of various substrates is illustrated by curves shown in Figure 4.3. These curves indicate that thermal conductivities of all substrate materials experience reduction as the operating temperature increases. The silicon curve shows that the thermal conductivity is about 1.55, 1.35, 1.12, and 0.88 W/cm °C at 20°C, 50°C, 100°C, and 200°C, respectively. It is important to mention that drastic variations in thermal conductivity of these substrates can have a significant impact on packaging density, thermal shock resistance, and CTE of the materials used in the MEMS packaging or housing. As stated earlier, the thermal conductivity of these materials will improve at cryogenic temperatures.

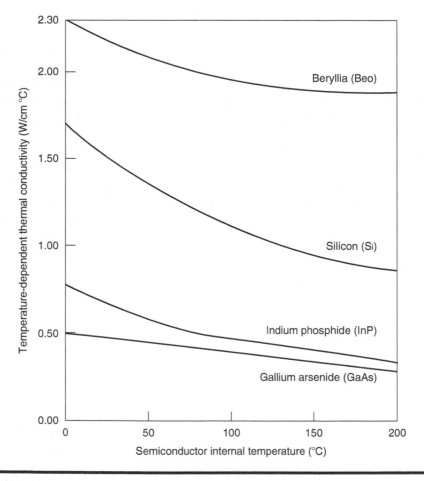

Figure 4.3 Impact of temperature on thermal conductivity of various semiconductor and substrate materials.

4.3.3 Special Alloys Best Suited for MEMS Applications

This section describes potential alloys with controlled expansion (CE) characteristics, which can significantly improve the mechanical and thermal performance of MEMS devices. The advantages and unique properties of CE-alloys [4] can be exploited in designing the future generation of MEMS devices. The ability to tailor the CTE values of these CE-alloys to ceramic boards and components operating at high frequencies, combined with their lightness, high thermal conductivity, dimensional stability, and manufacturability with minimum cost, have made them most attractive for RF/microwave packages, carriers, and heat sinks. The additional capability of high stiffness has made the CE-alloys with lower CTEs particularly suitable for optical and electro-optic housings. Furthermore, the higher thermal expansion of the CE-alloys are widely used in carrier plates for laminate PCBs, guided bar for circuit boards, and fixtures in semiconductor processing equipment and soldering ovens.

Currently, the CE-alloys are being considered for RF-MEMS and optical-MEMS devices because of their unique characteristics. These lightweight alloys are composed mostly of silicon and aluminum contents, and their CTEs can be controlled to a specific value between 5 and 20 ppm/°C through appropriate adjustment of the two constituents' properties. Brief studies performed by the author indicate that the new CE-alloys are compatible with many common microelectronic devices and substrates widely used in MEMS devices.

CE-alloys shown in Table 4.8 are typically three to six times lighter than conventional packaging and base plate materials widely used in RF/microwave devices. The excellent thermal conductivity of these alloys will significantly increase the power-handling capability of the future generation of MEMS devices, which is

Table 4.8   Physical Properties of CE-Alloys

CE-Alloy	Composition (Percent)	Coefficient of Thermal Expansion (CTE) (ppm/°C)	Thermal Conductivity K (W/cm °C)	Tensile Stress (MPa)	Density (g/cm³)
CE-11	Si-50 percent Al	11.0	1.49	125	2.50
CE-9	Si-40 percent Al	9.0	1.29	134	2.45
CE-7	Si-30 percent Al	7.4	1.20	100	2.40
CE-5	Si-15 percent Al	5.0	1.18	82	2.35

currently lagging. In addition, the high stiffness of these alloys are most amenable to standard machine operations such as drilling and milling.

Comprehensive examination of CE-alloy properties compared to other standard materials for RF/microwave devices or packaging indicate that CE-alloys will significantly improve the mechanical and thermal performance of MEMS devices because of their excellent thermal conductivity and thermal shock resistance characteristics. For example, the CTE of CE-7 alloy offers 33 percent reduction in weight, seven times improvement in thermal conductivity improvement at elevated temperatures, machining and electroplating improvement by 85 percent over tungsten and Kovar. Furthermore, the CE-alloys are nontoxic, because they contain silicon and aluminum materials. They are environmentally friendly and easy to handle and do not present any disposal problems It is interesting to note that machining operations do not produce burrs on the higher silicon CE-alloys, in contrast to most other metals. These alloys can be readily plated without any expensive operations. Heat is distributed evenly in these alloys, which makes it easy for soldering of feedthroughs in MEMS devices.

4.3.3.1 Benefits of CE-Alloys in RF/Microwave MEMS Packaging

CE-9 and CE-7 alloys are best suited for RF/microwave packaging for aerospace applications involving operational frequencies as high as 40 GHz, because of their close CTE match to circuit boards, high thermal conductivity, low density, dimensional stability, and ease of machining. CE-11 alloy is best suited for substrates for RF and microwave circuits and currently is widely used for the housing of a transmit/receive module operating in the 3–30 GHz range. CE-7 alloy is best suited in the fabrication of high-speed MEMS switching device, comprising an array of these modules widely used in aerospace applications. CE-7 provides low CTE, reduction in weight, enhanced stiffness, and excellent dimensional stability.

4.3.3.2 Benefits of CE-Alloys for Thermal Backing Plates

Thermal backing plates are used to laminate printed circuit boards (PCBs) formulated for high frequency MEMS applications. Sheets of CE-9 and CE-11 alloys can be filled with ceramic in the boards to produce a tight control of dielectric constant and improved thermal stability in the dielectric medium. Addition of woven glass reinforcement will provide additional stiffness in the *xy*-plane. CE-alloy backing plates are available in sizes up to 20×20 in.^{2} with thickness ranging from 0.5 to 3 mm. CE-alloys provide backing plates, which are capable of providing good CTE match, lightweight assembly, high thermal conductivity, and high specific stiffness. Its high stiffness permits the laminate substrate to be extremely thin, while maintaining flat surface with reduced microwave and RF dispersion losses at mm-wave MEMS devices.

4.3.3.3 Benefits of CE-Alloys in Integrated Circuit Assemblies

The controlled thermal expansion of CE-alloys at elevated operating temperatures is accurately predictable, which helps in precise positioning and soldering balls in IC assembly lines, leading to substantial savings in manufacturing costs of MEMS devices. CE-alloy resistance to chipping and ease of machining with tight tolerances will be found most useful in designing precision MEMS devices incorporating special materials like graphite needed for enhanced service life and uniform temperature distribution in the IC assembly. In summary, CE-alloys offer application solutions, which will provide enhanced product performance, improved reliability, enhanced functionality-to-weight ratio, and flexibility in manufacturing of diversified MEMS products with minimum cost.

4.3.4 Bulk Materials Best Suited for Mechanical Design of MEMS Devices

Properties of bulk materials widely used in the design of MEMS and NT devices will be described with particular emphasis on mechanical design to ensure high mechanical integrity and improved reliability under harsh operating environments. Research studies indicate that the bulk materials such as aluminum, nickel, silicon, and silicon nitride are best suited for fabrication of MEMS components, because they are readily available in markets at reasonable cost. Table 4.9 summarized the properties of bulk materials best suited for mechanical design of MEMS device components.

It is important to mention that tensile fracture strength is the property of a structure, whereas the fracture toughness is a material property, which is most relevant to the design of microsystems. Therefore, to use fracture toughness values in microsystem or MEMS design, it is necessary to evaluate the flaw sizes in

Table 4.9   Mechanical Properties of Bulk Materials for MEMS Devices

Bulk Material	Young's Modulus E (GPa)	Tensile Fracture (MPa)
Aluminum (Al)	69	200
Copper (Cu)	124	400
Nickel (Ni)	214	425
Polysilicon	130	2000
Silicon (Si)	180	3950
Silicon nitride (Si_3N_4)	280 and 310	5000 and 8000

Note: One pound per square inch (1 psi) is equal to 1430 MPa.

microscale structures using nondestructive techniques, which is extremely hard in actual practice. Thus, the tensile fracture strengths of representative structures are generally used in place of fracture toughness.

4.4 Material Requirements for Electrostatic Actuator Components

This section will describe the material requirements for the critical elements of the ES actuator of the MEMS device. Unique material properties are required for the critical components of a MEMS actuator. ES actuators are widely used by MEMS devices, because of reliable performance with minimum cost and high design flexibility. The critical components or elements include the flexible cantilever beams contact pads, wafers, and membranes. Materials needed in fabricating the cantilever beams must be capable of providing high rigidity, improved mechanical strengths, high structural reliability, and dimensional stability with low actuation voltage, while operating under harsh thermal and mechanical environments. In the cases of thermal, bio-, and trimorph actuators, most critical requirements include Young's modulus, CTE, and thermal conductivity (K) of the materials to be used. Material properties of various components used by the ES actuator are summarized in Table 4.10.

Table 4.10 Properties of Materials Widely Used for MEMS Actuator Components

Material	Young's Modulus E (GPa)	Properties		Application
		Coefficient of Thermal Expansion (CTE) (ppm/K)	Thermal Conductivity K (W/m K)	
Nickel (Ni) (bulk/plated)	208/135	12.7	90.5	Contact pad
Silicon nitride (Si₃N₄)	290	2.9	19.4	Mask
Silicon (Si) (crystal)	130	2.3	156	Wafer
Silicon (porous)	75	2.3	80	Beam
Palladium (Pd)	121	11.8	70	Membrane
Silver (Ag)	83	18.9	408	Membrane

4.4.1 Material Properties for MEMS Membranes

Material requirements for MEMS membranes are very unique. High mechanical strength across the membrane length is of critical importance. Mechanical properties of the materials must satisfy this requirement, regardless of the operating environments. Predicting the mechanical strength of the membrane is very complicated, because it is dependent on several factors such as membrane design configuration, membrane physical dimensions (length, width, and thickness), lithographic process, and properties of bulk materials involved.

The micro-fabricated membrane module is generally composed of several smaller membranes acting in parallel. Each smaller membrane is formed using palladium–silver (Pd–Ag) alloy. The mechanical strength of the silicon support that surrounds the Pd–Ag membrane elements is much higher than that of a single membrane. This means the strength of the whole membrane module will be determined by a single membrane element. Both Young's modulus (E) and yield stress (σ_{yield}) are temperature dependent and both parameters values will experience reduction at higher temperatures. The material properties of an alloy can be interpolated from the properties of the individual metals as shown in Table 4.11.

It is important to mention that the rupture strength membrane made from Pd–Ag alloy appears to be reasonably good. The Pd–Ag membrane does not break at a pressure difference of 4 bars or 4 dyn/cm^2 (cgs unit) over the membrane. However, there is possibility of breaking, if the pressure difference exceeds 5 bars. The micro-fabricated membrane is considered mechanically strong at room temperature and is able to maintain its operational capability under the desired pressure gradient.

4.4.2 Sacrificial Material Requirements for MEMS Devices

Sacrificial material requirements are dependent on the types of structural materials used in the fabrication of MEMS devices. Structural and sacrificial materials best suited for RF-MEMS and optical-MEMS devices [1] are summarized in Table 4.12.

Table 4.11 Mechanical Properties of Metals Best Suited for MEMS Membranes

Metal	Mechanical Properties		
	Young's Modulus E (GPa)	Yield Stress σ_{yield} (MPa)	Ultimate Stress $\sigma_{ultimate}$ (MPa)
Silver (Ag)	83	175–330	175–330
Palladium (Pd)	121	35–205	172–335
Pd–Ag (alloy)	150	80	Not available

Table 4.12 Sacrificial Material Requirements for MEMS Devices

Structural Material	Appropriate Sacrificial Material
Aluminum (Al)	Silicon (single-crystal)
Copper (Cu)	Chrome
Nickel (Ni)	Chrome
Polysilicon	Silicon dioxide
Polyimide	Aluminum
Silicon dioxide (SiO_2)	Polysilicon
Silicon nitride (Si_3N_4)	Undoped polysilicon

4.4.3 Three-Dimensional Freely Movable Mechanical Structure Requirements

MEMS-integrated systems can be designed with minimum cost, if three-dimensional (3-D) freely movable mechanical structures are integrated in the same wafer along with the electronics. The MEMS-integrated system designed this way can produce highly functional systems with outstanding performance with minimum size, cost, and power consumption, which is not possible otherwise. However, such design is strictly dependent on right materials selected for movable structures. The 3-D movable structures can be actuated using a variety of actuation mechanisms, including ES, electrothermal, electromagnetic, electrodynamics, or piezoelectric. Regardless of actuation mechanism selected, two pairs of electrodes are needed; one pair for the dc bias and the other pair for switching electrodes or contacts across which the RF signal exists. A sacrificial layer of high resistivity is required between the two pairs to provide sufficient isolation between the electrodes. If a MEMS device uses polysilicon as a structural material, then silicon dioxide must be used as a sacrificial material and hydrofluoride as an etchant for optimum results.

4.5 Substrate Materials Best Suited for Various MEMS Devices

This section will focus on the substrate requirements for possible applications in MEMS devices. Substrates best suited for RF/microwave MEMS devices will be identified and their important characteristics will be summarized with emphasis on insertion and dispersion losses. Electrical, mechanical, and thermal properties of substrates best suited for microwave MEMS devices and operating in mm-wave regions will be investigated as a function of environmental parameters. Major emphasis will be placed on material properties, which might have significant impact on RF performance under harsh operating environments. Parameters such as mechanical strength, strength-to-weight ratio, stiffness factor, thermal shock resistance, dimensional stability, and structural integrity must be given serious consideration, if reliable operation over extended periods is the principal requirement. Properties of soft, hard, and metallic substrates best suited for MEMS applications will be summarized with emphasis on fabrication cost, electrical performance, and device reliability under harsh operating conditions.

4.5.1 Soft Dielectric Substrates

Soft dielectric substrates include Teflon ($e_r = 2.20$), RT/DUROID 5880 ($e_r = 2.22$), RT/DUROID 6002 ($e_r = 2.94$), and RT/DUROID 6010 ($e_r = 10.5$). The RT/DUROID substrates manufactured by Roger Corp., Arizona, are classified as Teflon fiberglass laminates, which are generally referred as to (poly[tetrafluoroethylene]) PTFE/glass laminates. There are two types of such laminates: woven and unwoven (microfiber), and both meet the performance requirements of soft substrates.

4.5.2 Hard Dielectric Substrates

Sapphire, alumina ceramic, fused quartz, strontium titanate, and magnesium oxide are considered hard substrates. However, alumina ceramic and fused quartz are widely used in the fabrication of RF-MEMS devices because these substrates are readily available in market with minimum cost and rapid delivery. The most popular alumina ceramic is available with 99.6–99.9 percent purity. Fine-grained alumina has more random crystalline orientation than the course-grained alumina. Note uniform electrical properties are contingent on the grain size and random orientation. However, random orientation is dependent on the accuracy of the process control followed during the manufacturing phase of the dielectric material. It is important to mention that fine grain size in the substrate material reduces the microwave scattering effects.

It is critical to point out that the ceramic-PTFE substrate or RT/DUROID 6010 is a composite quasi-soft substrate and is best suited for RF/microwave MEMS devices and microelectronic circuits operating at 10 GHz and above. This substrate offers excellent performance even at cryogenic temperatures, precision control of permittivity, low moisture absorption, excellent dimensional stability, and nearly anisotropic properties. The dispersion effects in substrates at higher microwave and mm-wave frequencies can degrade the performance of MEMS devices, if the thickness of the substrate exceeds 0.010 in. Dispersion and surface mode losses will be much higher at mm-wave frequencies, if the substrate thickness is greater than 0.010 in. [6].

Alumina substrates are widely used in RF-MEMS switches, phase shifters, and coplanar waveguide (CPW) couplers operating at higher microwave and mm-wave frequencies. The adhesion of thin films on alumina substrates depends on the density of the grain boundary areas. Fine grain size produces a high-density alumina substrate, which yields minimum scattering and dispersion effects.

Another hard substrate known as fused silica has potential applications in mm-wave MEMS switches and phase shifters operating under harsh thermal and mechanical environmental conditions. This substrate has a CTE value close to 0.50 ppm/°C over a temperature range of 0°C–200°C as illustrated in Figure 4.4 [6].

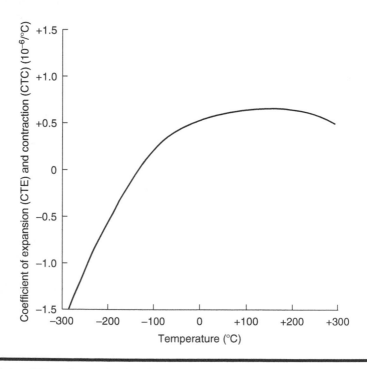

Figure 4.4   CTE and CTC for fused silica as a function of temperature.

4.5.3   *Electrical Properties of Soft and Hard Substrates*

Electrical properties of substrates are of critical importance, regardless whether the substrate is soft or hard. The most important properties of microwave substrates are dielectric constant or permittivity, and loss tangent or dissipation loss as a function of frequency and temperature. In addition, surface wave and dispersion losses at mm-wave frequencies must be taken into account. Finally, anisotropy property must be considered, if dimensional stability in the dielectric medium is of critical importance. The anisotropy parameter is the ratio of dielectric constant along the *xy*-plane to that in the *z*-plane (anisotropic constant $= \varepsilon_{xy}/e_z$) and its value close to unity is considered most desirable. Departure from unity value in the substrate occurs when a fill material such as ceramic or glass is added to achieve high dimensional stability. Note addition of fiberglass or ceramic to the basic substrate material will increase both the permittivity and the loss tangent, in addition to high dimensional stability as shown in Table 4.13. In brief, dimensional stability is achieved at the cost of high anisotropy. Uniform electrical properties of a substrate are possible when the anisotropy of the dielectric medium is close to unity. Important properties of various soft and quasi-soft substrates are summarized in Table 4.13.

It is important to point out that lowest values of dielectric constant and loss tangent are required to achieve minimum substrate losses at microwave and mm-wave frequencies. Furthermore, both the dielectric constant and loss tangent for soft and hard substrates increase as the temperature is elevated as shown in Figure 4.5a. Surface finish of substrates must be better than 2 μm to achieve optimum performance from the substrates such as lower insertion loss, reduced dispersion and surface mode losses, and uniform dielectric properties over wide temperature and frequency

Table 4.13   Important Properties of Microwave Soft and Quasi-Soft Substrates at 10 GHz Operations

Microwave Substrate	Relative Dielectric Constant (e_r)	e_{rxy}/e_{rz}	Dimensional Stability
Teflon poly(tetrafluoroethylene) (PTFE)	2.10	1.000	Excellent
Microfiber PTFE	2.20	1.025	Very good
Microfiber PTFE	2.35	1.041	Good
Woven PTFE	2.45	1.160	Worst
RT/DUROID ($t = 0.025$ in.)	10.5	1.021	Very good
RT/DUROID ($t = 0.050$ in.)	10.5	1.060	Fair

Note: Thinner substrate layers yield improved anisotropic properties as illustrated above.

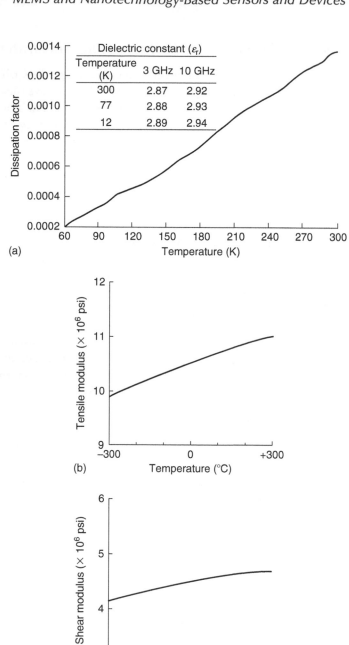

Figure 4.5 **Dielectric properties of RT/DUROID 6002. (a) Dissipation factor, (b) tensile modulus, and (c) shear modulus as a function of temperature.**

ranges. It is interesting to note that both the shear modulus and tensile modulus increase with temperature for hard substrates such as fused silica, as illustrated in Figure 4.5b and c. However, values of both parameters decrease at cryogenic temperatures [6]. In other words, mechanical strength of substrates deteriorates at cryogenic temperatures, regardless of substrate type or thickness.

Mechanical and thermal properties of alumina and fused silica are very critical, because they offer high mechanical integrity and instant heat removal capability from hot surfaces of MEMS devices are needed for device survival under harsh thermal environments. Preliminary studies performed by the author on fused silica substrate indicate that overall properties of this material is most suitable for mm-wave MEMS devices, namely, RF-MEMS switches and phase shifters operating over V-band (50–75 GHz) and W-band (75–110 GHz) frequencies and over wide temperature range under harsh thermal conditions. Its permittivity is close to 3.78 and dissipation factor is approximately equal to 0.0001 at 300 K. Its room temperature tensile strength is better than 85,600 psi and the compressive strength is in excess of 170,000 psi. It offers the lowest CTE of about 0.55 ppm/°C over a wide temperature range (0°C–200°C). Note low CTE value prevents solder and epoxy joints from cracking and deformations at cryogenic temperatures. The CTE values for the substrate and housing materials must not only be low but also as close as possible to maintain consistent performance, MEMS device reliability, and dimensional stability over extended periods.

Mechanical and thermal properties of hard substrates best suited for MEMS applications are summarized in Table 4.14.

Table 4.14 Thermal and Mechanical Properties of Substrates Ideal for MEMS

Substrate	Coefficient of Thermal Expansion (CTE) (ppm/°C)	Young's Modulus E ($\times 10^6$ psi)	Modulus for Rigidity G ($\times 10^6$ psi)	Knoop Hardness	Relative Dielectric Constant (e_r)
Silicon (Si)	3.4	15.7	8.4	950	11.80
Alumina (Al)	6.7	50.2	21.5	1370	9.82
Fused silica (or quartz)	0.5	10.6	4.5	741	3.78
Magnesium oxide (MgO)	5.9	36	22.4	1140	5.82

Note symbol E stands for Young's modulus and G stands for modulus of rigidity. Magnesium oxide substrates will be found most desirable for the MEMS devices operating at cryogenic temperatures. Sapphire substrate can be used in place of fused silica, which is also known as quartz. However, fused silica substrate offers the highest thermal shock resistance and is best suited for MEMS devices for aerospace and space applications, where thermal shock resistance is a critical requirement. Sapphire is a nonisotropic hard substrate with dielectric constant of 10.8 and dissipation factor or loss tangent of 0.00015 at 10 GHz. This substrate is most ideal for microwave MEMS tunable oscillators, where high unloaded Q better than ten million is the design requirement for minimum loss.

4.5.4 Glass–Ceramic Hybrid Substrate for MEMS

The advantage of glass–ceramic hybrid (GCH) substrate is the combination of both materials, which offers easy fabrication and outstanding mechanical and thermal properties. Because GCH substrates have zero porosity and submicrometer grains, they are mechanically stronger than many conventional glass or ceramic substrates, which may be best suited for some RF-MEMS devices. The dissipation factor or loss tangent of GCH substrate is less than 0.0003 at 10 GHz. The demand for miniaturization and enhanced reliability of microwave equipment and MEMS devices justifies the increased use of glass or ceramic substrates. GCH substrate is neither readily available nor cost effective. Note the ceramic substrates have good thermal conductivities. However, glazed ceramic substrates offer both surface smoothness, better than 0.25 microinch, and good thermal conductivity. Deployment of GCH substrates in MEMS devices operating in high microwave regions is recommended, if the material is readily available at a reasonable cost.

4.5.5 Para-Electronic Ceramic Substrates for MEMS Applications

Para-electronic ceramic (PEC) substrates may play a critical role in future generations of MEMS devices. These substrates have relatively high dielectric constants, extremely low values of CTE, and exceptionally high unloaded Q's (Q_{un}) and are best suited for lower microwave frequency operations. Dielectric properties of these substrates are mainly determined by the ionic polarization and the complex dielectric constant can be computed from the classical dielectric dispersion equations. Their important properties are summarized in Table 4.15.

It is important to point out that due to their higher dielectric constants, these substrates will be best suited for MEMS devices operating at 100 MHz and below. However, the insertion losses per unit length of the substrate will be higher due to very high permittivity.

Table 4.15   Properties of Para-Electronic Substrates at 100 MHz

Material	Dielectric Constant at 25°C	Loss Tangent at 25°C	Unloaded Quality Factor (Q) at 100 MHz
Titanium dioxide (TiO$_2$)	104	0.00025	40,000
Calcium titanate (CaTiO$_3$)	180	0.00082	7,000
Strontium titanate (SrTiO$_3$)	304	0.00014	3,500

4.5.6   Insulation and Passivation Layer Materials

Silicon nitride is a special form of nitride ceramic and is widely used as an insulation layer or passivation layer in the design and fabrication of RF-MEMS devices such as switches and phase shifters. This material has low value of CTE as well as low thermal conductivity. Silicon nitride has a dielectric constant of 6.7, which remains pretty constant over the 10–60 GHz frequency range. This material has very high resistively greater than 10^{12} Ω from room temperature to 200°C, and thus can be used as an insulation layer in MEMS switches. It has demonstrated a tensile strength greater than 25, 000 psi at elevated service temperatures and excellent thermal shock resistance. This material is best suited for MEMS contact pads. Studies performed by the author indicate that silicon nitride experiences very little volume change, thereby making it most desirable for a MEMS assembly where close dimensional tolerances are very critical. Pure silicon nitride is found in the form of a thin, dense insulation layer, which is widely used as a passivation layer in RF-MEMS switches.

Besides silicon nitride, Pyrex-glass and BCB materials are also available for insulation layer or passivation layer for applications in microwave and mm-wave MEMS devices. Dielectric properties and applications of these three materials are shown in Table 4.16.

Table 4.16   Dielectric Properties of Various Passivation and Insulation Materials

Material	Dielectric Constant	Application in Microelectromechanical Systems (MEMS) for
Silicon nitride (Si$_3$N$_4$)	6.7	Passivation layer
Pyrex-glass	4.3	Insulation layer
Benzocyclobutene (BCB)	4.2	Passivation layer

Note the BCB material can be also used for sealing and bonding applications. BCB has excellent RF property in addition to minimal outgassing and low moisture uptake. The sealing caps are made from low-loss borosilicate glass. BCB offers strong bond capability and shear strength greater than 10 MPa. However, true hermeticity cannot be expected from a polymer like BCB. Leak rates better than 10^{-8} mbar L/s for 100 μm wide sealing rings are considered most appropriate for MEMS devices.

4.5.7 Material Requirements for MEMS in Aerospace Systems

Material requirements for MEMS devices for aerospace applications are very stringent. High mechanical strength, high thermal conductivity, low CTE, and improved thermal shock resistance are the principal requirements for such materials. Important properties of aerospace materials are summarized in Table 4.17.

The room temperature thermal conductivity values will decrease by 25–30 percent at 125°C, where the tensile strength, comprehensive strength, and bending strength would experience a reduction approximately by 15–20 percent at 125°C. Still these materials will retain optimum mechanical and thermal properties under

Table 4.17   Properties of Materials Best Suited for Aerospace Applications

	Materials		
Properties	Aluminum Nitride (AlN)	Silicon Carbide (SiC) (6H-SiC)	Silicon (Si)
Hardness (kg/mm²)	1200	2,800	950
Tensile strength (psi)	29,000	28,250	25,500
Compressive strength (psi)	300,000	560,000	385,000
Bending strength (psi)	64,350	38,500	31,250
Coefficient of thermal expansion (CTE) (ppm/°C)	4.1	3.8	4.2
Thermal conductivity (W/cm °C) at 25°C	1.8	4.5	1.5

severe thermal conditions. These materials are most suited for MEMS devices for possible deployment in aerospace and satellite applications.

The silicon technology is fully matured and the material has been widely used in the design of solid-state devices, IC circuits, and hybrid circuit devices. This particular material is widely used as substrates for applications in RF-MEMS and optical-MEMS devices, because of its superior dielectric, mechanical, and thermal properties. MEMS devices using silicon substrates can operate reliably under severe thermal (250°C) and mechanical environments with no compromise in device performance. Silicon films can be thermally grown and selectively etched using standard photolithographic process. Note very precision and complex geometries in silicon films can be formed with minimum cost. Silicon material has been used in fabrication of cantilever beams for various RF-MEMS and optical-MEMS devices with minimum cost and complexity.

4.6 Summary

This chapter describes the properties of materials for various process applications involving fabrication and packaging of MEMS devices. Smart materials have been identified for possible applications of MEMS devices capable of operating in satellite and aerospace environments. Material requirements for packaging, sealing, electrode capping, contact pads, and passivation layer are summarized. Potential sealing materials and seal types are discussed with particular emphasis on limitation of differential expansion between glass, ceramic, or metal to avoid production of stresses in the seal. Material requirements for zero-level and first-level packaging are defined. Effects of temperature, pressure, and humidity on MEMS packaging and housing are identified. Packaging requirements for MEMS and NT devices operating under harsh thermal and mechanical environments are discussed in greater details. Wafer material requirements are summarized. Material properties for cantilever beams using ES and piezoelectric actuation mechanisms are discussed in great details. Effects of temperature on thermal conductivity and CTE on materials used by various elements of MEMS devices are identified. Electrical, thermal, and mechanical properties of substrate materials for RF-MEMS and optical-MEMS devices are discussed extensively. Properties of soft, quasi-soft, and hard substrates for possible applications in MEMS devices are summarized. Properties of alumina and fused silica best suited for RF-MEMS switches and phase shifters operating at mm-wave frequencies are discussed with emphasis on dimensional stability, efficient heat dissipation capability, structural integrity, and strength-to-weight ratio. RF substrate materials with low dissipation factor, reduced insertion loss, low dispersion, and minimum surface wave losses are identified. Glass, ceramic, and ceramic–glass substrates are discussed with emphasis on dimensional stability, thermal conductivity, and CTE parameters.

References

1. I.D. Wolfe, P. Czar et al., Influence of packaging environments on the functionality and reliability of capacitive RF-MEMS switches, *Microwave Journal*, December 2005, 102–116.

2. H.J.D.L. Santos, RF MEMS for ubiquitous wireless connectivity, *IEEE Microwave magazine*, December 2004, 45–46.

3. A.R. Jha, Technical report on smart skin materials, Jha Technical Consulting Services, Cerritos, California, September 28, 1991, pp. 17–19.

4. D.M. Jacobson, A.J.W. Ogilvy et al., Applications of CE alloys in defense, aerospace, telecommunications and other electronic markets, *Microwave Journal*, August 2006, 150–152.

5. R. Weast, (Ed.), *Handbook of Chemistry and Physics*, 51st ed., CRC Press, Boca Raton, Florida, 1988, p. F-96.

6. A.R. Jha, *Superconductor Technology: Applications to Microwave, Electro-Optics, Electrical Machines, and Propulsion Systems*, John Wiley and Sons, New York, 1998, pp. 46–48.

Chapter 5

RF-MEMS Switches Operating at Microwave and mm-Wave Frequencies

---

5.1 Introduction

This chapter focuses on the research, development and design aspects, and performance capabilities of radio frequency microelectromechanical systems (RF-MEMS) switches operating at microwave and mm-wave frequencies. Preliminary studies conducted by the author on the subject concerned indicate that RF-MEMS direct-contact switches offer significant advantages over the conventional semiconductor switches operating in microwave and mm-wave regions. Recent research and development activities have been directed toward development of RF-MEMS switches for wireless communications applications. The studies further indicate that RF-MEMS switches developed so far are capable of handling 1–2 W of RF power with switching times between 2 and 10 μs using electrostatic (ES) actuation pull-in voltages between 30 and 80 V. It is important to point out that the design of RF-MEMS devices is quite challenging, because their overall RF performance is affected by the simultaneous coupling of ES, electromagnetic, thermal, and mechanical phenomena, and selection of appropriate materials in fabrication of devices.

It is essential to define a design optimization process aimed to achieve three distinct objectives, namely, high power-handling capability, lower pull-in voltages between 10 and 40 V, and short switching times ranging between 1 and 5 μs. Impact of nonlinear effects generated by the MEMS capacitive switches will be investigated. Because most of the MEMS devices or circuits operate at power levels of 1–1000 mW, linearity aspect is a critical parameter to prevent distortions and interchannel interferences. Because RF-MEMS switches have potential applications in wireless communication systems, electronic warfare (EW) equipment, and missile tracking radar incorporating electronically steering-phased array antennas where phase distortion is not acceptable. A fixed-beam RF-MEMS shunt switch and a cantilever RF-MEMS series switch (Figure 5.1) will be selected to investigate the adverse effects of nonlinearity [1]. MEMS switches are characterized by an upstate capacitance, which varies as a function of bridge height. The cantilever beam can be moved using an ES force between the bridge and the bottom electrode, which is

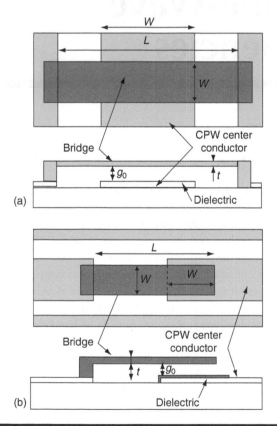

Figure 5.1 Plan and side views of (a) fixed-beam RF-MEMS shunt switch and (b) cantilever RF-MEMS series switch architecture. (Adapted from Figure 1 of Dussopt, L. and Rebeiz, G., *IEEE Trans. MTT*, 51, 1247, 2003.)

generated by the ES actuator. The upstate capacitance is affected by the nonlinear effects such as intermodulation (IM) products generated by the RF power. Data presented in latest publications reveals that these distortions are generally small and can be neglected. This is a distinct performance advantage of MEMS devices as compared to other switching devices using PIN-diodes, metal-semiconductor-field-effect transistor (MESFET), varactor diodes, or ferroelectric films.

5.2 Operating Principle and Critical Performance Parameters of MEMS Devices

It is in the best interests of readers to get familiarized with the operating principle, IM products generated by the capacitive elements, and critical performance parameters of RF-MEMS switches or devices. An RF-MEMS switch has several critical elements and their performance parameter requirements will be briefly discussed. These critical elements include bridge structure, cantilever beam, contact pads, substrate material, passivation layer, and actuating mechanism. These elements have specific material and geometric requirements to meet performance specifications of a MEMS device. Note both the MEMS capacitive element and the spring constant (k) when greater than 10 N/m can generate very low IM products and this could lead to a two-tone third-order (TTTO) IM intercept points (IIP3) exceeding +40 dBm over the resonance bandwidth of 3 to $5f_0$, where the f_0 represents the mechanical resonance frequency of the membrane or cantilever beam. The magnitude of IIP3 can increase to +80 dBm for a difference signal bandwidth even at 5 MHz. Both the IP and IIP3 parameters are dependent on the bridge position and its height, input power level, and parallel-plate (PP) area. In addition, phase noise and amplitude noise are generated in the RF switches, which are dependent on phase delay. Engineers and scientist working on RF-MEMS switches indicate that these switches generate very low IM products, even when the difference signal frequency exceeds 200 kHz. Because of very low IM products, RF-MEMS switches are best suited for phase shifter applications in tracking radar and communications systems, where low IM products, low phase errors, and fast switching are the critical requirements.

5.2.1 Critical Performance Parameters Affected by Environments

Insertion loss (IL), isolation, switching speed, linearity, and power consumption are the most critical performance requirements, regardless whether the switch has a shunt or series configuration. An RF-MEMS switch with tuned design configuration offers two distinct advantages, namely, reduced reflection loss in the upstate operation and enhanced isolation in downstate operation over wide frequency bandwidth. A two-bridge-tuned switch configuration and a four-bridge "cross" configuration will be described with emphasis on isolation and IL. An RF switch design with tuned-bridge configuration offers compact packaging, cheaper fabrication process, and higher reliability. Impact of switch bridge capacitance, physical dimensions (length, thickness, and width) of the bridge, and gap height on switch performance will be investigated. It is important to mention that an inductively tuned-bridge configuration has demonstrated much higher isolation over a capacitive-tuned bridge at microwave and mm-wave frequencies.

Effects of packaging process and environment on the functionality and reliability of capacitive RF-MEMS switches need to be discussed. Limited test data available in open literatures indicates that the packaging temperature profile can change the mechanical stress in the suspended metallic bridge of the switch. The data further indicates that decreasing the environmental pressure in the cavity or housing strongly influences the switching speed of the switch, which can cause vibration effects if it becomes too low. As far as the impact of environments is concerned, the lifetime of the capacitive-based RF switches will be much higher in a nitrogen environment than in an ambient air environment.

5.2.2 Two Distinct Configurations of RF-MEMS Switches and Design Aspects

Reflective MEMS switches using ES actuation mechanisms are best suited for low-loss microwave and mm-wave applications. RF-MEMS switches are fabricated with thin metal membrane, known as cantilever beam, which can be electrostatically actuated in an RF transmission line using a (direct current) dc bias voltage called actuation voltage. There are two classes of MEMS switches, namely, RF-MEMS series switch involving metal-to-metal contacts and an RF-MEMS shunt switch incorporating shunt capacitance. The MEMS series switch consists of a flexible S-shaped film actuator with a switching contact, which moves between the top and bottom electrodes under the influence of ES touch-mode actuation. This particular series switch design allows a low actuation voltage independent of the contact distance in the off-state condition. However, a high off-state isolation is only possible with large contact distance and large overlapping contact areas. In addition, the RF transmission line and the MEMS part of the series switch are fabricated on separate wafers, thereby requiring two different RF substrate materials. The series switch design provides the possibility of full switch assembly on a full wafer bonding, leading to a near-hermetic switch package. Typical actuation voltage requirement is 12 V to open and 16 V to close the contacts. The switch requires large switching areas ranging from 4,000 to 18,000 μm^2 with contact forces in excess of 100 μN, approximately. Its major advantages include low actuation voltage and high design flexibility.

The shunt-capacitive MEMS switch have several advantages such as very little power consumption, very high downstate-to-upstate capacitance ratio, low IL, and high isolation over an RF-MEMS series switch. The major drawbacks of the shunt switch are slow switching speed and higher actuation voltages. A shunt-capacitive MEMS switch consists of a metallic bridge suspended over the center conductor of a coplanar waveguide (CPW) or microstrip line and fixed at both ends to the ground conductors. A dielectric layer known as passivation layer is used to isolate the switch from the CPW center conductor. When the state of the switch is up, the switch presents small shunt capacitance and when the switch is pulled down to the center conductor, the shunt capacitance increases by a factor of 25 to 125, presenting an RF short-circuit condition [2]. Note the shunt-capacitive MEMS switches have demonstrated operations well above 110 GHz with lower IL and higher isolation over wide bandwidths.

Regardless of RF-MEMS switch types, the overall performance of an RF-MEMS switch is much better than that from a conventional RF switch using solid-state elements such as PIN-diodes, MESFETs, or high electron mobility transistors (HEMTs) based on the cutoff frequency criteria. Performance comparison between various RF switches is based on the cutoff frequency parameters of the devices summarized in Table 5.1.

It is important to mention that the cutoff frequency of the switching element or device occurs when the ratio of off-impedance to on-impedance degrades to unity. The cutoff frequency can be expressed as

$$F_c = \left[\frac{1}{2\pi R_{on} C_{off}} \right] \tag{5.1}$$

where the parameters R_{on} and C_{off} indicate the on-state resistance and off-state capacitance of the device, respectively. The data summarized in Table 5.1 indicates that a MEMS switch can operate with minimum power consumption and lowest IL, because of its lower parasitic. Because of impressive RF performance of shunt switches, the majority of the discussion will be limited to this particular switch type.

Table 5.1 Performance Comparison between Various RF Switches

Device Type	Parameter		Figure of Merit (Cutoff Frequency) (GHz)
	R_{on} (Ω)	C_{off} (fF)	
Gallium arsenide–metal-semiconductor-field-effect transistor (GaAs-MESFET)	2.2	250	285
GaAs-high electron mobility transistor (HEMT)	4.4	80	432
GaAs-PIN diode	5.6	42	755
MEMS switch (with metal membrane)	(a) 0.45	35	10,105
	(b) 0.15	35	30,315

5.3 Performance Capabilities and Design Aspects of RF-MEMS Shunt Switches

A MEMS shunt switch can be fabricated either using CPW or microstrip transmission-line configuration. Cross section of a MEMS shunt switch using CPW transmission line is shown in Figure 5.2 [2] along with its equivalent circuit. The switch is fabricated on a high-resistivity silicon substrate using CPW transmission line with typical dimensions of G, W, and G equal to 60, 100, and 60 μm assuming a 50 Ω design for a dc-40 GHz shunt switch configuration. The symbol G represents the gap between the two transmission-line segments and W is the center conductor

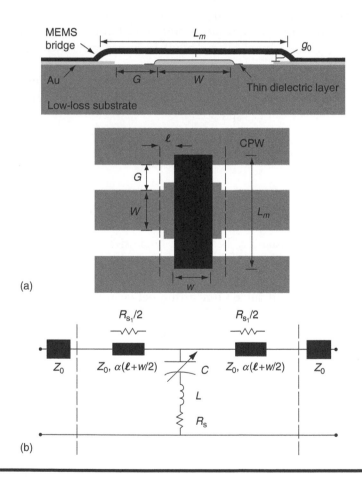

Figure 5.2   Capacitive-based RF-MEMS shunt switch using CPW transmission-line section (a) upstate position and top view and (b) equivalent circuit of the switch. (Adapted from Figure 1 of Muldavin, J.B. and Rebeiz, G., *IEEE Trans. MTT*, 48, 1045, 2000. With permission.)

width of the CPW line. The switch membrane length (L_m) can vary between 200 and 400 μm and the membrane width (w) can vary between 20 and 150 μm to achieve different switch inductance and capacitance values. The membrane height (h) varies from 1.5 to 2.5 μm, while the membrane thickness (t) is typically 2 μm.

5.3.1 Electrostatic Actuation Requirements for the Shunt Switch Using Membranes

ES actuation mechanism provides actuation force with minimum cost and optimum reliability. The center conductor of the CPW transmission line (Figure 5.2) requires dc bias with respect to ground to provide actuation force. The ES force pulls the membrane toward the center conductor using appropriate magnitude of pull-down voltage (V_{pull}), which is written as

$$V_{pull} = 0.544 \left[\frac{kh^3}{\varepsilon_0 Ww} \right]^{0.5} \tag{5.2}$$

where
W is the center conductor width of the CPW line
h is the membrane height
w is the membrane width
ε_0 is the free-space permittivity (8.85×10^{-12} F/m)
k is the spring constant of the membrane

This spring constant of the membrane can be written as

$$k = \left[\frac{32\, Ewt^3}{L_m^3} + \frac{8\sigma(1-v)tw}{L_m} \right] \tag{5.3}$$

where E is Young's modulus of the membrane material (80 GPa) and v is the Poison's ratio (equal to 0.42) for electroplated gold membrane, residual tensile stress (σ) equal to 0 for pull-down voltage of 29.2 V and equal to 20 MPa for pull-down voltage of 39.4 V. It is important to mention that once the dc bias is withdrawn, the mechanical stress in the membrane will overcome the sticking forces and pull the membrane away from the dielectric layer, thereby returning to its original position. It takes time for the membrane to pull itself up from the dielectric layer.

Fabrication of a MEMS switch requires specific materials and processes for membrane, CPW line sections, high-resistivity silicon substrate, and sacrificial layers. Fabrication process flow and materials required are shown in Figure 5.3 [3]. The electroplating gold thickness is approximately 2–3 μm. It is important to mention that the gold-plated membrane switches have a low compressive residual tensile stress. The sacrificial layers are removed using appropriate etchants to remove the seed layers and acetone to remove the resist layer. The pull-down voltage for the

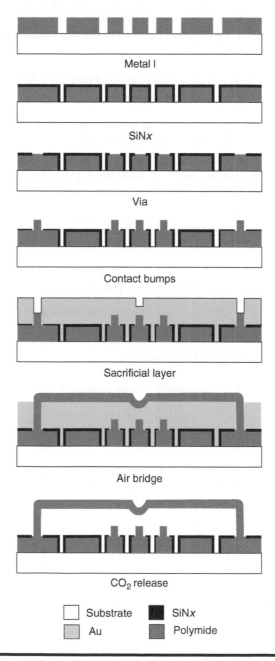

Metal I

SiN*x*

Via

Contact bumps

Sacrificial layer

Air bridge

CO_2 release

☐ Substrate ■ SiN*x*
▨ Au ▨ Polymide

Figure 5.3 **Fabrication process flow and materials required in the design of an RF-MEMS switch actuator.**

shunt switches varies from 12 to 20 V, depending on the membrane thickness, resist profile, electroplating rate, and stress gradients.

5.3.1.1 Sample Calculations for Spring Constant and Pull-in Voltage

Assuming a membrane width of 20 μm, membrane thickness of 2 μm, membrane length, residual tensile stress (σ) of 0 and 20 MPa, modulus of elasticity of 80 GPa, Poisson's ratio of 0.42 and inserting these parameters in Equation 5.3 the spring constant comes to

$$k = \left[\frac{32 \times 80 \times 10^9 \times 8 \times 20}{300^3 \times 10^6} \right]$$

$$= [15.17] \text{ N/m (for residual stress} = 0)$$

$$k = [15.17] + \left[\frac{8 \times 20 \times 10^6 \times (1 - 0.42) \times 2 \times 20}{10^6} \right]$$

$$= [15.17] + [12.36]$$

$$= [27.53] \text{ N/m (for residual stress} = 20 \text{ MPa)}$$

Assuming the membrane height (h) of 1.5 μm, center conductor width of 100 μm for the CPW transmission line, membrane width of 20 μm, spring constant (k) values as computed above and inserting these parameters in Equation 5.2, the computed values of pull-down voltage come to

$$V_{\text{pull}} = 0.544 \left[\frac{15.17 \times 1.5^3}{8.85 \times 10^{-12} \times 100 \times 20} \right]^{0.5} = [0.544 \times 53.78]$$

$$= [29.2 \text{ V}] \text{ (for } \sigma = 0)$$

$$V_{\text{pull}} = 0.544 \left[\frac{1.5^3 \times 27.53}{8.85 \times 10^{-12} \times 100 \times 20} \right]^{0.5} = [0.544 \times 72.45]$$

$$= [39.4 \text{ V}] \text{ (for } \sigma = 20)$$

These sample calculations provide useful information to readers regarding the magnitudes of spring constant (k) and pull-down voltage (V_{pull}) under various residual tensile stress conditions.

5.3.2 Computer Modeling Parameters for MEMS Shunt Switch

A MEMS shunt switch can be modeled by considering two short sections of CPW transmission line and a lumped (Common Language Runtime) CLR Model of the bridge involving capacitance for its upstate and downstate values. It is evident from

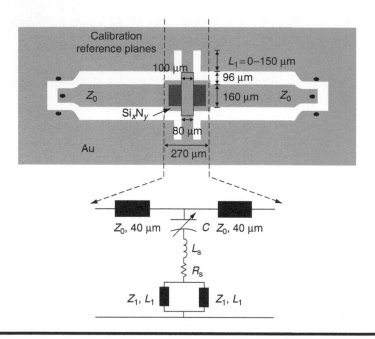

Figure 5.4 **RF-MEMS shunt switch configuration using CPW transmission-line structure and its equivalent circuit. (Adapted from Figure 6 of Muldavin, J.B. and Rebeiz, G., *IEEE Trans. MTT*, 48, 1053, 2000. With permission.)**

Figure 5.4 that the transmission-line section [4] has a length of $(w/[2 + l])$, where l is a distance of about 35–45 μm from the reference plane to the edge of the bridge. Typical values of upstate and downstate capacitances are about 35 fF and 2.8 pF, respectively, an inductance value is about 7.5 pH, and a series resistance value is close to 0.25 Ω for an mm-wave MEMS shunt switch. For an X-band MEMS shunt switch, typical corresponding values are roughly 100 fF, 8 pF, 4.5 pH, and 0.15 Ω. For the switch configuration shown in Figure 5.4, the shunt impedance can be written as

$$Z_s = [R_s + j\omega L + 1/j\omega C] \tag{5.4}$$

where subscript "s" stands for shunt and C stands for upstate or downstate capacitance.

The resonance frequency of the shunt switch can be written as

$$F_0 = \left[\frac{1}{2\pi(LC)^{0.5}}\right] \tag{5.5}$$

The CLR Model of this switch acts like a capacitance below the resonance frequency and like an inductor above the resonance frequency. At resonance, the impedance is reduced to series resistance of the MEMS's bridge. Using the above-mentioned values for an mm-wave shunt switch, the resonance frequency comes to 321 GHz in the upstate position of the switch and 35 GHz in the downstate position. One can notice from the above equations that the inductance of the bridge plays no role in the upstate position for operations below 100 GHz. However, the inductance plays an important role in the downstate position of the switch. It is important to mention that the shunt switch inductance limits the downstate RF performance at a much lower frequency than the resonance frequency. However, higher values of resonance frequency in downstate position offers improved isolation up to twice the resonance frequency.

5.3.2.1 Computation of Upstate and Downstate Capacitances

The upstate position capacitance can be written as

$$C_{\text{up}} = \left[\left(\frac{\varepsilon_0 W w}{h_b} \right) + \left(\frac{t_d}{e_r} \right) \right] = \left[\frac{\varepsilon_0 A_b}{h_b + (t_d/e_r)} \right] \tag{5.6}$$

where

h_b is the bridge height (typically 4 μm)
t_d is the dielectric layer thickness (typically 150 nm)
A_b is the area of the bridge and is equal to product of CPW center conductor and width of the bridge
e_r is the relative dielectric constant of the dielectric layer (typically 7.6 for a silicon nitride [Si$_3$N$_4$] layer)

Note the dielectric layer must be thick enough to withstand the actuation voltage between 20 and 40 V without dielectric breakdown. The contact area must be about 50 percent of the bridge area. Under these guidelines, one will find the fringing capacitance as small as possible.

The downstate position capacitance can be written as

$$C_{\text{down}} = \left[\frac{\varepsilon_0 e_r A}{t_d} \right] \tag{5.7}$$

Because, the dielectric thickness (typically 100–150 nm) is so small that the fringing capacitance can be neglected. The expression for the downstate to upstate capacitance ratio can be written as

$$\frac{C_{\text{down}}}{C_{\text{up}}} = \left[\frac{A}{t_d} \right] \left[e_r h_b + \left(\frac{t_d}{W w} \right) \right] \tag{5.8}$$

For optimum capacitance ratio, it is important to have a dielectric layer not exceeding 150 nm and the surface roughness of the MEMS bridge and that of the dielectric layer be kept below 1.5 nm. Assuming the parameters values and contact area percentage as mentioned above, one gets the value of this capacitance ratio of 77 for 1.5 μm gap height and 122 for the gap height of 2.0 μm.

As mentioned previously, the CPW transmission-line configuration is best suited for the MEMS bridges. Complete MEMS bridge structure using CPW line is shown in Figure 5.4 along with the lumped-element CPW transmission-line parameters. The MEMS bridge parameters such as the bridge capacitance, total capacitance, and inductance per unit length are identified for rapid calculations.

5.3.2.2 Current Distribution and Series Resistance of the MEMS Bridge Structure

Precise estimation of current distribution in the bridge is extremely difficult. However, it is possible to obtain normalized current distribution using the Sonnet software program. Note that there is no RF current on the bridge portion, which is above the CPW transmission line. Furthermore, the RF current is concentrated on the edge of the bridge over the gap of the CPW line. According to Sonnet simulations, the current distribution remains constant, hugging the edges of MEMS bridge, if the bridge width exceeds 40 μm. It is important to mention that the upstate inductance and resistance of the MEMS switch are different from the downstate values. However, the inductance in downstate position is of critical importance, because it affects the performance of the switch particularly at mm-wave operations.

As far as the series resistance of the MEMS switch is concerned, it has two components. The first component is due to the transmission loss including the ohmic and dielectric losses and the second is due to the MEMS bridge structure. For a CPW transmission line on high-resistivity silicon substrate with a 0.8 μm thick gold-plated center conductor and 2 μm thick gold ground plane, the measured attenuation loss is approximately 1.7 dB/cm at K-band frequencies centered at 35 GHz. A switch length of 160 μm shows a loss of 0.028 dB. For a 0.8 μm thick gold-plated CPW line with $G/W/G$ values of 30/100/30 μm on quartz substrate ($e_r = 3.78$), the attenuation is about 0.4 dB/cm at 35 GHz, because of lower dielectric constant compared to silicon ($e_r = 11.8$).

The resistance of the MEMS's bridge is extremely difficult to predict or to measure because of the current distributions in up and downstates. The bridge resistance can be predicted or estimated based on the skin depth phenomenon, which is defined as

$$\delta = \left[\frac{1}{(\pi f \ \mu \sigma)^{0.5}} \right] \tag{5.9}$$

where
   f is the frequency
   μ is the permeability
   σ is the electrical conductivity of the bridge material

**Table 5.2 Computed Values of Skin
Depth for MEMS Bridge Materials (μm)**

Frequency (GHz)	Gold	Aluminum
10	0.73	0.84
20	0.52	0.59
30	0.42	0.48
60	0.30	0.33
90	0.24	0.28

As shown in Table 5.2, the skin depths for the gold and aluminum metals are 0.73 and 0.84 μm at 10 GHz and 0.42 and 0.48 μm at 30 GHz, respectively. This data indicates that if the bridge thickness is smaller than two skin depths, the MEMS switch resistance is constant for all frequencies. For MEMS bridges with gold thickness greater than 1.5 μm, the resistance will vary as a function square root of the frequency above 30 GHz due to skin depth effect and will remain constant at frequencies lower than 30 GHz. However, for bridges made of aluminum with thickness less than 1 μm, the bridge resistance remains constant at all operating frequencies.

5.3.2.3 Estimates of Switch Inductance and Capacitance Parameters

The most accurate method for modeling the switch inductance is to assume that the MEMS switch capacitance is large enough to provide the short-circuit condition. The modeling can be accomplished using Sonnet or HFSS software. Accurate results are possible because the dielectric layer is ignored. The model provides the S-parameters from dc to 60 GHz using the CRL Model from which the values of series resistance and inductance can be extracted.

Both the MEMS switch inductance and capacitance can be determined using the above-mentioned software within ±3 percent accuracy. Because a majority of the RF current flows on the edge of the CPW center conductor and ground planes, the portions of the bridge over the CPW center conductor (Figure 5.5) [2] of width G and ground planes contribute very little to the bridge inductance. Thus, the bridge inductance is dominated by the portion of the bridge over the CPW gaps of width G. The switch inductance remains constant at 10 pH, even as the width of the bridge over the center conductor is varied.

Bridge width parameter (w) has an impact on the inductance and capacitance values. It is interesting to note that the fitted capacitance varies from 1 to 1.8 pF as the bridge width (w) increases. As far as the bridge capacitance is concerned, it can

$(L_m = 300\ \mu m,\ t = 2\ \mu m)$

$w \times W$ [μm^2]	g [μm]	C_t [fF]	C_{pp} [fF]	C_f [fF]	C_f/C_{pp} [Percent]
40 × 100	4	13.8	8.8	5	57
80 × 100	4	23.4	17.6	5.8	33
140 × 160	4	59.2	49.3	9.9	20
40 × 100	1.5	29.8	23.2	6.5	28
80 × 100	1.5	55.6	46.6	9	19

(a)

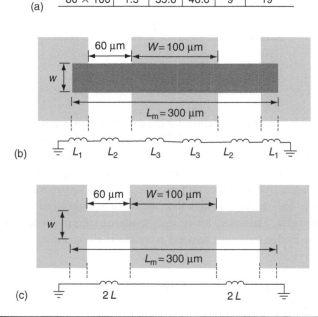

(b)

(c)

Figure 5.5 RF-MEMS shunt switch bridge. (a) MEMS bridge inductance for several bridges with various widths ($L_m = 300\ \mu m$, $t = 2\ \mu m$), (b) physical layout of an RF-MEMS bridge, and (c) equivalent circuit in the downstate position of the MEMS switch.

be estimated accurately if the PP capacitance is known. Note the downstate capacitance is roughly 0.5–0.8 times the PP capacitance, which can be easily computed from the readily available relevant parameters. Modeling parameters indicate that the bridge inductance changes by a factor of 3 if the bridge width (w) is increased from 20 to 140 μm. This indicates that the RF current is concentrated on the first edge of the bridge and is weakly dependent on the width of the bridge.

5.3.2.4 Insertion Loss in a MEMS Switch

The IL in a MEMS switch can be estimated from the S-parameters which can be written as

$$\text{Insertion loss (IL)} = \left[1 - S_{11}^2 - S_{21}^2\right] \qquad (5.10)$$

where S_{11} is the reflection coefficient and S_{21} is the loss coefficient. In the upstate position, when S_{11} is less than -14 dB and the series impedance Z_s is much less than the characteristic impedance Z_0, the expression for switch loss is reduced to

$$\text{IL} = \left[w^2 C_u^2 R_s Z_0 \right] \tag{5.11}$$

where

    w is the width of the transmission-line widths at both sides of the switch, which is typically between 40 and 80 μm

    C_u is the upstate capacitance (typically 30–40 fF)

    R_s is the series resistance including the resistance of the CPW transmission-line sections (typically 0.25 Ω)

In the downstate position when $S_{11} > -10$ dB and series impedance is less than characteristic impedance, the switch loss is approximately given as

$$\text{IL} = \left[\frac{4R_s}{Z_0} \right] \tag{5.12}$$

The IL is dominated by the transmission loss in the upstate position due to use of high-resistivity silicon substrate. In the upstate position, the MEMS switch loss can vary between 0.25 dB and 1.5 dB/cm at 35 GHz. In the downstate position, the IL is about 0.18 dB for a series resistance less than 0.25 Ω and less than 0.1 dB for a series resistance of 0.1 Ω at operating frequencies greater than 10 GHz. It is extremely difficult to measure IL less than 0.12 dB, particularly in upstate position.

5.3.2.5 Estimation of Series Resistance of the Bridge and Impact of Switch Inductance on the Isolation

Most accurate estimate for the series resistance is possible at resonance conditions. Under such condition, the series resistance is equal to series inductance. This means the S-parameter at resonance can be approximately written as

$$[S_{21}] = \left[\frac{2R_s}{Z_0 + R_s} \right] \tag{5.13}$$

For a gold-plated MEMS bridge with 40 μm width, the series resistance is about 0.072 Ω, inductance (L) is close to 7.8 pH, and the downstate capacitance is about 2.6 pF. The resonance frequency for these inductance and capacitance parameters comes to 35 GHz. The LC resonance frequency for most MEMS shunt switches ranges approximately from 25 to 60 GHz, depending on the dimensions of the switch. Furthermore, the switch isolation can be significantly improved by 20–25 dB by selecting appropriate resonance frequency, which is strictly dependent on the L and C values. It is important to mention that the inductance in the downstate position of the switch has a significant impact on the isolation, which can actually improve the switch performance by 20–25 dB at resonance. It is critical to point out that as the series resistance gets smaller, the resonance peak in S-parameter S_{21} gets deeper, which means the series resistance is 0.07, 02.5, and 0.5 Ω at the resonance peak of −48, −40, and −34 dB, respectively. This indicates IL measurement equipment must have excellent calibration to −50 dB to measure a resistance of 0.1 Ω with high accuracy. Note the S-parameters for a MEMS switch must be measured at the resonance frequency if accuracy of the measurement is of critical importance. Isolation improvement due to inductance tuning of a MEMS shunt switch is possible by choosing appropriate values of inductance. In this case, the downstate capacitance is kept constant at 2.7 pF. Both the MEMS inductance and capacitance are affected by the bridge width (w) and thickness (t). Total capacitance and the PP capacitance as a function of width and thickness are summarized in Figure 5.5. However, the combined effect of bridge with downstate capacitance and inductance as a function of frequency is evident from the curves shown in Figure 5.6 [2].

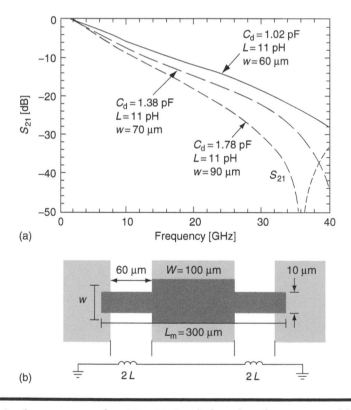

Figure 5.6 *S*-parameters of an RF-MEMS switch (a) in a downstate position where width of the CPW section over the center conductor is varied and (b) schematic and equivalent circuit for the inductance.

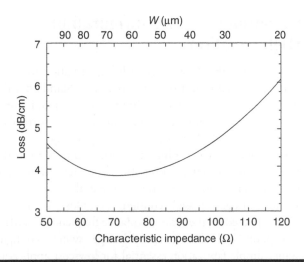

Figure 5.7 Simulated insertion loss (IL) at 90 GHz for a CPW transmission line as a function of center conductor width (*W*) and characteristic impedance of the CPW transmission-line structure.

5.3.2.6 Typical Upstate and Downstate Insertion Losses in a MEMS Shunt Switch

As stated earlier, it is very difficult to measure low IL as low as 0.1 dB with high accuracy. However, rough estimates of losses can be obtained from the curve shown in Figure 5.7 [2]. ILs in upstates and downstates as a function of frequency at 40 μm from the edge of a MEMS switch can be estimated using the CPW line parameters shown in Figure 5.7. Estimated losses are summarized in Table 5.3.

Table 5.3 Upstate and Downstate Losses in a MEMS Shunt Switch (dB)

	Upstate Loss		Downstate Loss	
Frequency (GHz)	$R_s = 0.1\ \Omega$	$R_s = 0.25\ \Omega$	$R_s = 0.1\ \Omega$	$R_s = 0.25\ \Omega$
10	0.000	0.007	0.030	0.080
20	0.001	0.011	0.030	0.084
30	0.003	0.013	0.30	0.086
40	0.005	0.018	0.035	0.090

Note: These losses assume C_{up} and C_{down} capacitance values as 35 and 2.8 fF, respectively.

5.4 MEMS Shunt Switch Configuration for High Isolation

Some applications require MEMS switches with high isolation to obtain high signal spectral purity and minimum electronic interference. Standard MEMS switches discussed so far will not be able to meet high isolation requirements over wide frequency bandwidth and at mm-wave frequencies. This particular section will focus on the designs of MEMS shunt switches incorporating CPW lines and tuned multibridge configurations to achieve high isolation and minimum IL at mm-wave frequencies. The tuned switch configuration offers lower reflection loss and wideband operation in upstate position and improved isolation in the downstate. A two-bridge-tuned switch or a four-bridge-"crossed" configuration illustrated in Figure 5.8 offers improved isolation [4] based on the series resonance frequency of a MEMS switch. It is important to point out that a tuned-bridge switch configuration offers compact packaging, simple fabrication essential for low-cost applications, lower IL,

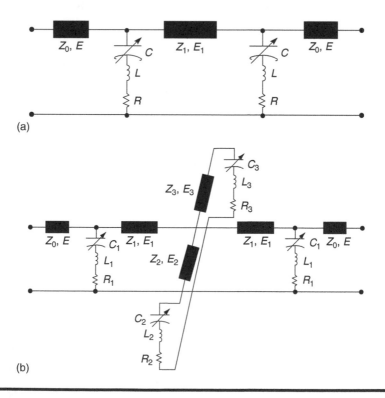

Figure 5.8 Equivalent circuit showing the electrical parameters. (a) Equivalent circuit for the two-bridge MEMS switch and (b) equivalent circuit for the tuned-cross MEMS switch.

and higher isolation over wide frequency bandwidths. This particular design technique can be applied to MEMS shunt switches capable of operating up to 100 GHZ or higher.

5.4.1 Tuned Two-Bridge Design and Its Performance Capabilities

The simplest structure of the tuned-bridge switch is shown in Figure 5.8, which consists of two single MEMS shunt switches separated by a short high-impedance CPW transmission line. The midsection transmission-line length is selected such that the reflection from the first membrane and the reflection from the second membrane cancel each other at the input port, when the MEMS switch is in the upstate position. Because of the reflection null of the tuned switch, the upstate capacitance can be increased. It is evident from Equation 5.6 that the capacitance can be increased either by increasing the area of the bridge or by lowering the height of the bridge or a combination of both parameters. It is important to mention that increasing the area will increase the isolation during the downstate position and lowering the membrane gap height will lower the pull-down voltage as evident from Equation 5.2, which will be at the expense of smaller downstate to upstate capacitance ratio [2]. By lowering the membrane gap height from 3.5 to 1.5 μm, the pull-down voltage can be reduced from 50 to 15 V for a 300 μm long gold-plated bridge and same mechanical spring constant. Using these equations, one will find that the downstate to upstate capacitance ratio is close to 42.5 and 85 for a membrane gap height of 1.5 and 3.5 μm, respectively, using the dielectric (silicon nitride) layer thickness of 0.2 μm and permittivity of 7.6.

The midsection transmission-line section length can be reduced if the midsection line impedance (Z_m) is equal to characteristic impedance of the switch port (Z_0). Now the electrical length of the midsection CPW transmission line can be given as

$$\text{Tan}\,\beta l = \left[\frac{2}{C_b Z_0 \omega_c}\right] \qquad (5.14)$$

where
β is the phase constant
C_b is the bridge capacitance
ω_c is the cancellation frequency to be selected

Assuming the upstate capacitance of each bridge between 55 and 60 fF, the midsection line impedance of 61 Ω with a line length of about 400 μm, typical bridge inductance of 4.8 pH, and bridge thickness of 3 μm, the downstate to upstate capacitance ratio will be 15–19, which is not high for improved isolation. Higher isolation would require this ratio better than 40:1. The upstate IL is estimated between 0.2 and 0.4 dB over 20–40 GHz bandwidth. Computer simulations undertaken by various switch designers indicate that the downstate capacitance around 2.2 pF would result in improved isolation over 20–40 GHz frequency range, while still maintaining a pull-down voltage of 15 V.

5.4.2 Design Aspects and Performance Capabilities of Four-Bridge Cross Switch

This cross switch design offers increase in bandwidth and downstate isolation of the tuned switch physical implementation. The physical implementation and equivalent circuit for the cross switch are shown in Figure 5.8. It is evident from this figure that the in-line of the switch consists of two tuned bridges separated by high-impedance transmission lines. However, the shunt sections of the switch are open-ended stubs, which are loaded with smaller switches at the ends. The in-line and shunt sections are responsible for production of two reflection nulls.

In the upstate operation, the CPW transmission lines and the electrical lengths can be optimized to achieve minimum return loss over a wide frequency bandwidth. In the downstate position of the switch, the reflection nulls shifted to lower frequency and the RF short-circuit conditions presented by the capacitively loaded studs provide higher isolation. Single bias voltage can be used to pull four MEMS switches to the downstate position. It is also important to mention that the parasitics of the cross junction may present design problems.

The characteristic impedances of CPW lines are dependent on frequency, effective dielectric constant, and elliptic integrals (Figure 5.9). The midsection CPW transmission line of length 350 μm has an impedance (Z_1) of about 66 Ω, whereas the shunt section line of 170 μm has an impedance (Z_2) of 50 Ω. The bridge height of 1.5 μm is needed to keep the pull-down voltage to 15 V. The upstate capacitance of the in-line membrane elements with a width of 64 μm is about 0.065 pF and 0.28 fF in the shunt sections.

Regarding the performance capabilities of a tuned-cross switch, its return loss is better than −22 dB from 23 to 39 GHz with an estimated IL of 0.32 and 0.62 dB, respectively. The upstate return loss is less than −16 dB over dc to 40 GHz. The cross switch isolation (S_{21}) is greater than 40 dB from 16 to 40 GHz, with capacitance and inductance of 1.5 pF and 9 pH, and 0.65 pF and 12 pH, for in-line membranes 1 and 2, respectively. Accuracy for computations of these parameters above 35 GHz is strictly dependent on the substrate properties and thickness and the coupling effects between the CPW transmission-line sections. The major advantage of the MEMS cross switch is its capability to provide high isolation over 10–20 GHz range even with a downstate capacitance ratio as low as 22, compared to a single MEMS switch. An isolation greater than 50 dB is possible over 10–40 GHz frequency range if the cross switch is designed with a downstate capacitance ratio in excess of 46.

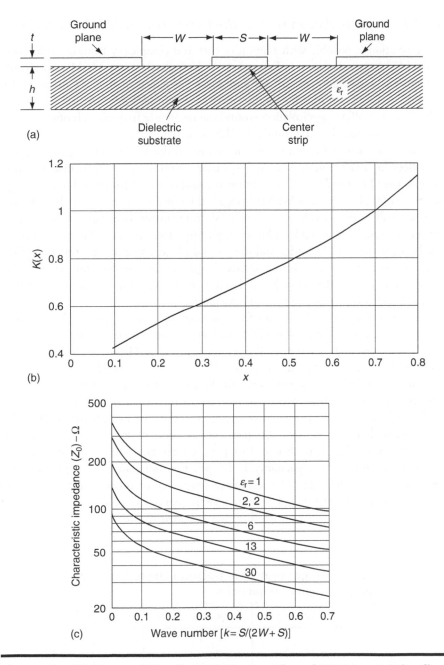

(a)

(b)

(c)

Figure 5.9   CPW transmission line. (a) Cross section of CPW transmission line, (b) plot of elliptic integral of first-order $K(x)$, and (c) characteristic impedance of CPW line as a function dielectric constant.

5.4.2.1 High Isolation with Inductively Tuned MEMS Switches

High isolation is possible with minimum cost and complexity from an inductively tuned MEMS switch shown in Figure 5.4. An inductively tuned MEMS switch with a downstate capacitance of 2.7 pF and series resistance of 0.35 Ω offers higher isolation as a function of bridge inductance, as shown in Table 5.4. Because such a MEMS switch offers high isolation around the resonance frequency, it offers limited bandwidth compared to the standard MEMS switch.

These values indicate that inductance values in the 45–70 pH range are needed to achieve high isolation at X-band frequencies. This is true at X-band frequencies because the Q of the *LC*-resonant circuit is equal to wL/R_s for an inductance range of 45–70 pH. However, the MEMS bridge inductance is limited to 15–20 pH, even with a small bridge width over the CPW transmission-line gaps. A large series inductance can be achieved by adding a high-impedance section of the transmission line between the MEMS bridge and the ground plane. By selecting proper length of this line, one can lower the resonance frequency in the X-band region. This technique offers the design of X-band standard MEMS shunt switches with high isolation and with minimum cost and complexity.

Just to demonstrate the design feasibility, an inductively tuned MEMS shunt switch was fabricated using a 50 Ω CPW transmission-line implementation (*G/W/G* = 96/160/96) on high-resistivity silicon substrate ($e_r = 11.8$). A 300 μm long bridge with electroplated gold (thickness of 0.6 μm) was suspended over the center conductor. The metallization underneath the bridge consists of a layer of 0.2 μm refractory metal, which is covered with silicon nitride layer ($e_r = 7.6$) of 0.21 μm thickness. The width of the capacitive section over the center conductor of the CPW transmission line was chosen 100 μm. The length of the inductive section of the transmission line was chosen 150 μm. Under no inductive tuning and in downstate position, the capacitance is 4.8 pF, the inductance is 12 pH, and the series resistance is 0.31 Ω. The resonance frequency of tuned circuit with these parameters comes to 21 GHz. The resonance frequency can be shifted to 12 GHz if a 150 μm long inductive transmission line is added to the switch. This 150 μm

Table 5.4 Improvement in Isolation due to Inductive Tuning at Higher Frequencies

Bridge Inductance (pH)	Frequency (GHz)	Isolation (dB)
64	12	38
32	18	38
16	25	38
8	34	38

inductive transmission line provides an equivalent series inductance equal to 37 pH at 12 GHz. The inductively tuned bridge is achieved by taking the L and C values of the MEMS bridge and adding two short sections of 56 Ω CPW transmission line. The IL for the tuned X-band MEMS switch is less than 0.2 dB and the return loss was better than -15 dB. In summary, an inductively tuned MEMS switch offers high isolation in the downstate and improved return loss in the upstate position, regardless of operating band. Furthermore, such MEMS switches are best suited for V-band and W-band operations, where high isolation and low IL are difficult to achieve.

5.4.3 MEMS Shunt Switches for Higher mm-Wave Frequencies

Discussion in the previous section is limited to MEMS series and shunt switches operating over dc to 40 GHz frequencies. It is important to note that capacitive MEMS shunt switches are equally good for higher mm-wave frequencies covering V- and W-bands. However, discussion and design methodology will be limited to a W-band (75–110 GHz) MEMS capacitive shunt switch using the CPW transmission-line structure. To achieve lower IL and higher isolation, switch design [5] must deploy quartz substrate ($e_r = 3.78$ and $\tan \delta = 0.00025$ at 25 GHz and beyond) and T-match and π-match circuit configurations [5].

5.4.3.1 W-Band MEMS Shunt-Capacitive Switch

Structural features of a typical W-band MEMS shunt switch using CPW transmission-line configuration and quartz substrate ($e_r = 3.78$) are shown in Figure 5.8 along with its equivalent circuit. The switch is suspended above the center conductor of the CPW line and has a length, width, and thickness of 280, 40, and 0.8 μm, respectively. The switch incorporates a T-matching network with gap/center conductor width/gap $(G/W/G)$ dimensions equal to 20/80/20. But the transmission line is tapered to $G/W/G$ dimensions of 35/80/35 underneath the bridge structure to achieve lower upstate capacitance. For a pull-down electrode with area equal to 2000 μm^2, the PP upstate capacitance C_{upp} $(\varepsilon_0 A/h_b)$ comes to 11.8 fF and the total upstate capacitance with 30 percent fringing capacitance is about 15.3 fF. The silicon nitride substrate of 0.15 μm thickness is used underneath the bridge, which offers a downstate capacitance $(C_{down} = \varepsilon_0 A e_r/h_b + t)$ of 897 fF. Considering the dielectric roughness between the bridge membrane and the silicon nitride layer [5], the downstate capacitance is reduced to approximately 403 fF (by 45 percent). Note the substrate thickness is not critical in the designs of mm-wave switches using CPW transmission lines. The switch can be biased using bias tee with a pull-down voltage ranging from 35 to 45 V. The corresponding

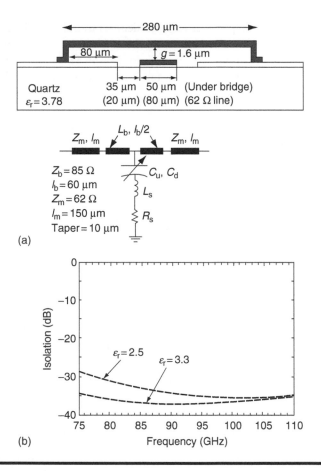

Figure 5.10 **RF-MEMS shunt design. (a) Physical layout and equivalent circuit for an RF-MEMS shunt switch using CPW transmission-line structure and (b) improvement in isolation over W-band as a function of substrate permittivity.**

spring constant comes to 22 N/m at 40 V bias level and the mechanical resonance frequency comes to 103 kHz. Note the high-impedance line of 62 Ω (Figure 5.10) on either sides of the bridge along with a small section underneath the bridge presents a series inductance, which acts like a T-matching network. The return loss is better than 26 dB over the entire W-band frequencies and the IL is about 0.2 dB. Minimum IL occurs when the metal conductivity is changed to match the reference line characteristics impedance of 52 Ω. The downstate the switch isolation is better than −30 dB at 100 GHz and above, but maintains a minimum isolation better than −22 dB over the frequency range from 75 to 110 GHz.

In the downstate position of the switch, the inductance is roughly 4 pH, the capacitance is 403 fF, and the bridge series resistance (R_s) is around 0.8 Ω. The above capacitance and inductance values increase the downstate resonance frequency to 125 GHz, which provides higher isolation over the entire W-band frequencies. Of course, higher switch RF performance can be achieved using a T-matching network.

Studies performed by the author indicate the isolation performance can be further improved using the π-matching network in the upstate position of the switch. The high-impedance line between the bridges has $G/W/G$ dimensions as 30/60/30 μm to achieve membrane impedance (Z_m) of 76 Ω with a length of 150 μm. The reference planes are 470 μm apart. With these physical parameters and using π-matching network, the IL is slightly higher around 0.4 dB, the return loss is better than -25 dB over the entire W-band, and the isolation can be improved better than -35 dB over the limited band. Using a dielectric interlayer with permittivity of 2.5, a downstate capacitance can be reduced to 295 fF, which improved the downstate resonance frequency to 146 GHz leading to much higher isolation in the downstate. Impact of permittivity of the interlayer dielectric medium on switch isolation is evident in Figure 5.10. The isolation of a capacitive shunt switch can be computed using the following equations:

$$\text{Isolation} = 20 \log \left[\frac{4}{\omega^2 Z_0^2 C_d^2} \right], \quad \text{if } f \ll f_{res}$$

$$= 20 \log \left[\frac{4 R_s^2}{Z_0^2} \right], \quad \text{if } f = f_{res}$$

$$= 20 \log \left[\frac{4 \omega^2 L_s^2}{Z_0^2} \right], \quad \text{if } f \gg f_{res} \tag{5.15}$$

$$f_{res} = \frac{1}{2\pi (C_d L_s)^{0.5}} \tag{5.16}$$

where
C_d is the downstate capacitance
L_s is the series resistance
f_{res} is the resonance frequency of the switch in downstate position

5.4.3.2 Switching Speed of MEMS Shunt Switches

Switching speed is the most critical requirement for the MEMS phase shifters and switches deployed by the satellite communications systems and missile tracking radar. The switching speed is dependent on the pull-down voltage (V_{pull}), actuation or supply voltage (V_s), and the mechanical resonance frequency (f_o).

Higher mechanical frequency and lower pull-down voltage are needed to achieve lower switching time or higher switching speed. However, higher mechanical resonance frequency can bring mechanical instability of the bridge or cantilever beam structure. Also higher actuation voltage is not compatible with microelectronic circuits. Therefore, careful and judicial selection of operating parameters is absolutely essential, when a higher speed or lower switching time is the principal design requirement for the MEMS switches. The switching time of a MEMS switch is given as

$$T_s = 3.674 \left[\frac{V_{pull}}{V_s \omega_{mo}} \right] \tag{5.17}$$

Assuming a pull-down voltage of 40 V, actuation or supply voltage of 55 V, and a mechanical resonance frequency of 103 kHz, and on inserting these values in Equation 5.17, one gets

$$T_s = 3.674 \left[\frac{40}{55 \times 2\pi \times 103,000} \right]$$

$$= \left[\frac{(0.5847)(0.72727)}{103,000} \right] = 4.129 \times 10^{-6} s$$

$$= 4.13 \ \mu s$$

Computed values of switching times as a function of various assumed parameters are summarized in Table 5.5.

Table 5.5   Computed Values of Switching Times

Pull-Down Voltage (V)	Actuation Voltage (V)	Resonance Frequency (kHz)	Switching Time (μs)
45	60	80	5.48
40	55	103	4.13
30	45	116	3.83
30	50	120	2.92
25	58	95	2.65
20	55	110	1.93

5.5 MEMS Switches Using Metallic Membranes

5.5.1 Introduction

This section identifies potential advantages of micromechanical (MM) membrane for MEMS switches and its wide applications in various disciplines. MM switches have deployed electrostatically actuation using cantilever, rotary, and membrane topologies. Designers of these switches claim that the moving low-loss metal contacts possess low parasitics at microwave and mm-wave frequencies because the contact and spreading resistance associated with ohmic contacts are eliminated. High conductivity films are used to fabricate metallic structures to carry RF currents with minimum losses. These switches offer low on-state resistance and high on-state capacitance. Significant improvements have been incorporated in the design of metal membrane RF switches, which can operate with ultralow losses at higher microwave frequencies, increased switching speeds, and negligible power consumption [6].

5.5.2 Operating Principle and Design Aspects of Capacitive Membrane Switches

An MM switch deploys the mechanical connection of two metallic surfaces to actuate a low resistance electrical connection. The capacitive membrane of an RF-MEMS switch generally suffers from problems, namely, lower switching speeds, unpredictable stiction forces, and microwelding. However, these problems have been solved using capacitive connection [6]. Specific details on the geometry of a capacitive metal membrane switch with low resistance connection are shown in Figure 5.11. This capacitive membrane switch consists of a thin metallic membrane suspended over a low-loss dielectric and a thin film is deposited on top of the bottom electrode as illustrated in Figure 5.11. When an ES actuation voltage is applied between the membrane and the bottom electrode, the electrostatically generated force pulls down the metallic membrane down onto the bottom electrode covered with dielectric. The dielectric film prevents the stiction between the two metallic surfaces.

When the membrane is not actuated or the switch is in off-state position or downstate position, the capacitance between the contacts is very low and can be written as

$$C_{\text{off}} = \left[\frac{1}{(t_\text{d}/e_\text{c} A) + (h_\text{a}/\varepsilon_0 A)} \right] \tag{5.18}$$

where

  t_d is the dielectric film thickness
  A is the overlap area between the bottom electrode and the membrane
  h_a is the height of the air gap between the membrane and the dielectric film
    or layer when switch is in upstate position

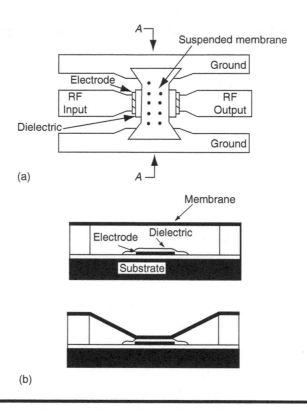

Figure 5.11 Capacitive metallic membrane-based shunt switch. (a) Top view of an RF-MEMS switch showing the switch critical elements and (b) cross section of the switch showing up and down position of the membrane.

All other parameters are well known and have been defined previously. The off-state capacitance varies between 20 and 50 fF.

When the switch is in upstate position or in on-state position or it is actuated, the capacitance of the metal–dielectric–metal sandwich structure can be written as

$$C_{on} = \left[\frac{\varepsilon_0 A}{t_d} \right] \qquad (5.19)$$

This on-position under the actuation conditions the capacitance is typically between 3 and 6 pF. The on-state to off-state impedance ratio is equal to the on-state to off-state capacitance ratio. This means that Z_{on}/Z_{off} is equal to C_{on}/C_{off}. For switching RF signal at microwave or mm-wave frequencies, the on-state to off-state capacitance must be greater than 100 for good quality signal.

A well-designed metallic membrane switch must have minimum IL and maximized on/off ratio of the device. The geometry and parameters of the electrode and membrane must be selected to minimize the electrical path resistance and to achieve an off-state capacitance that can be matched in impedance to frequencies exceeding 40 GHz. The characteristics of the dielectric film, properties of electrodes and membrane, and gap geometry must be set to achieve the on/off ratio in excess of 100. Note the switching pull-down voltage, dielectric breakdown, and switching speed contribute to trade-offs that need to be considered for the RF performance of the switch. Computer-generated plots on the on/off ratio or the switching ratio as a function of overlap area (A), air gap height (h_a), permittivity of the dielectric layer (ε_d), and dielectric passivation layer thickness (t_d) are shown in Figures 5.12a through d and 5.12e.

Close examination of these plots indicates that with overlap area of 14,000 μm^2 and passivation layer thickness of 0.1 μm and permittivity of 7.6, on/off ratios greater than 100 are possible with air gaps not exceeding 2 μm. With overlap area of 4000 μm^2 and passivation layer thickness of 0.2 μm, ratios greater than 100 can be achieved with air gaps less than 3 μm. However, the on/off ratio greater than 100 is only possible with air gap close to 5 μm, when the overlap area is 12,000 μm^2 and the passivation layer thickness and permittivity are 0.3 and 6.7 μm, respectively. It is important to mention that smaller air gaps (2 or 3 μm) must be selected to keep the actuation voltages well below 20 V essential for higher reliability. These plots present multiple options to the MEMS designers for the selection of passivation layer parameters and overlap areas to achieve optimum membrane MEMS shunt switch design with high performance and enhanced reliability.

The switch can be fabricated using CPW transmission-line sections which have an impedance of 50 Ω using the CPW line parameters shown in Figure 5.9. This impedance value matches the impedance of the switch. The spacing between the ground lines and signal line is about 80 μm. The transmission lines are 120 μm in width, when fabricated on high-resistivity silicon wafer. The lines can also be fabricated on high-resistivity sapphire or gallium arsenide wafers, whichever is easily available with minimum cost. A thin layer of silicon nitride between 0.1 and 0.2 μm can be used to insulate the electrodes to avoid stiction problem. Tungsten film with gold electroplating must be deposited on electrodes to provide high reliability under high current and contact pressure operating environments.

Fabrication requirements for this switch are relatively simple. The switches can be fabricated on high-resistivity wafer materials such as silicon, sapphire, or gallium arsenide. These devices can be designed with loose geometrical tolerances to achieve high yield. However, using high-resistivity silicon wafers allow high integration compatibility with complementary metal-oxide semiconductor (CMOS) or micro-electronic circuits for multifunction assemblies. Some precautions must be observed during fabrication process. All transmission lines must be oversized by 5 μm or so to compensate for the decrease in the line width during wet etching process. The sacrificial spacer layer must be removed by oxygen plasma etch to release the membrane. The gap between the membrane or the top electrode and the bottom electrode must be as small as possible to actuate with low actuation voltage.

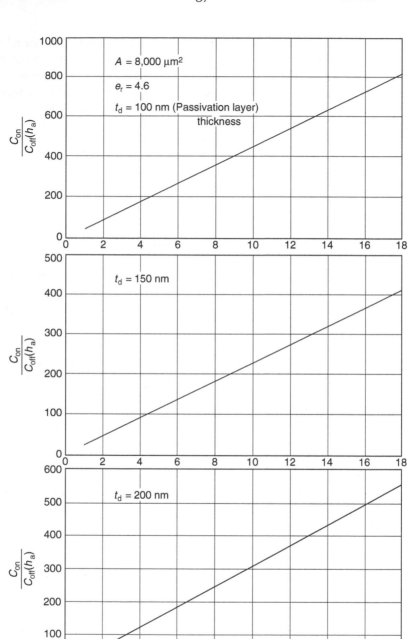

Figure 5.12 **(a) On/off capacitance ratio as a function of air gap,**

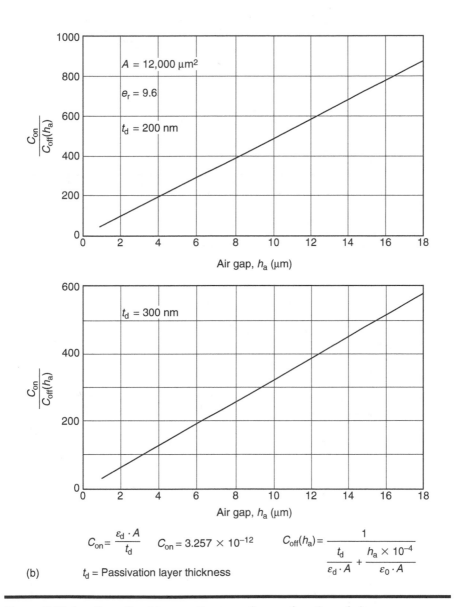

$$C_{on} = \frac{\varepsilon_d \cdot A}{t_d} \qquad C_{on} = 3.257 \times 10^{-12} \qquad C_{off}(h_a) = \frac{1}{\dfrac{t_d}{\varepsilon_d \cdot A} + \dfrac{h_a \times 10^{-4}}{\varepsilon_0 \cdot A}}$$

(b) t_d = Passivation layer thickness

Figure 5.12 (continued)   (b) capacitance ratio as a function of air gap,

(continued)

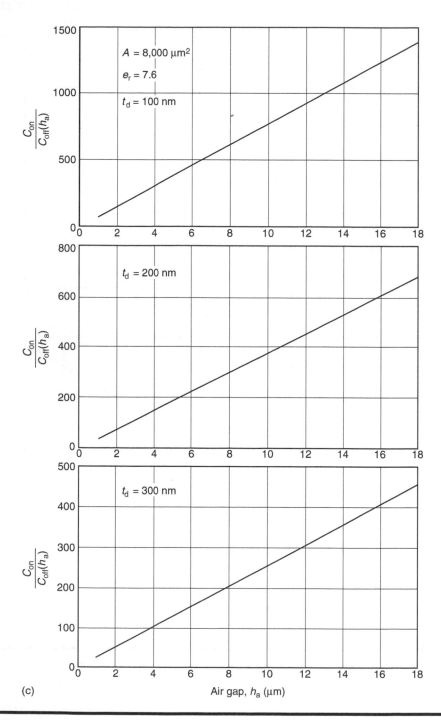

Figure 5.12 (continued) **(c) capacitance ratio as a function of air gap,**

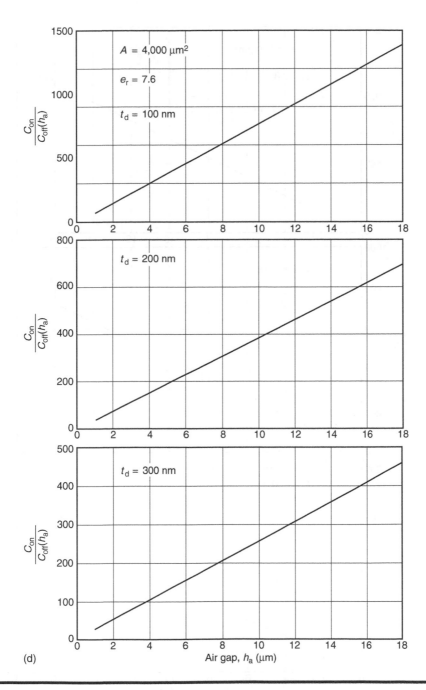

(d)

Figure 5.12 (continued)   (d) capacitance ratio as a function of air gap,

(continued)

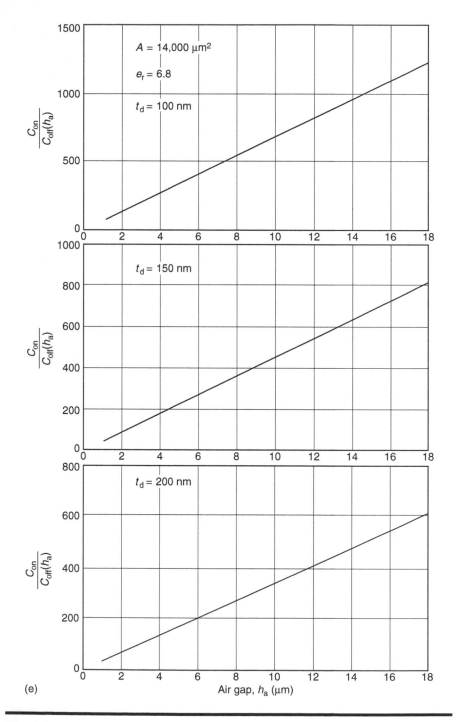

Figure 5.12 (continued) (e) capacitance ratio as a function of air gap.

A thin layer of chromium may be deposited on the top of aluminum to achieve a smooth surface. Electrode surfaces must be very smooth for low-loss contact. It is important to mention that surface roughness has significant impact on the on-capacitance of the metallic membrane shunt switch. Compressive and tensile stresses on the suspended membrane must be avoided to prevent buckling of the membrane and to avoid adverse switch operation during the pull-down operation. The stresses on the membrane depend on the disposition conditions and the materials used for the spacer layer and the membrane. Engineering material research undertaken by the author reveals that the latest aluminum alloys, tungsten, and aluminum/silicon/aluminum sandwich are best suited for the membrane. A new type of aluminum alloy known as controlled expansion (CE)-alloy (referred to as CE-5 alloy) contains 85 percent silicon and 15 percent aluminum and is best suited for MEMS bridges, cantilever beams, and membranes. This particular alloy has a coefficient of thermal expansion (CTE) less than 5 ppm/°C over the temperature range from 25°C to 125°C, density of 2.35 g/cc, thermal conductivity of 1.18 W/cm °C, Young's modulus of 135 GPa, and bending stress of 148 MPa.

5.5.3 RF Performance Parameters of Membrane Shunt Switch

When the switch is unactuated, the RF signal passes underneath the membrane with negligible attenuation. When the switch is actuated through the ES actuation, the metal/dielectric/metal sandwich provides a low impedance path to the surrounding CPW ground lines, thereby preventing the RF signal from traversing beyond the switch structure. The IL of the switch shown in Figure 5.11 is well below 0.18 and 0.30 dB at 10 and 35 GHz, respectively, compared to typical loss of 1 dB in a field-effect transistor (FET) or PIN-diode RF switches at the same frequencies. For time-delay phase shifters, which may require several such switches, the total IL is still significantly lower over other electronic switches.

Due to excellent impedance match to 50 Ω, these switches offer return loss better than 15 dB up to 40 GHz in the unactuation state, because of low off-capacitance of the switch ranging from 20 to 50 fF. Whether the actuation mechanism is off or on, there is a possibility of high electric field across the thin dielectric layer and becoming trapped, thereby requiring higher actuation voltages than the voltages allocated for the switch operation.

As far as the switching performance of these switches is concerned, the estimated switching speeds vary between 1.5 and 5.0 μs. A nonlinear dynamic model may be required, which will include the effects of ESs, residual compressive and tensile stresses, membrane deformation due to buckling, inertia, damping, and electrode parameters. However, as a first-order approximation, the switching speed can be determined by the mechanical resonance frequency of the MEMS switch, which can vary anywhere between 50 and 150 kHz. The switch will change the state between up and down positions in one-fourth of a cycle of the natural resonance frequency.

This switch has specific actuation voltage requirements, which are dependent on the air gap height (h_a) between the membrane and bottom electrode, dielectric medium permittivity, and the actuation area. Assuming the modulus of elasticity of 70 GPa for the bulk aluminum and residual stress of 120 MPa, the estimated pull-down voltage is roughly 56 V. The pull-down voltage can also be computed from the following equation:

$$V_{pull} = 0.544 \left[\frac{K h_a^3}{e_r A} \right] \tag{5.20}$$

Assuming an area of 6000 μm^2, dielectric constant of 6.7, and initial air gap height of 3 μm, one gets the pull-down voltage equal to 54.6 V.

5.6 RF-MEMS Switches with Low-Actuation Voltage

5.6.1 Introduction

Most of the MEMS switches described here before need actuation voltages between 30 and 75 V, thereby making these switching devices impractical for mobile wireless communications and reconfigurable circuit applications that normally operate between 3 and 12 V. The low-actuation MEMS switches can be directly integrated with current monolithic microwave integrated circuit (MMIC) devices without using the bulky and costly voltage up-converter circuits. Current research and development activities are being directed to development of MEMS shunt switch design with metal-to-metal contact, because such switch configuration provides ultralow loss in the upstate position and excellent RF performance at low actuation voltages over wide frequency band due to metal-to-metal contact [7].

It is important to point out that the lifetimes of a capacitive shunt MEMS switch with metal-to-metal contact significantly improve as the actuation voltage is reduced. Tests performed by Rockwell, Raytheon, and Hughes Research Laboratories on such switches demonstrated an exponential relation between the lifetime and actuation voltages with lifetimes between 10^4 at 65 V and 10^8 at 30 V. The test results indicate that lifetime improves on the order of a decade with every 5–7 V decrease in actuation voltage. These test results predict that to achieve lifetime in excess of 10^{10} or ten billion cycles, an actuation voltage less than 12 V is required. Furthermore, the root cause of sticking problem becomes somewhat of least concern. These tests have demonstrated that a low-voltage RF switch can achieve seven billion cold switching cycles [7].

5.6.2 Fabrication Process Steps and Critical Elements of the Switch

Critical elements of the shunt MEMS switch with metal-to-metal contact are shown in Figure 5.12 along with fabrication process steps. This particular switch incorporates a metal bridge made of gold, which spans a CPW transmission line of appropriate impedance. In the upstate position, the bridge is suspended at 3 μm above the signal line and in the downstate position, the switch is pulled into direct metal-to-metal contact with the signal transmission line, thereby creating a short circuit from signal line to ground line. The ES force generated upon the application of actuation voltage pulls down the switch and the restoring force pulls up the switch back into its original position. The dc applied voltage or the actuation voltage can be applied to the actuation pads beneath the bridge structure.

The pull-in voltage requirements are strictly dependent on the spring constant (k) of the switch, mechanical structure strength parameters, actuation area, and the air gap (d) between the switch bridge and the actuation pad. Assuming the ES force equal to the restoring force of a mechanical spring, the pull-in voltage can be written as denoted by Equation 5.20. Computer-generated plots of pull-in voltages as a function of spring constant (k), air gap (d), dielectric constant (e_r), and actuation pad area (A) can be obtained using Equation 5.20. One can select the optimized values of air gap and actuation pad area to achieve a desired actuation voltage. With an air gap of 2 μm, one can select an actuation voltage less than 15 V for spring constants ranging from 1 to 10 N/m.

5.6.3 Reliability Problems and Failure Mechanisms in the Shunt MEMS Switches

Although the metallic shunt MEMS switch offers the lowest actuation voltage capability, there are some reliability problems which cannot be ignored. Some of these problems include metal stress bowing, which comes from the residual stress in the switch electrode. The gold bridge tends to exhibit compressive strains and bowing above the silicon wafer surface. A switch with a low spring constant will have a larger residual stress in the bridge structure causing the bridge to bow more. Any increase in the cantilever length corresponds to a lower spring constant. For a cantilever with 8 μm width, the air gap increases to from 2 to 6.2 μm as the cantilever length increases from 90 to 450 μm. These statements clearly show the trade-off that exists between the air gap (d) and the spring constant (k), which must be undertaken before designing such MEMS switches.

Incorporating a dimple in the bridge structure will reduce the air gap between the switch electrode and the signal line, thereby ensuring a good electrical contact. It is critical to mention that a good contact between the electrode and signal line is absolutely necessary for adequate isolation between the input and output terminal of the switch. Furthermore, if the electrode does not make a contact, the switch structure behaves like a variable capacitor. If the stress in the metallic bridge is significant, the pad arches over the signal line and most likely it will not make contact with the signal line when the actuation voltage is applied. Addition of a dimple improves the contact with the signal line and at the same time it reduces the air gap. Optimization of the cantilever and bridge geometries is required to achieve lower actuation voltage from 15 to 10 V, while maintaining good contact between the signal line and the ground plane. A MEMS switch with an actuation voltage of 10 V can meet the goal of ten billion cycles of switch operation [7].

Sticking presents the most serious reliability problem in the RF-MEMS switches. Moving parts have a tendency to stick together during the fabrication process, particularly during the release of sacrificial layers. These parts can also stick during the operational phase. Sticking problem during the fabrication phase can be solved, but it is difficult, if not impossible, during the operational phase. MEMS switch designers feel that surface roughness, molecular force, and electronic charge force contribute to the sticking problem. But the sticking issue can be isolated and resolved by conducting hot and cold switching tests [7]. Some MEMS switch designers propose a noble design that prevents the switch electrode from contacting the sacrificial through the use of strategically located separation posts. These posts typically stand about 1 μm above the dielectric sacrificial layer, so when the switch is pulled down upon actuation, it comes to rest on the separation posts instead of the sacrificial layer. This new gap reduces the magnitude of the electric field across the dielectric layer, thereby eliminating any accumulation of electric charge. This technique essentially eliminates the sticking problem due to charge accumulation.

When the actuation voltage is applied to the switch, the greatest amount of contact and highest electric fields occur between the metallic bridge and the bottom actuation pad. However, by positioning separation posts at each corner and within the actuation pad, contact between the suspended metal pad and the dielectric layer is eliminated. Thus, implementation of separation posts will significantly improve the lifetime because failures due to accumulation of charge are eliminated. Tests conducted by Rockwell and Raytheon [7] have achieved lifetimes in excess of seven billion cycles, which are at least three orders of magnitude greater than the lifetimes achieved by MEMS switches without deployment of separation posts.

Three dominant failure mechanisms include electric charge accumulation, contact degradation, and contact deformation due to high RF current. The most serious failure mechanism is due to contact degradation. A rapid deterioration that usually occurs over the last few 1000 cycles can deposit thin insulation film on the contact. This will increase the resistance and decrease the isolation. Reduction of actuation voltage by optimizing the bridge geometry will improve the reliability. With lower actuation voltages, the charge accumulation problem is minimized because the electric field across the dielectric layer is very low, leading to significant reduction in sticking problems. Furthermore, lower actuation voltages require small gaps between the switch and the actuation pad, which further improves the switch reliability, thereby eliminating any internal stress caused by the motion of the cantilever. In addition, MEMS switches yield higher reliability with shorter cantilevers. Note the geometry of the separation post and separation between the two posts play key role in achieving lower actuation voltage. To achieve low actuation voltages, the separation post area must be minimized for the largest possible actuation pad area if high reliability and mechanical stability are the principal design requirements.

5.6.4 RF Performance Capabilities

The switch performance is very impressive from dc to 40 GHz. The IL in the upstate is less than 0.1 dB and isolation in the downstate is better than 25 dB from dc to 40 GHz. Isolation better than 20 dB up to 40 GHz indicates the existence of effective short distance between the signal and ground. The switching speed is greater than 20 μs with transition time less than 3 μs, when the switch operates at 15 V. The actuation voltage with minimum fluctuations is necessary if an accurate measurement of RF performance parameters is desired. Optimum switch geometry and strategically located separation posts will provide low-voltage operation with highest reliability.

5.7 RF-MEMS Series Switches

5.7.1 Introduction

In previous sections, the discussions were limited to RF-MEMS shunt switches and their performance capabilities and limitations. The focus of this section is exclusively on RF-MEMS series switches. Design aspects, performance capabilities and limitations, fabrication process steps, and potential applications of the MEMS series switch will be discussed in greater detail. Its major benefits and shortcomings will be described. Failure mechanisms, if any, will be identified.

5.7.2 Description and Design Aspects of the MEMS Series Switch

The MEMS series switch opens and closes the metal contacts and is most suitable for dc to RF applications such as automatic test equipment (ATE). The classical MEMS switch uses ES actuation mechanism and metal contacts. The switch design (Figure 5.11) configuration consists of a cantilever beam or membrane structure with a switching metallic contact. This contact closes the gap between the membrane and the CPW transmission line when an actuation voltage is applied between the cantilever beam or membrane structure and an electrode embedded in the substrate. A large distance or separation between the beam and the substrate is required to achieve a high isolation in the off-state position of the switch. But a large separation or gap exceeding 5–6 μm would require higher actuation voltage ranging from 25 to 60 V. To maintain high isolation at low separations, the overlapping switching contact area is limited not to exceed 500 μm^2 to minimize the capacitive coupling. Switches with metal contacts require substantial force to open the switch to avoid a risk due to contact microwelding. The restoring force to open the switch is provided by the spring energy stored in the cantilever beam, thereby indicating that the membrane structure or cantilever beam requires the right amount of stiffness to increase the actuation voltage. An active force is necessary to open the switch using a top electrode [8].

The actuator used for this switch consists of a membrane, whose one end is attached to the top electrode and the other end to a bottom electrode. The membrane moves between these electrodes, namely, bottom and top electrodes. Note the thin and flexible membrane is the most critical and delicate part of the S-shaped film actuator and is shown in Figure 5.11 along with top and bottom electrodes with sacrificial layer. Various fabrication steps for this switch are also shown in this figure [8].

Both for opening and closing the switch, the electrodes operate like a "touch-mode" or "zipper-like" actuation mechanism. Note electrodes are separated only by few micrometers ranging from 2 to 4 μm. Even at low actuation voltages, high ES forces can be developed. The actuator allows the displacement distance of the switching contact, which defines the off-state isolation independent of the effective actuation distance or separation. The independence of the contact distance and the actuation voltage is an essential feature of the electrostatically actuated RF-MEMS switches. Because the switch can also be opened by ES force generated between the top and bottom electrodes, spring energy stored in the mechanical-based membrane structure is not required to open the switch. This can make the moving structure very thin and flexible, thereby further lowering the actuation voltage requirements.

Due to large separation between the contacts in the off-state position of the switch, the overlapping switching contact area (*A*) can be designed significantly larger without decreasing the isolation. A large overlapping contact does not reduce the contact resistance, which is strictly dependent on the actuation force, the hardness of the contact surfaces, the surface roughness of the contact material, and contact surface contamination. Note large contact areas are best suited for switching higher RF currents. Geometrical dimensions of the CPW transmission line can be designed to reduce the power dissipation in the switch, thereby improving the heat distribution in the contact interface at high current densities. Note large contact area with soft material like gold will increase the risk of contact stiction during the off-state position of the switch, which must be avoided at all costs. The qualitative performance criteria correlated to features of the switch actuator are summarized in Table 5.6.

Table 5.6 Performance Criteria Dependent on Switch Features

Performance Parameters Criteria	Design Features
High isolation requirements	Large contact separation in off-state
Low actuation voltage requirements	Thin and flexible membrane structure
High power-handling capability	Wide signal lines with large contact areas
Packaging requirements	Assembly comprising of two parts
Integration with RF circuits	RF and MEMS on two separate wafers provides easy integration with RF substrate
High-speed switching requirement	Gas displacement around rolling membrane

5.7.3 Fabrication Process Steps and Switch Operational Requirements

Fabrication must be achieved using two separate wafer substrates: one containing RF circuits and the other containing MEMS mechanical part, to achieve maximum fabrication flexibility and highest production yield. The major advantage of two-wafer fabrication is that a MEMS switch can be fabricated with minimum time, lowest cost, and highest yield. In addition, switch can be integrated with RF substrates of different materials and physical dimensions. All parts can be housed in a 100 μm wide ring-shaped cavity with polymer walls with thickness ranging from 15 to 20 μm. The cavity size can be as small as 1000×1000 μm^2, approximately. The major emphasis is on large off-state contact distance achieving high isolation and low actuation voltage. Using a pattern adhesive layer, one can achieve a near-hermetic package-integrated MEMS switch.

The thin, flexible S-shaped actuator is about 1000 μm long and 1 μm thick using polyimide as a sacrificial layer to release the switch membrane. A sacrificial layer is selected to avoid stiction of the flexible membrane to the substrate [8]. The electroplated gold switching contact is about 200 μm long with a closing and opening gap of 120 μm in the CPW signal line, thereby providing an overlap of about 35 μm with each end of the signal line. This forms two large switching contact areas of 3500 μm^2 each. An actuation voltage of 15 V provides the contact force of slightly greater than 100 μN for each contact between the S-shaped membrane and the bottom electrodes as shown in Figure 5.13. This figure shows four electrodes needed for MEMS switch operation, namely, the top electrodes, the membrane electrodes with clamping electrodes, the bottom electrodes, and bottom

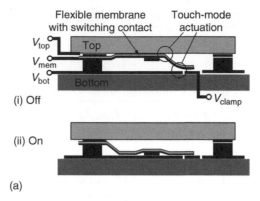

(a)

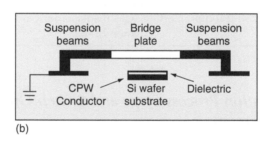

(b)

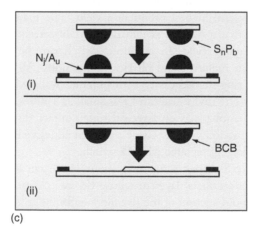

(c)

Figure 5.13   RF-MEMS switch design. (a) Actuation mechanism with S-shaped membrane showing on- and off-states of an RF-MEMS switch, (b) schematic cross section showing critical elements of a MEMS switch, and (c) sealing concepts for the switch (i) metal/solder seal, (ii) benzocyclobutene (BCB) seal.

clamping electrodes. The function of the clamping electrodes is to keep the membrane secured in touch-mode known as zippy-mode with the bottom electrodes during the off-state position when the membrane is pulled up toward the top electrode positions. Without the clamping electrodes, the switching membrane will completely be pulled up toward the top electrodes in the off-state position of the switch, and the switch will not be closed again in case the membrane should stick to the top electrodes even after releasing the top electrode actuation voltage.

5.7.4   RF Design Aspects

The moving membrane structure consists of 0.2 μm thick silicon nitride layer, which provides the needed isolation between the top and membrane electrodes. A 2 μm thick electroplated gold finite ground CPW is selected as an RF transmission line on the bottom of the switch. The overall length of the line is about 3400 μm. The signal line width is 100 μm, the distance between the signal line and ground line is 55 μm, and the ground line width is 200 μm. These values are needed to achieve a nominal characteristic impedance of 50 Ω. The vertical traveling distance of the switching contacts from open to close positions is 14 μm and with large switching contact areas it is good enough for high isolation in the off-state position of the switch. As stated earlier, the switch fabrication must use two separate wafers to keep the fabrication cost to minimum and to achieve high production yield. The bottom part of the switch is fabricated on a 100 μm diameter, 500 μm thick high-resistivity silicon substrate, while the top part of the switch is fabricated on a 100 μm diameter, 500 μm thick Pyrex 7740 glass wafer to keep fabrication cost low. Because the top of the switch is about 21 μm away from the CPW transmission line, the properties of Pyrex substrate have no impact on the RF performance of the switch [8].

5.7.5   RF Performance Parameters of the Switch

The actuation voltage between the membrane and top electrode requires 12 V to open the switch with an application clamping voltage of 20 V. The switch requires 16 V between the membrane electrode and the bottom electrode to close it. Any voltage variation within 0.7 V will not have any impact on the switch performance or reliability. The S-shaped thin-film actuation design offers high isolation at low actuation voltages and with large overlapping switch contact areas. The dc resistance of the switch at a closed position is about 0.65 Ω and each contact resistance is 0.275 Ω. One will experience higher contact resistance if the contact surface is not clean or the contact pressure exceeds 100 μN. Contact contamination is possible due to outgassing of the benzocyclobutene (BCB) polymer used during the fabrication process, which will further increase the contact resistance. The off-state isolation is about 65 dB at dc, 35 dB at 10 GHz, 28 dB at 20 GHz, 25 dB at 35 GHz, and 20 dB at 40 GHz, which is due to large overlapping contact areas. The upstate IL is less than 0.25 dB from dc to 40 GHz and can be further reduced by decreasing the contact resistance [8].

5.8 Effects of Packaging Environments on the Functionality and Reliability of the MEMS Switches

5.8.1 Introduction

The material presented in this section focuses on the impact of packaging environments on the functionality, performance, and reliability of the RF-MEMS switches. The first positive impact on the MEMS switch is the packaging temperature profile. Temperature profiles higher than any step seen by the switch before packaging can have detrimental effects. These adverse effects include outgassing of the packaging material and deterioration of contact surface leading to higher contact resistance. On the other hand, optimum packaging parameters must be used to control the pressure in the switch cavity, which can affect the switching speed. Packaging will protect the MEMS switches from the external environments such as dust, humidity, and winds [9].

5.8.2 Impact of Temperature on the Functionality and Reliability

Note when the temperature is increased in excess of 200°C, the stress curve of the materials starts deviating from its linear characteristic at a certain critical temperature. Furthermore, the stress changes during the increase in operating temperature due to plastic deformation in the materials used in the fabrication of the MEMS switch. However, upon cooling, one will notice higher tensile stress compared to that at room temperature. As a matter of fact, cooling down from any temperature higher than the critical temperature (T_c), which varies from material to material, would result in higher tensile stress. Therefore, if the packaging temperature profile exceeds the T_c for certain materials, the tensile stress in the MEMS switch bridge will change, leading to some deformation in the bridge and consequently, failure of the MEMS switch. In the case of a very small deformation of the bridge in "good direction" after zero-level packaging of a MEMS switch, slight degradation in RF performance can be expected, but no catastrophic failure of the switch. Exact impact of packaging environments on switch reliability and performance can be determined using very sophisticated instrumentation capable of measuring the fringe contrast images and associated temperature profile images through the glass cap of a zero-level packaged MEMS switching device.

5.8.3 Impact of Pressure on Switch Reliability and RF Performance

The adverse effect of pressure on the switch functionality, reliability, and RF performance can be determined by investigating two different bridge design configurations: one mechanically stiff bridge with small holes and the other with larger holes in the structure. Electrical life tests should be performed to monitor through optical probes the capacitance changes in the RF-MEMS switches as a function of time and nitrogen pressure. These tests must be performed in a closed chamber, where it is possible to achieve tight control of pressure, temperature, and gas composition inside the test chamber. Tests conducted by few MEMS switch designers [9] were able to observe the capacitance changes during the actuation and release of a capacitive RF-MEMS switch. These tests reveal that when the bridge is actuated at voltage higher than the pull-in voltage, the switch capacitance abruptly increases, leading to downward motion of the bridge and touching the dielectric medium. Upon the release of the actuation voltage, the bridge structure moves up and the capacitance decreases. From this test, one can estimate the pull-out time of the switch. The switching speed can be increased if the nitrogen pressure in the switch housing or cavity is reduced to 2×10^{-4} mbar. In addition, the pull-in and pull-out times are much shorter. It is important to mention that a long settling time is normally observed under large overshoot condition, when the bridge returns to the upstate position or low capacitance situation. Under large overshoot conditions or critical damping situations, the switch bridge can start vibrating with large amplitudes, leading to uncontrollable mechanical instability.

The change in capacitance is a function of bridge mechanical strength and stiffness and operating pressure in the preliminary tests performed by switch designers indicate that decrease in pull-in time occurs with decreasing pressure. The stiff bridge does not show overshoot even at a pressure as high as 125 mbar, but no overshoot appears in vacuum (0.0002 mbar). With a less stiff switch and lower surface area, overshot exists even at a pressure of 250 mbar. Both the internal pressure in the package and the gases inside the cavity affect the switch performance. The change in capacitance during the upstate and downstate positions of the switch is strictly dependent on the number of cycles. It is interesting to mention that MEMS switches using silicon oxide, titanium oxide, and silicon nitride-insulating materials exhibited constant change in capacitance for a certain number of cycles regardless of air or nitrogen gas environments, and then displayed sudden drops in capacitance. However, the change in capacitance is more in nitrogen environment than the change in air. Under high actuation voltages, high electrical fields are induced causing charge trapping. Under these conditions, the bridge sticks to the insulator and remains in the downstate position. This is considered as a failure mode in capacitive-based RF-MEMS switches, which is caused by the charge that builds up in the insulator when the bridge touches the dielectric. The lifetime is roughly 100 times longer in the nitrogen gas environments than the air environments, regardless of actuation voltages or insulating materials used.

Under humid environments, charge trapping increases, leading to a faster buildup in addition to a shorter lifetime of the capacitive MEMS switches. The MEMS packaging must provide protection to the device against external environmental parameters such as pressure, temperature, humidity, chemicals, and atmospheric particles known as aerosols. Encapsulation at wafer level is necessary to prevent exposure of the microstructures to particles and debris produced during the dicing and handling processes. MEMS switches must be packaged on wafer level using glass capping and BCB sealing techniques capable of providing electrical isolation, but not hermetic seal.

5.8.4 Effects of Zero-Level Packaging on MEMS Switch Performance

This section will summarize the effects of environmental factors or parameters on the RF performance of MEMS switches. Adverse effects from temperature, pressure, and gases are summarized. Temperature profiles required for the packaging process are discussed, which can affect the mechanical stress in the MEMS's bridge. Improper packaging techniques can impact the functionality and performance of the switch. Selecting the right nitrogen pressure and a bridge with optimum stiffness will certainly optimize the switching speed of the MEMS switch. However, choosing a very low gas pressure can produce some vibrations in the switch structure. In brief, a zero-level packaging with constant, well-controlled nitrogen gas pressure inside the packaging will provide optimum RF performance and high reliability. Requirements for zero-level packaging are quite different from those required for a complete hermetic package.

5.9 Packaging Material Requirements for MEMS Switches

This section is dedicated to packaging materials required for RF-MEMS switches. Suitable metals, alloys, and substrate materials best suited for MEMS switches are identified. Important properties of these materials such as thermal, electrical, and mechanical are summarized. Major emphasis will be placed on the physical properties and manufacturing characteristics of CE- alloys [10], which are considered most suitable for RF-MEMS applications.

5.9.1 Properties and Applications of CE-Alloys for RF-MEMS Devices and Sensors

Although CE-alloys were discussed in detail in Chapter 4, nevertheless, their potential applications in the design of RF-MEMS and optical-MEMS devices and sensors are carefully identified in this section. The ability to tailor the CTE values of certain CE-alloys to ceramic and high-performance dielectric hard substrates is impressive. The major benefits of applications in the design and development in MEMS devices are discussed. These CE-alloys have demonstrated [10] high thermal conductivity, excellent dimensional stability, enhanced rigidity, improved reliability under harsh thermal and mechanical environments, and low-cost manufacturability. Their outstanding characteristics have made them most attractive for RF/microwave/optical-MEMS packaging and miniaturized heat sinks. Certain CE-alloys, namely, CE-11, CE-9, CE-7, and CE-5 have demonstrated remarkable combination of properties that are most ideal for RF-/optical-MEMS devices (Table 5.7).

Note the CE-5 alloy with 85 percent silicon and 15 percent aluminum composition offers the lowest CTE of 5 ppm/°C over the temperature range from 25°C to 125°C. In general, the CE-alloys are three to six times lighter than standard packaging and base plate materials used in RF/microwave MEMS devices. These alloys possess high thermal conductivity and specific stiffness and are amenable to standard machining operations such as milling and drilling. Furthermore, the CE-alloys are composed of silicon and aluminum, which are environmentally friendly materials. These alloys do not belong to strategic material category and, hence, are readily available with no restriction. Due to unusual combination of high thermal conductivity and low thermal capacity, these alloys can provide uniform heat distribution, thereby making them most suitable for easy soldering of RF feedthroughs. In addition, these alloys can be widely used as packaging materials in MEMS devices for spacecraft and aerospace sensors and systems operating at frequencies as high as 40 GHz.

Table 5.7 Physical Properties of CE-Alloys

CE-Alloy	Si_xAl_{1-x} (Percent Content)	Coefficient of Thermal Expansion (CTE) (ppm/°C)	Deflection Profile δ (g/cc)	Thermal Conductivity K (W/°C cm)	Young's Modulus E (GPa)	Yield Strength (MPa)
CE-11	50/50	11.2	2.50	1.40	121	125
CE-9	60/40	9.1	2.45	1.22	124	135
CE-7	70/30	7.4	2.40	1.20	129	101
CE-5	85/15	5.0	2.35	1.18	138	75

Note the CTE of CE-11 alloy is sufficiently close to that of aluminum, thereby making it most suitable for RF-/optical-MEMS devices, where hermetic sealing is the principal design requirement. CE-7 alloys are best suited for MEMS optical housing, which forms a part of a high-speed RF-MEMS switching device comprising of an array of these modules. These arrays have potential applications in aerospace systems. As mentioned earlier, CE-5 alloy provides the lowest CTE, reduced weight, high stiffness and rigidity, improved reliability, and high modulus of elasticity. A housing made of CE-5 alloy can be electroplated with gold over nickel to reduce RF losses. Furthermore, such a housing can be precisely and manually tuned through adjustment of the micromirrors, which are fixed in the position by soldering the supporting areas to the optical housing. Both the CE-7 and CE-5 alloys are dimensionally stable over a wide temperature range, which makes them most attractive for unpredictable thermal environments. CE-alloy is especially most suitable for RF-MEMS devices operating at mm-wave frequencies, where superior machinability and close tolerance are of critical importance. In summary, the CE-alloys offer MEMS application solutions that enhance MEMS device performance, increase the functionality-to-weight ratio, and yield most cost-effective design [10].

It is important to mention that most CE-alloys are best suited for RF/microwave MEMS packaging and housing because they provide close CTE match to circuit boards and components (CE-9 and CE-7), high thermal conductivity, high specific stiffness or high stiffness-to-weight ratio, close expansion match to lens glass (CE-9), improved hermetic capability (CE-11), excellent dimensional stability, and burr-free machining.

5.10 Summary

This chapter has summarized design concepts, performance capabilities, and limitations of an RF-MEMS shunt, and series switches operating at microwave and mm-wave frequencies.

Operating principles and critical performance parameters, namely, the IL and isolation of both switches are identified. ES actuation voltages for shunt and series MEMS switch are specified with emphasis on pull-in or pull-down voltages. Design requirements of an RF-MEMS shunt switch with high isolation and minimum IL at mm-wave frequencies are described. Design parameters for a W-band (75–110 GHz) shunt switch are optimized for low IL and high isolation. Sample calculations for upstate and downstate losses, isolation in a MEMS shunt switch, switching speed, and on/off capacitance ratio are provided for rapid comprehension by the readers. Design aspects and material requirements for critical switch elements such as membrane, cantilever bridge, and electrode contacts are described for optimum RF performance and high reliability. Benefits of low actuation mechanisms are summarized. Reliability problems and failure mechanisms of critical switch elements are identified and remedies to eliminate them are recommended. Material requirements for fabrication and packaging of RF-MEMS switches are defined. Zero-level packaging requirements are summarized. Design aspects and parameters for an mm-wave RF-MEMS series switch with high isolation and lowest IL are identified. Requirements for high isolation, high power-handling capability, optimum switching speed, and high reliability are discussed. Effects of packaging environments on the functionality and performance of RF-MEMS switches are summarized with particular emphasis on reliability.

References

1. L. Dussopt and G. Rebeiz, Intermodulation distortion and high power handling capability of RF-MEMS switches, varactors and tuning filters, *IEEE Transactions on MTT*, 51(4), April 2003, 1247–1252.
2. J.B. Muldavin and G. Rebeiz, High isolation CPW MEMS shunt switches—Part I: Modeling, *IEEE Transactions on MTT*, 48(6), June 2000, 1045–1046. [used Fig. 1 on p. 1045].
3. H. Toshiyoshi and M. Mita, A MEMS piggyback actuator for hard disk-drives, *Journal of MEMS*, 11(6), December 2002, 648–652. [used Fig. 2 on p. 649].
4. J.B. Muldavin and G. Rebeiz, High isolation CPW MEMS shunt switches—Part II: Design *IEEE Transactions on MTT*, 48(6), June 2000, 1053–1056. [used Fig. 3 from p. 1054].
5. J.B. Rizic and G.M. Rebeiz, W-band CPW RF-MEMS switch on quartz substrate ($e_r = 3.78$), *IEEE Transactions on MTT*, 51(7), July 2003, 1256–1262.
6. J. Jamie, S. Chen et al., Micromachined low-loss microwave shunt switches, *Journal of MEMS*, 8(2), June 1999, 128–133. [used Fig. 1 on p. 129 and Fig. 2 on p. 130].
7. R. Chen, R. Lesmict et al., Low-actuation voltage RF MEMS shunt switch with cold switching lifetime of seven billions cycles, *Journal of MEMS*, 12(5), October 2003, 713–714.
8. I.D. Pioter, C. Necki et al., The influence of package environments on the functionality and reliability of capacitive RF MEMS switches, *Microwave Journal*, December 2005, 102–104.
9. J. Oberhammer and G. Stemme, Design and fabrication aspects of an S-shaped film actuator based on DC to RF MEMS series switch, *Journal of MEMS*, 13(3), June 2004, 421–427.
10. D.M. Jacobson, A.J.W. Ogilvy et al., Applications of CE-alloys in defense, aerospace, telecommunication and other electronic markets, *Microwave Journal*, August 2006, 150–163.

Chapter 6

RF/Microwave MEMS Phase Shifter

6.1   Introduction

This chapter is exclusively dedicated to radio frequency (RF)/microwave microelectromechanical system (MEMS) phase shifters. Phase shifters can be classified into two categories: absorption types and reflection types. An absorption-type phase shifter is widely used for RF and microwave frequencies, whereas a reflection-type phase shifter is best suited for higher microwave and mm-wave frequency applications. A phase shifter is the most critical component of an electronically scanning-phased array antenna, widely deployed in electronic warfare (EW) systems, missile tracking radar, forwarding looking radar used by airborne fighter/bomber aircraft, communications systems, and space-based surveillance and reconnaissance sensors. MM-wave phase shifters using conventional field-effect transistors (FETs) and PIN-diodes suffer from excessive insertion loss (IL), high power consumption, and poor reliability. It is important to mention that the dominant IL of a phase shifter demands excessive power consumption, which is not easily available in airborne systems or satellite-based sensors because of stringent requirements for power consumption, weight, and size. Current research and development activities on MEMS and nanotechnologies (NTs) reveal that integration of MEMS-based switches, tuning capacitors and air gap 3 dB coplanar waveguide (CPW) couplers will lead to successful design and development of MEMS phase shifters operating at microwave and mm-wave frequencies with low IL, high switching speeds, minimum power consumption, higher reliability, and life cycles exceeding ten billions. The activities further reveal that mm-wave phase shifters using MEMS technology will not only offer low IL and minimum power consumption but also provide ultrawide bandwidth and reduced intermodulation (IM) distortion compared to conventional devices using FETs or PIN-diodes. In summary, a bulk micromachined MEMS phase shifter fabricated on a high-resistivity silicon wafer offers minimum IL, reduced surface and dispersion losses, and high reliability over ultrawide bands at mm-wave frequencies. Furthermore, use of a micromachined mm-wave phase shifter yields a distributed MEMS true-time-delay (TTD) phase shifter capable of providing high phase shift per decibel loss, large phase shift per unit length, and lowest power consumption at mm-wave frequencies compared to conventional phase shifter using FETs or PIN-diodes. Such a phase shifter has demonstrated lowest IL and highest phase shift per decibel from direct current (dc) to 60 GHz. Furthermore, the distributed MEMS TTD phase shifter design is fully compatible with monolithic technology, which will further reduce the losses and power consumption in distributed MEMS TTD phase shifters operating at mm-wave frequencies.

Note that design of MEMS phase shifters involves MEMS bridges, metal–insulator–metal (MIM) capacitances, MEMS switches, MEMS capacitors, 3 dB CPW couplers, unloaded inductances and capacitances, and transmission-line segments. MEMS phase shifters make massive use of CPW transmission-line technology [1]. Therefore, it is appropriate to provide useful information on the relevant parameters of the CPW transmission lines, namely, the characteristic impedance (Z_0), center conductor width (W), and thickness (t) of the CPW transmission line, and substrate thickness as a function of frequency and dielectric constant of substrate deployed.

6.2 Properties and Parameters of CPW Transmission Lines

Cross section of the CPW transmission line depicted in Figure 6.1a shows the center conductor width (W), conductor thickness (t), gap or spacing on either side of center conductor (G), substrate height (h), and the dielectric constant of the substrate (e_r). Characteristic impedance (Z_0) plots for the CPW transmission lines with zero thicknesses ($t = 0$) and as a function of wave number ($k = W/[W + 2G]$) and dielectric constant are shown in Figure 6.1b. For more accurate characteristic impedance calculations, values of effective dielectric constant (e_{eff}) must be used. Plot of complete elliptic integral of first-kind $K(k)$ is shown in Figure 6.2, which must be used in computations of CPW line impedances as a function of wave number (k) and effective dielectric constant of the substrate materials.

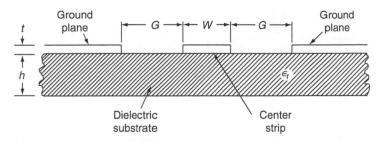

W = center conductor width of CPW transmission line, t = conductor thickness
(a)   G = gap or spacing on either side of the center conductor

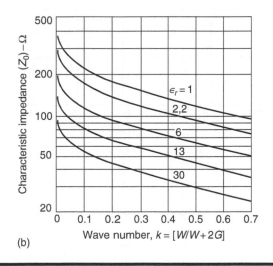

(b)

Figure 6.1 **CPW transmission-line structure showing (a) cross section of CPW transmission line and (b) characteristic impedance of CPW line as a function dielectric constant.**

6.2.1 Computations of CPW Line Parameters

The characteristic impedance of the CPW transmission line can be written as

$$Z_0 = \left[\frac{94.25}{\sqrt{e_{\text{eff}}}}\right]\left[\frac{K(k')}{K(k)}\right] \quad (6.1)$$

where e_{eff} is the effective dielectric constant of the substrate and $K(k)$ and $K(k')$ are the complete elliptic integrals of first kind.

$$e_{\text{eff}} = \left[\frac{(e_r + 1)}{2}\right] + \left[\frac{(e_r - 1)(0.5)}{\sqrt{1 + 10(t/h)}}\right] \quad (6.2)$$

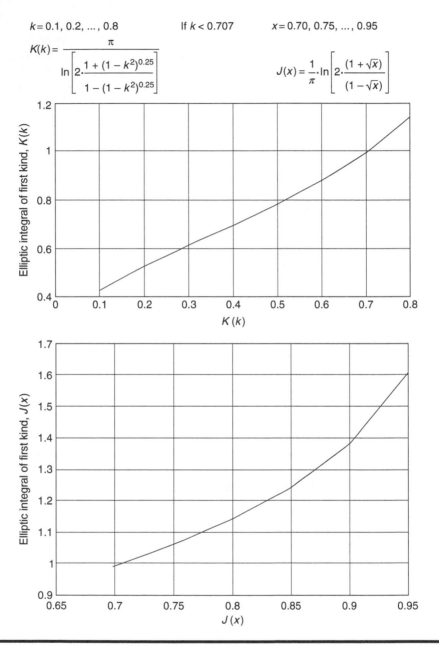

$k = 0.1, 0.2, \ldots, 0.8$ If $k < 0.707$ $x = 0.70, 0.75, \ldots, 0.95$

$$K(k) = \cfrac{\pi}{\ln\left[2 \cdot \cfrac{1 + (1 - k^2)^{0.25}}{1 - (1 - k^2)^{0.25}}\right]}$$

$$J(x) = \frac{1}{\pi} \cdot \ln\left[2 \cdot \frac{(1 + \sqrt{x})}{(1 - \sqrt{x})}\right]$$

Figure 6.2 Plots of elliptic integral of first kind (k less than 0.707 and x less than 1.0).

where

   e_r is the relative dielectric constant of the substrate
   t is the conductor thickness
   h is the substrate height or thickness

$$k' = \left[\frac{(W + \Delta)}{(W + 2G - \Delta)} \right] \qquad (6.3)$$

where

   W is the center conductor width,
   G is the gap on either side of the center conductor as shown in Figure 6.1a, and
      the parameter Δ is given as

$$\Delta = \left[\frac{1.25t}{\pi} \right] \left[1 + \ln \left(\frac{4\pi W}{t} \right) \right] \qquad (6.4)$$

When the conductor thickness is zero, the parameter Δ becomes zero and Equation 6.3 is modified and is written as

$$k = \left[\frac{W}{W} + 2G \right] \qquad (6.5)$$

$$k' = \left[\sqrt{1 - k^2} \right] \qquad (6.5a)$$

The parameter defined by this equation is known as the wave number. The characteristic impedance plots for the CPW transmission lines using the above equations and as a function of dielectric constant and wave number, which is a function of W and G dimensions, are shown in Figure 6.1b.

The dielectric loss of the CPW line can be defined as

$$L_d = 0.2729 \left[\frac{\tan \delta}{\lambda_g} \right] \left[\frac{e_r}{e_{eff}} \right] \left[\frac{(e_{eff} - 1)}{(e_r - 1)} \right] \text{ dB/cm} \qquad (6.6)$$

where $\tan \delta$ is the loss tangent of the dielectric medium or substrate. The expression for the conductor loss can be written as

$$L_c = \frac{\pi}{\lambda_g} [1 - (Z_1 - Z_\delta)] \qquad (6.7)$$

where Z_1 is the characteristic impedance of the line when the losses are neglected and Z_δ is the impedance when the dimensions W, G, and t are replaced by W', G', and t' such that

$$W' = [W - \delta] \qquad (6.7a)$$

Table 6.1 Skin Depths for Copper and Gold Conductors as a Function of Frequency (µin/nm)

Frequency (GHz)	Copper	Gold
10	26/662	30.9/785
20	18/469	22/559
30	15.2/386	17.8/452
40	13/331	15.5/393
60	10.6/269	12.6/320
80	9.2/234	10.9/277
90	8.7/221	10.3/262

$$G' = [G + \delta] \tag{6.7b}$$

$$t' = [t - \delta] \tag{6.7c}$$

where δ is the skin depth of the conductor. Calculated values of skin depth in copper and gold conductors as a function of frequency are summarized in Table 6.1.

Sample Calculations for Characteristic Impedance

Assuming 100 µm for conductor width, 30 µm for gap or spacing, conductor thickness to substrate thickness ratio (t/h) of 0.04, conductor thickness of 2 µm, and relative dielectric constant of 6, and inserting these values in above equations, one gets

$$e_{\text{eff}} = \left[\frac{6+1}{2}\right] + \left[\frac{6-1}{2}\right]\left[\frac{1}{\sqrt{1 + 10 \times 0.04}}\right]$$

$$= [3.5] + [2.5]\left[\frac{1}{\sqrt{1.4}}\right] = [3.5 + 2.11]$$

$$= 5.61$$

$$\Delta = \left[\frac{1.25 \times 2}{\pi}\right]\left[1 + \text{In}\left(\frac{4\pi \times 100}{2}\right)\right] = [0.7958][1 + 6.44]$$

$$= 5.92 \text{ µm}$$

$$= 0.625$$

$$K(0.625) = 0.88$$

$$k' = \left[\sqrt{1 - 0.625^2}\right] = \left[\sqrt{1 - 0.3906}\right] = \left[\sqrt{0.6094}\right]$$

$$= 0.7806$$

$$K(0.7806) = 1.168$$

$$k_e = \left[\frac{100 + 5.92}{160 - 5.92}\right] = \left[\frac{105.92}{154.08}\right]$$

$$= 0.6874 \text{ (from Figure 6.1)}$$

$$K(0.6874) = 0.975$$

Now inserting these calculated parameters in Equation 6.1, one gets the value of characteristic impedance as

$$Z_0 = \left[\frac{94.25}{\sqrt{5.61}}\right]\left[\frac{1.168}{0.880}\right] = [39.8][1.3273]$$

$$= 52.83 \ \Omega \text{ (calculated value)}$$

$$= 54 \ \Omega \text{ (from Figure 6.1)}$$

Thus, these calculations verify the authenticity and accuracy of the impedance plots as illustrated in Figure 6.1. Therefore, one can use these curves to estimate the characteristic impedance of the CPW transmission lines with minimum time as a function of wave number, which is dependent on the center conductor width and gap parameters.

It is important to mention that the parameters k and k_e are strictly dependent on the center conductor width (W), gap (G), and the conductor thickness (t). Values of these parameters as a function of these physical dimensions are summarized in Table 6.2.

Table 6.2   Computed Values of Parameters k and k_e

	Thickness, t (μm)			
	0	*1*	*2*	*5*
W/G (μm)	k	k_e	k_e	k_e
100/30	0.625	0.658	0.687	0.761
100/40	0.555	0.584	0.608	0.676
100/50	0.500	0.525	0.546	0.604
100/60	0.454	0.476	0.495	0.546
100/80	0.384	0.402	0.417	0.457

Table 6.3   Properties of Substrates Best Suited for MEMS Applications

Substrate	Permittivity	Tan δ at 10 GHz	Thermal Conductivity (W/°C cm)
Alumina (Al)	9.75	0.0002	0.30
Gallium arsenide (GaAs)	12.91	0.0016	0.31
Quartz	3.78	0.0001	0.02
Sapphire	11.72	0.0001	0.41
Silicon (Si)	11.82	0.0052	0.92

Using the values summarized in this table as argument, the magnitude of the elliptic function of first-kind $K(k)$ and $K(k')$ can be determined from the curve shown in Figure 6.2. The contents of this table and the curves shown in the above figure can be used to determine the physical dimensions of CPW transmission line for a given impedance and substrate's permittivity.

Various substrate materials are being used in the fabrication of MEMS devices. Dielectric properties and thermal conductivity of potential substrates are summarized in Table 6.3, which can provide useful data to MEMS designers and scientists.

6.3   Distributed MEMS Transmission-Line Phase Shifters

6.3.1   Introduction

Distributed MEMS transmission-line (DMTL) circuits and elements are widely used in the design and development of filters, traveling-wave amplifiers, nonlinear transmission-line segments, matching circuits, and phase shifters. Deployment of periodic-loaded transmission line offers very wideband operations in microwave and mm-wave spectral regions [1]. For the DMTL configuration, transmission-line geometrical parameters and physical dimensions can be defined to achieve a periodic transmission line with a characteristic impedance of 50 Ω. It is critical to mention that the DMTL is a high-impedance line with the characteristic impedance (Z_0) greater than 50 Ω with CPW center conductor width (W) capacitively loaded by the periodic installation of MEMS bridges [1] with specific width (w) and spacing (s) as illustrated in Figure 6.3a. The DMTL is connected to 50 Ω feed lines and test pads needed to obtain relevant performance parameters. The transmission-line model of the DMTL along with MEMS bridge is shown in Figure 6.3b, identifying the bridge capacitance (C_b) and the transmission-line inductance (L_t) and capacitance (C_t)

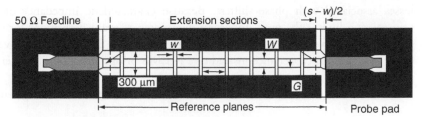

W = Center conductor width of CPW line

G = Gap or spacing on either side of center conductor

W/W + 2G = Total width of the CPW transmission line

w = Width of MEMS bridge

(a) s = Spacing between the MEMS bridges

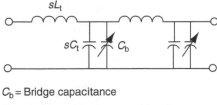

C_b = Bridge capacitance

C_t = Line capacitance per unit length

L_t = Line inductance per unit length

(b)

Figure 6.3 Physical layout of the DMTL structure. (a) Layout of a DMTL using CPW line and (b) lumped-element model for the DMTL structure including the bridge. (From Barker, N.S. and Rebeiz, G.M., *IEEE Trans. MTT*, 48, 1957, 2000. With permission.)

per unit length. The periodic unloaded transmission-line elements can be easily implemented using CPW transmission-line segments. The top view of the DMTL using CPW lines is shown in Figure 6.3 identifying the bridge dimensions, namely, width (w), and the length l equal to $W + 2G$, where G represents the gap or spacing on either side of the center conductor width (W). The bridge structure has a thickness equal to t. The periodic spacing between the two consecutive bridges is equal to s as illustrated in the above figure. One can use a varactor diode instead of a MEMS bridge to obtain variable capacitances, which has high series resistance ranging between 2 and 5 Ω at mm-wave frequencies.

6.3.2   Computations of DMTL Parameters

Sample calculations are performed to provide readers rough estimates of magnitudes for Bragg frequency, phase shift, phase delay, loaded line impedance, and various losses associated with phase shifters. Because characteristic impedance of CPW transmission lines is the most important parameter, relevant derivations for various parameters involved will be first discussed in great detail. DMTL can use a varactor diode as a variable capacitance instead of a MEMS bridge structure, if higher losses are acceptable.

The characteristic impedance of the CPW transmission line can be given as

$$Z_0 = 95.25 \left[\frac{K(k')}{(\sqrt{e_{\text{eff}}})K(k)} \right] \tag{6.8}$$

where $K(k')$ and $K(k)$ are the elliptic integrals of first kind and e_{eff} is the effective dielectric constant of the substrate.

$$k = \left[\frac{W}{W + 2G} \right] \tag{6.9}$$

$$k' = \sqrt{1 - k_2} \tag{6.9a}$$

$$e_{\text{eff}} = \left[\frac{(e_r + 1)}{2} \right] + \left[\frac{(e_r - 1)}{2} \right] \left[\sqrt{1 + 10 \left(\frac{t}{h} \right)} \right] \tag{6.9b}$$

where
    W is the center conductor width
    G is the spacing on either side of center conductor
    e_r is the relative permittivity of substrate
    t is the conductor thickness
    h is the substrate height or thickness as shown in Figure 6.1.

6.3.2.1 Bragg Frequency Computations

The Bragg frequency plays a key role in designing the MEMS phase shifters, which can be written in quadratic form as

$$f_b = 0.159 \left[\left(b - \sqrt{\frac{b^2 - 4ac}{2a}} \right) \right]^{0.5} \tag{6.10}$$

$$a = \left[s^2 L_t C_t L_b C_b \right] \tag{6.10a}$$

$$b = \left[s^2 L_t C_t \right] + \left[s L_t C_b \right] + \left[4 L_b C_b \right] \tag{6.10b}$$

$$c = 4 \tag{6.10c}$$

where

L_t stands for inductance

C_t stands for capacitance per unit length of transmission line

L_b stands for bridge inductance

C_b stands for bridge capacitance

s stands for periodic spacing between the bridges

The periodic spacing decreases as the operating frequency increases.

$$L_t = \left[C_t Z_0^2 \right] \tag{6.11a}$$

$$C_t = \left[\frac{\sqrt{e_{\text{eff}}}}{V_l Z_0} \right] \tag{6.11b}$$

where V_l is the velocity of light, 3×10^{10} cm/s.

Assuming $s = 200 \, \mu m$, $Z_0 = 100 \, \Omega$, $e_{\text{eff}} = 2.5$, $L_b = 10 \, pH$, and $C_b = 40 \, fF$, one gets the values of variable parameters as

$$C_t = 0.527 \times 10^{-12} \text{ F/cm}$$

$$L_t = 0.527 \times 10^{-8} \text{ H/cm}$$

$$a = 0.444 \times 10^{-48}$$

$$b = 6.92 \times 10^{-24}$$

Now inserting these parameters in Equation 6.10, one gets

$$f_b = 0.159 \left[6.927 \times 10^{-24} - \left(\frac{\sqrt{47.98 \times 10^{-48} - 7.104 \times 10^{-48}}}{0.888 \times 10^{-48}} \right) \right]^{0.5}$$

$$= \left[0.159 \times 775 \times 10^9 \right]$$

$$= 123.225 \text{ GHz}$$

$$= 123.23 \text{ GHz}$$

Assuming various values of bridge inductance, calculated values of Bragg frequency are summarized in Table 6.4.

It is important to mention that the Bragg frequency must be set slightly higher than the calculated values to achieve return loss better than -15 dB. Furthermore, the total width of the CPW transmission line S, which is equal to $W + 2G$, must be chosen close to $\lambda_d/8$ at the maximum operating frequency. It is interesting to note that at smaller center conductor width and higher CPW line impedance, higher phase shift (degree per centimeter) is possible regardless of the operating frequency, which is evident from Figure 6.4. Note the Bragg frequency must be at least three times the maximum operating frequency, if return loss of better than -15 dB is

Table 6.4 Calculated Values of Bragg Frequency, F_b (GHz)

Bridge Inductance (pH)	Bragg Frequency (GHz)
0	137.92
10	123.23
20	112.14
30	103.25
40	95.57
50	89.55

Note: Variables assumed: $Z_0 = 100\ \Omega$, $C_b = 40$ fF, $e_{eff} = 2.5$, and bridge spacing $s = 200\ \mu m$.

desired over the band of interest. The Bragg frequency requirements increase with the increase in operating frequency and in center conductor characteristic impedance, as illustrated in Figure 6.5. Furthermore, higher phase shift can be achieved at higher capacitance ratio, as shown in Figure 6.5.

6.3.2.2 Computations of Bridge Impedance (Z_B) and Phase Velocity (V_p)

It is important to mention that the creation of periodic structures is dependent on the cutoff frequency or the Bragg frequency at a point where the guided wavelength approaches the periodic spacing (s) of the discrete RF component. This cutoff frequency can be selected in such a way that will not limit the performance of a MEMS device. In the case of the DMTL structures, the self-resonant frequency of a MEMS bridge is about 210 GHz [1], and therefore, the device operation is limited by the bridge frequency of the transmission line. On the basis of the above statement, the characteristic impedance of a MEMS bridge can be written as a function of shunt capacitance as

$$Z_B = \left[\frac{\sqrt{sL_t}}{sC_t + C_b}\right]\left[\sqrt{1 - \left(\frac{f}{f_b}\right)^2}\right] \tag{6.12}$$

where f is the operating frequency.

The phase velocity of a lossless loaded transmission line as function of parameters shown in Figure 6.2 can be written as

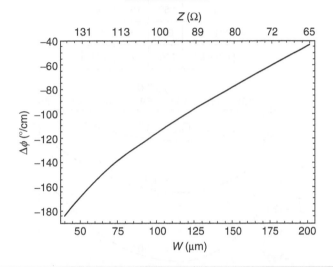

DMTL phase-shifter specifications

ϵ_r	3.8
Z_{lu}	48 Ω
f_B	120 GHz
C_r	1.2

Figure 6.4 **DMTL phase shifter performance specifications and phase shift per centimeter at 40 GHz as a function of characteristic impedance and center conductor width of the CPW transmission line with total CPW line width of 300 μm. (From Barker, N.S. and Rebeiz, G.M.,** *IEEE Trans. MTT,* **48, 1957, 2000. With permission.)**

$$V_p = \left[\frac{s}{\sqrt{sL_t(sC_t + C_b)}} \right] \left[1 + \left(\frac{1}{6}\right)\left(\frac{f}{f_b}\right)^2 \right] \qquad (6.13)$$

If the operating frequency is well below the Bragg frequency, then Equation 6.13 is reduced to

$$V_p = \left[\frac{s}{\sqrt{Lc}} \right] \qquad (6.14)$$

where $L = sL_t$ and $C = sC_t + C_b$. It is evident from Equation 6.14 that by varying the bridge capacitance, one can vary the phase velocity of the transmission line leading to formation of a variable delay line or a TTD phase shifter.

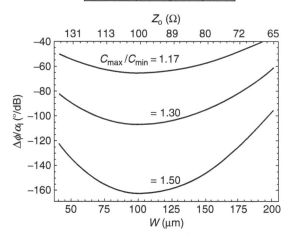

Dimensions of the DMTLs used to verify the optimization
the Bragg frequency is calculated using $L_b = pH\ 20$.
$R_b = 0.15\ \Omega$ at 30 GHz

$W\ (\mu m)$	50	100	150
$Z_u\ (\Omega)$	131	100	80
$C_{bo}\ (fF)$	37.9	33.8	28.6
$s\ (\mu m)$	150	197	249
$w\ (\mu m)$	80	35	17
$f_B\ (GHz)$	119	122	127

Figure 6.5 **Parameters required for optimization of Bragg frequency of DMTL and phase shift per decibel loss at 40 GHz as a function of capacitance ratio. (From Barker, N.S. and Rebeiz, G.M., *IEEE Trans. MTT*, 48, 1957, 2000.)**

6.3.2.3 Insertion Loss in the DMTL Section

The IL of the loaded CPW transmission line must include the resistance in series with the line inductance L, which is equal to sL_t and the bridge resistance (R_b) in series with the bridge capacitance C_b. Combining the losses in these resistors, total IL per section of the distributed transmission line can be written as

$$IL = \left[\frac{R_s}{2Z_l}\right] + \left[\frac{\omega^2 C_b^2 R_b Z_l}{2}\right] \tag{6.15}$$

where Z_l is the impedance of the loaded line and is different from the unloaded impedance of the transmission line. The impedance of a loaded line is about 50 Ω,

Table 6.5 Loaded Transmission-Line Loss as a Function of Bridge Resistance (dB/cm)

Frequency (GHz)	Bridge Resistance (Ω)		
	0.15	0.30	0.60
20	0.01	0.03	0.15
40	0.12	0.23	0.42
60	0.25	0.45	1.12
80	0.45	0.92	1.83
100	0.72	1.45	3.55

Note: These computed values have assumed center conductor width of 100 μm, total width of 300 μm, Bragg frequency of 180 GHz, periodic spacing of 110 μm, and bridge capacitance of 21 fF. Note the bridge resistance varies as $\sqrt{f}$ at mm-wave frequencies between 30 and 70 GHz with errors not exceeding 2 percent.

where as the impedance of an unloaded line is about 100 Ω. Preliminary studies performed on distributed transmission lines indicate that IL due to MEMS bridge resistance is about 0.07 dB and 0.28 dB/cm at 30 GHz and 60 GHz, respectively, although the IL due to transmission line is about 1.6 dB and 2.5 dB/cm at 30 GHz and 60 GHz, respectively. These IL values are possible for a distributed transmission line fabricated on a quartz substrate with an effective dielectric constant of 2.5, bridge capacitance of 40 fF, and a periodic spacing of 200 μm. The transmission-line losses will increase with higher dielectric constant and loss tangents of the substrates.

Loaded transmission-line loss is a function of center conductor width (W), total width ($W + 2G$), periodic spacing (s), Bragg frequency (f_b), bridge capacitance (C_b), and bridge resistance (R_b), which can vary from 0.15 to 0.7 Ω. Loaded transmission-line losses as a function of bridge resistance and operating frequency are summarized in Table 6.5.

6.4 Design Aspects and DMTL Parameter Requirements for TTD Phase Shifters Operating at mm-Wave Frequencies

Completion of the design of a DMTL-based phased shifter is dependent on Bragg frequency, permittivity of the dielectric constant of the substrate, unloaded impedance, and loaded impedance. These parameters are needed to compute the zero-bias bridge capacitance (C_{bo}) and the periodic spacing (s).

$$C_{bo} = s \left[\frac{L_t}{Z_{ul}^2 - C_t} \right] \qquad (6.16)$$

where Z_{ul} is the impedance of unloaded transmission line, which is close to 100 Ω. The periodic spacing (s) between the bridges can be written as

$$s = \left[\frac{Z_{ul}}{\pi f_b} \right] \left[\frac{1}{\sqrt{L_t \left[C_r L_t - (C_r - 1) C_t Z_{ul}^2 \right]}} \right] \qquad (6.17)$$

where C_r is the ratio of maximum bridge capacitance to minimum bridge capacitance and is equal to C_b / C_{bo}, where C_{bo} is the minimum bridge capacitance. It is important to point out that the minimum bridge capacitance occurs at zero bias, whereas the maximum capacitance occurs at reverse bias voltage around of 12–14 V in the downstate position of a MEMS bridge. The bridge capacitance at zero bias can vary from 34 to 36 fF, and theoretically, the capacitance ratio can vary anywhere from 1.2 to 1.5. But, in actual practice, this ratio can be achieved between 1.15 and 1.30. All variables appearing in Equation 6.17 have been defined earlier, except the capacitance ratio.

6.4.1 Computations of Unloaded Line Impedance (Z_{ul}), Line Inductance, and Capacitance per Unit Length of the Transmission Line

Assuming center conductor impedance of 100 Ω and effective dielectric constant of 2.4 for the quartz substrate (relative dielectric constant $e_r = 3.78$ and loss tangent $= 0.0001$ at 95 GHz and up) and inserting these values in Equations 6.11 and 6.11a one gets

$$C_t = \left[\frac{1.5492}{3 \times 10^{10} \times 100} \right] = 0.516 \text{ pF/cm}$$

$$L_t = \left[0.516 \times 10^{-12} \times 100 \times 100 \right] = 0.512 \times 10^{-8} \text{ H/cm}$$

Note that the unloaded line impedance (Z_{ul}) close to 48 Ω seems to offer maximum change in the zero-bias bridge capacitance and can be computed using Equation 6.16. Inserting 40 fF for the zero-bias capacitance, 200 μm for bridge spacing (s), and the computed values of inductance and capacitance per unit length in this equation, one gets

$$[40 \times 10^{-15}] = 0.020 \left[\frac{0.516 \times 10^{-8}}{Z_{ul}^2 - 0.516 \times 10^{-12}} \right]$$

$$Z_{ul}^2 = \left[\frac{10,000}{4.876} \right] = 2050$$

$$Z_{ul} = 45 \ \Omega$$

Table 6.6   Computed Values of Zero-Bias Bridge Capacitance (C_{bo}), fF

Spacing (μm)	Loaded Line Impedance (Ω)			
	48	47	46	45
100	17.3	18.2	19.2	20.0
150	25.9	27.3	25.1	30.5
200	34.5	36.4	38.4	40.6
300	51.8	54.6	49.9	60.9

Computed values of zero-bias capacitance as a function of loaded line impedance and periodic spacing are summarized in Table 6.6 for a CPW line impedance of 100 Ω.

These computed values indicate that there is a difference of about 1 fF at 100 μm spacing, 2 fF at 150 and 200 μm spacings, and 3 fF at 300 μm spacing in the zero-bias bridge capacitance, when the loaded line impedance varies from 48 to 45 Ω. From these values, one can select optimum periodic spacing to meet a specific zero-bias bridge capacitance.

6.4.2   Digital MEMS Distributed X-Band Phase Shifter

Before embarking on mm-wave distributed MEMS phase shifters, it will be very helpful to get familiarized with the design concept for the X-band distributed MEMS phase shifter. The design is based on the distributed MEMS transmission line (DMTL) cascaded with high-capacitance ratio varactors. These high-capacitance varactors are fabricated using a series combination of MEMS bridge variable capacitors and fixed-value MIM capacitors, as depicted in Figure 6.6. The phase shifter design involves a periodic loaded line with MEMS bridges in series with lumped element capacitors as shown in above figure. It is important to mention that the variable capacitance is provided by the MEMS bridge suspended over the CPW line, whereas the fixed-value capacitance is provided by the MIM capacitors. The MIM capacitor has a series resistance less than 0.15 Ω. Combination of two capacitors produces a high-capacitance ratio (1.5–2.5), resulting in a large loading on the CPW line, which offers large phase shift with minimum loss. This phase shifter when fabricated on a low-loss substrate such as quartz with e_r of 3.78 and loss tangent less than 0.0001 can produce an X-band, TTD phase shifter with reflection coefficient or return loss better than −15 dB and IL less than 0.5 dB over a wide band at X-band frequencies. This design concept can be converted into distributed MEMS phase shifter into a digital version, which is not sensitive to electrical noise and at the same time provides high-capacitance ratio for larger phase shift.

The transmission-line length depends on the number of MEMS bridges and the number of bits involved in the design of DMTL phase shifter. Typically, a 1-bit

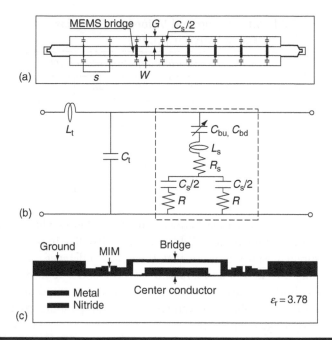

Figure 6.6 CPW-based phase shifter showing (a) bridge and CPW line parameters, (b) distributed lumped parameters of the bridge and CPW line, and (c) cross section of the device showing the center conductor, bridge, and MIM capacitance on the quartz substrate ($e_r = 3.78$). (From Hayden, J.Y. and Rebeiz, G.M., *IEEE Microw. Guided Wave Lett.*, 10, 142, 2000.)

phase shifter may involve four MEMS bridges, a 2-bit phase shifter may involve eight MEMS bridges, a 3-bit phase shifter may involve 16 MEMS bridges, and so on. The bias for the bridge ranges from 12 to 14 V in the downstate position and 30 to 50 V in the upstate position. It is possible to cascade such DMTL sections to design a 2-bit, 3-bit, or 4-bit digital phase shifters with excellent wideband performance at X-band frequencies. The performance of the DMTL phase shifter with MEMS bridge alone is limited by the capacitance ratio of 1.2 due to the bridge's vertical movement and its mechanical instability. The digital version of DMTL phase shifter offers minimum loss and large phase shift per centimeter length of the CPW transmission line.

6.4.3 Design Procedure for mm-Wave DMTL Phase Shifters

Optimum design of a DMTL phase shifter requires comprehensive examination of the critical design parameters and parametric analysis data. For example, the total CPW transmission-line width must be selected close to $\lambda_d/8$ at the maximum operating frequency, if optimum performance is the principal requirement.

In addition, the periodic spacing and other parameters appearing in the Bragg frequency expression must be selected to achieve higher Bragg frequency to obtain return loss better than -15 dB on the entire operating bandwidth. The MEMS bridge parameters must be chosen to achieve maximum phase shift with minimum IL. The unloaded line impedance (Z_{ul}) must be selected close to 48 Ω to maximize the change in MEMS bridge capacitance. The Bragg frequency computations must include the bridge inductance. A constant bridge inductance of 20 pH can be used regardless of the bridge width (w), because the bridge inductance is a weak function of the bridge width. The bridge resistance is dependent on the bridge width (w) and bridge height (h). Because the bridge resistance does not vary much with the width, and the bridge height typically of 1.5 μm is very small compared to bridge width, the bridge resistance practically remains constant at 0.15 Ω at 30 GHz. However, it will increase with increase in the frequency. CPW dimensional parameters, bridge parameters, and the Bragg frequencies of three 30 GHz optimized DMTL phase shifter designs are shown in Figure 6.5. Computed values of these parameters have assumed a bridge resistance of 0.15 Ω and an inductance of 20 pH. Note the three unloaded impedances (Z_{ul}) of 131, 100, and 80 Ω are calculated assuming the CPW center conductor widths of 50, 100, and 150 μm, respectively.

The author has performed comprehensive trade-off studies for optimum design of DMTL phase shifters operating at mm-wave frequencies and the results of these studies involving several design parameters are summarized in Table 6.7.

These parametric analysis results will provide unique guidance and useful design tips to the MEMS engineers and designers, who are actively involved in the design and development of mm-wave DMTL phase shifters.

Table 6.7   Optimized Design Parameters for mm-Wave DMTL Phase Shifters

Design Parameter	40 GHz	60 GHz	90 GHz
Total CPW line width (S), μm	593	395	263
CPW line impedance (Z_0), Ω	100	100	100
Periodic spacing (s), μm	300	200	150
Bridge capacitance ratio (C_r)	1.2	1.2	1.2
$\varepsilon_{r,eff}$ for quartz ($\varepsilon_r = 3.78$)	2.4	2.4	2.4
Zero-bias capacitance (C_{bo}), fF	40	40	40
Loaded line impedance (Z_{ul}), Ω	48/47/46	48/47/46	48/47/46
Bragg frequency (f_b), GHz	103/101/99	154/151/147	206/202/197

6.4.4 Expression for Phase Shift

The phase shift can be written in terms of loaded and unloaded impedances as

$$\Delta\Phi = \left[\frac{\omega Z_0 \sqrt{\varepsilon_{r,\text{eff}}}}{c}\right]\left[\frac{1}{Z_{ul}} - \frac{1}{Z_{L1}}\right] \text{rad/m} \qquad (6.18)$$

where c is the velocity of light (meter per second), and Z_{ul} and Z_{L1} are the DMTL characteristic impedances for the low and high bridge capacitance states, respectively [1]. Note that the characteristic impedance of the center conductor as the line width decreases (Figure 6.2), which means larger phase shifts are possible with narrow center conductor width. It is evident from Equation 6.18 that large loading capacitance per unit length (C_{bo}/s) is required to reduce the impedance close to 50 Ω. This will increase the zero-bias bridge capacitance (C_{bo}) by 20 percent approximately, thereby showing a larger effect on the phase velocity. Calculated phase shift (degree per centimeter) as a function of center conductor width (W) at 40 GHz is shown in Figure 6.4, assuming the loaded line impedance of 48 Ω, dielectric constant of quartz as 3.78, Bragg frequency of 120 GHz, and total center conductor ($W + 2G$) width of 300 μm. It is important to mention that the Bragg frequency must be selected greater than three times the maximum operating frequency for minimum IL and maximum phase shift per decibel loss for the DMTL phase shifter operating at mm-wave frequencies exceeding 30 GHz.

Numerical example for phase shift
Assuming the DMTL phase shifter design frequency of 40 GHz, CPW center conductor width of 100 μm, total CPW linewidth of 300 μm, and loaded line impedance of 48 Ω, and inserting these parameters in Equation 6.18, one gets the phase shift as

$$[\Delta\Phi]^{100\,\Omega} = \left[\frac{0.25 \times 10^{12} \times 100 \times \sqrt{2.4}}{3 \times 10^{10}}\right]\left[\frac{1}{48} - \frac{1}{44.5}\right] \text{rad/cm}$$

$$= 57.3[0.13 \times 10{,}000][0.02083 - 0.02198] \text{ rad/cm}$$

$$= 122^\circ/\text{cm} \quad \text{for } Z_{dl} = 44.5 \ \Omega \text{ and } W = 100 \ \mu\text{m}$$

$$[\Delta\Phi]^{131\,\Omega} = 57.3[0.13 \times 1.31 \times 10{,}000]\left[\frac{1}{48} - \frac{1}{44.4}\right]$$

$$= [9.758 \times 16.92]$$

$$= 165^\circ/\text{cm} \quad \text{for } Z_{dl} = 44.4 \ \Omega \text{ and } W = 50 \ \mu\text{m}$$

$$[\Delta\Phi]^{80\,\Omega} = 57.3[0.13 \times 0.80 \times 10{,}000]\left[\frac{1}{48} - \frac{1}{45.1}\right]$$

$$= [5.959 \times 13.43]$$

$$= 80^\circ/\text{cm} \quad \text{for } Z_{dl} = 45.1 \ \Omega \text{ and } W = 150 \ \mu\text{m}$$

Table 6.8 Phase Shift as a Function of Center Conductor Width and Unloaded CPW Transmission-Line Impedance (°/cm) Using the Quartz Substrate

W (μm)	Z_0 (Ω)	40 GHz	60 GHz	90 GHz
150	80	80	120	180
100	100	122	183	274
50	131	165	274	371

Note: The insertion loss (IL) in the loaded CPW transmission line is about 0.85, 1.15, and 1.45 dB/cm at 40, 60, and 90 GHz, respectively. The unloaded impedance can be computed using Equation 6.1 or from the curve shown in Figure 6.4 using the dielectric constant of 3.78 for the quartz substrate. When using this curve, assume the value of parameter k equal to 0.333 for 100 μm CPW line width, 0.500 for CPW line width of 150 μm, and 0.166 for the CPW line with of 50 μm. When using Equation 6.1, one must use the computed values of elliptic functions shown in Figure 6.2.

These computed values at 40 GHz design agree with the phase shift magnitudes shown in Figure 6.4. Computed phase shifts as a function of center conductor width and unloaded CPW line impedance for 40, 60, and 90 GHz DMTL phase shifters are summarized in Table 6.8.

It is important to point out that the phase shift as well as the IL increases with the increasing operating frequency. One might expect much higher losses at 90 GHz and up due to dispersion and surface wave losses, which have been neglected in the above calculation. As a rule of thumb, one must consider least 15 percent additional loss due to these two sources, if a realistic assessment of total IL is desired.

6.4.5 Optimized Design Parameters for a W-Band DMTL Phase Shifter

So far the discussion was limited mostly to phase shifters designed to operate from 10 to 60 GHz discrete frequencies. Now the focus will be on design concept and parameters, which will lead to optimum design of a W-band DMTL phase shifter capable of operating over the entire W-band from 75 to 110 GHz. There are additional effects; operating conditions and device parameters must be taken into consideration at W-band frequencies such as dispersion effects in the substrate, low phase shift per decibel loss, lower capacitance ratio due to higher levels of compressive stress within the bridge structures, and increased nonuniformity within the dielectric wafer and higher IL due to bridge resistance, which is quite significant at W-band frequencies. At these frequencies, the operating bias voltage or the pull-down voltage is much higher. Still higher bias voltage will be required, if the bridge height (h) is slightly increased to 1.5 μm from a nominal value of 1.2 μm.

As stated earlier, the MEMS bridge resistance can vary from 0.15 to 0.6 Ω at 30 GHz. The loss will increase with frequency as the square root of the operating frequency when the bridge resistance is less than 0.3 Ω. However, the bridge loss starts to dominate when the bridge resistance approaches 0.6 Ω. The bridge resistance can vary from 0.26 to 1.04 Ω at 90 GHz. In addition, the CPW transmission loss due to conductor and the substrate will be much higher at higher frequencies. These losses can be calculated using expressions given in textbooks dealing with mm-wave techniques. For a 32-bridge DMTL fabricated on a substrate with effective dielectric constant of 2.4, total width of 300 and gold-plated (0.8 μm) center conductor width of 100 μm, the characteristic impedance is 100 Ω, which can be loaded to 48 Ω at a Bragg frequency close to 180 GHz and at capacitance ratio of 1.15. This capacitance ratio is possible at a pull-down or a bias voltage of 25 V. It is important to mention that bridge height determines the bias voltage. An increase in bridge height from 1.2 to 1.5 μm (25 percent increase) will require double the pull-down voltage. The zero-bias load impedance is 48 Ω, whereas the unloaded CPW line impedance is 100 Ω. For an optimized W-band DMTL phase shifter, the phase shift per decibel is maximum (110°/dB) at 100 GHz when the bridge resistance is zero. This phase shift value drops to below 52°/dB for a bridge resistance of 0.3 Ω. These parameters are valid when the zero-bias bridge capacitance is 19 fF, bridge inductance is 20 pH, and the periodic spacing is 110 μm. Note the Bragg frequency is 186 GHz when the bridge capacitance is 20.7 fF, which is possible with a bridge width of 25 μm and height of 1.5 μm. However, the Bragg frequency can be increased to 192 GHz when the bridge inductance of 20 pH is chosen, which will yield better return loss and lower phase shifter loss. The phase shift for the W-band (75–110 GHz) phase shifter is fairly constant around 72°/dB or 5 dB for a 360° phase shift over the entire band of the device. It is interesting to note that the performance of W-band device comprising of 24-MEMS bridges can be improved to 136°/dB and 224°/dB at the capacitance ratio of 1.3 and 1.5, respectively, which comes to an IL of 2.6 and 1.6 dB, respectively at 100 GHz.

The design equations used for the W-band unit can be used to compute the electrical and geometrical parameters for the U-band unit operating over a 40–60 GHz spectrum. The U-band phase shifter design offers 70°/dB at 40 GHz and 90°/dB at 60 GHz with a change in loading capacitance of only 15 percent [1]. The bridge series resistance of 0.15 dB is assumed for the U-band device. The performance of the DMTL mm-wave phase shifters can be significantly improved if the bridge capacitance can be increased from 30 to 50 percent, which can be accomplished by using the bridge capacitance in the upstate and downstate positions of the bridge. Note the increase in bridge capacitance would introduce a large change in loaded impedance. However, if the MEMS bridge capacitance is placed in series with a fixed-value MIM capacitor, then the resulting capacitance can be limited to 1.5–2.0 to avoid bridge structure instability. The bridge height must be selected between 1 and 1.2 μm to keep the ES actuation voltage or pull-down voltage between 25 and 45 V.

6.5   Two-Bit MEMS DMTL Phase Shifter Designs

A wideband DMTL phase shifter can be designed based on the DMTL structure loaded with MEMS bridges and MIM capacitors as illustrated in Figure 6.6. As stated earlier, the distributed loading can lower an unloaded impedance say from 100 Ω to 45–48 Ω depending on the loading intensity or the density of the transmission line. The change in line impedance will result in change of phase velocity leading to a phase shift in the CPW line [2]. The phase shift expression can be written as

$$\Delta\Phi = 57.3 \left[\frac{wZ_0\sqrt{e_{r,\text{eff}}}}{c} \right] \left[\frac{1}{Z_{\text{ul}}} - \frac{1}{Z_{\text{l}}} \right] °/\text{cm} \qquad (6.19)$$

where
    w is the operating frequency
    c is the velocity of light
    Z_{ul} and Z_{l} are the unloaded and loaded impedances of the CPW line

Numerical example for phase shift in a 2-bit, X-band DMTL phase shifter
Assuming a center frequency (f_{o}) of 10 GHz, a Bragg frequency (f_{b}) of 30 GHz (three times the center frequency to achieve a reflection coefficient better than −15 dB), number of MEMS bridge section equal to 16, unloaded CPW line impedance of 100 Ω, loaded impedance of 53.2 Ω, and effective dielectric constant of 2.4 for quartz substrate, and inserting these parameters in Equation 6.19, one get the phase shift (°/cm) as

$$\Delta\Phi = 57.3 \left[\frac{2 \times \pi \times 10^{10} \times 100 \times \sqrt{2.4}}{3 \times 10^{10}} \right] \left[\frac{1}{100} - \frac{1}{53.2} \right]$$

$$= 18589[0.01 - 0.0187997] = [18589 \times 0.008796]$$

$$= 163.53°/\text{cm}$$

for phase shifter consisting of 16 MEMS bridge sections with periodic spacing of 906 μm, approximately. This phase shift of 163.53 will increase to 183.97°/cm for the phase shifter comprising of 18 MEMS bridge sections. This indicates that higher the number of bridge sections, larger will be the phase shift per centimeter.

6.5.1 Design Parameters and Performance Capabilities of 2-Bit, X-Band Phase Shifter

It is important to mention that the bridge capacitance in the upstate position (C_{bu}) is smaller than the total lumped capacitance C_s under bias-free operation and the effective capacitance of the CPW line C_{ul} is equal to bridge capacitance in upstate position. When a bias is applied to the line and the MEMS bridge is in the downstate position, the bridge capacitance C_{bd} is 40–70 times higher than the C_s [2].

As stated earlier, the impedance change of the DMTL determines both the phase change and the reflection coefficient or the return loss of the phase shifter. However, a wider swing in the impedance change will result in more phase shift, but at the expense of higher return loss. Preliminary studies performed by the author indicate that twice the phase shift can be achieved, if the maximum reflection of -10 dB is acceptable rather than -15 dB. Because a 2-bit phase shifter contains two sections, one for 90° and other for 180° phase shift, the reflection coefficient for the cascaded device will be close to -10 dB.

The X-band, 2-bit DMTL phase shifter [2] consists of a CPW line with a total width of 900 μm, center conductor width (W) of 300 μm to minimize the line loss, bridge spacing (s) of 906 μm, MEMS bridge structure of 16 sections for the 180° bit and eight sections for the 90° bit. The MEMS bridge structure has a thickness of 0.8 μm, length of 350 μm and width of 60 μm and is suspended at a height of 1 μm above the quartz substrate to keep the pull-down voltage for the MEMS bridge alone to 13 V. The spring constant (k) of the bridge is 72 N/m and the residual stress is 72 MPa. Note in the upstate position, the ES actuation voltage in the pull-down operation will be much higher, which could be in the 30–50 V range depending on the height of the suspended bridge and the series fixed-value MIM capacitor to the ground. The downstate-loaded impedance can vary anywhere from 46 to 53 Ω depending on the maximum allowable reflection coefficient and acceptable IL over the band of interest.

The overall performance of the phase shifter is dependent on the Bragg frequency, number of MEMS bridge sections, MIM capacitor rating, spacing between the bridges, loading capacitances, and maximum acceptable reflection coefficient over the band of interest. As mention earlier, the 2-bit DMTL phase shifter consists of a 90° bit comprising of eight MEMS bridge sections and a 180° bit comprising of 16 sections. The overall performance is equal to bridge length times the number of sections plus spacing between the bridges times the number of spacing. In this particular case, the bridge length ($350 \times 16 + 906 \times 18$) comes to 21.9 mm for the 16-MEMS bridge sections capable of providing a phase shift of 163.5°/cm. It is important to point out that an even number of MEMS bridge sections must be used for the 180° bit so that the 90° bit results in an integer number of MEMS sections. The four phase shifter states comprising of 0°, 90°, 180°, and 270° can have phase error of ±3° (Figure 6.7) due to phase ripples, impedance variations, bias fluctuations, and mechanical instability of the MEMS bridge.

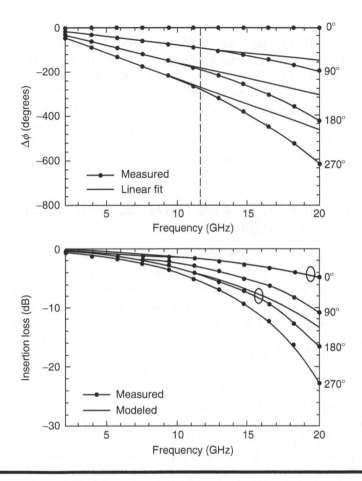

Figure 6.7 **Phase shift and IL of a 2-bit MEMS phase shifter in its four bits (0°, 90°, 180°, and 270°) as a function of frequency.**

6.5.2 *Insertion Loss in a DMTL Phase Shifter*

The IL in a DMTL phase shifter is a function of series resistance, loaded impedance, loaded capacitance, and operating frequency. The IL can be expressed as

$$IL = \left[\frac{Z_l R_s C_l^2 \omega^2}{4} \right] \qquad (6.20)$$

where ω, which equals $2\pi f$, is the design frequency and subscripts s and l stand for series and loaded conditions, respectively. The overall phase shifter loss includes the

IL per section contributed by the MEMS bridge resistance and MIM resistance in series with a MEMS bridge. But the MIM resistance is strictly contingent on the Q of the capacitor, which is typically around 15 at 10 GHz. This is equivalent to a series resistance of 10.7 Ω of the MIM capacitor. The low Q is due to thin metallization layer, which is much less than the skin depth. If the metallization thickness is increased to yield a Q of 65 or more for the MIM capacitor, the IL will be less than 0.8 dB at 10 GHz for a 180° DMTL phase shifter. Typical phase shift and IL for a phase DMTL-based phase shifter capable of operating over X-band and K_u-band regions are shown in Figure 6.7.

6.5.3   Digital Version of the DMTL Phase Shifter with 360° Phase Capability

This particular digital-distributed MEMS phase shifter consists of a high-impedance CPW line capacitive loaded by the periodic installations of discrete varactors, which are formed by the cascaded MEMS bridge sections. This phase shifter design offers significant change in the distributed capacitance with application of small bias voltage (about 12 V) leading to a compact phase shifter with 360° phase capability. The high-capacitance ratio Schottky diode varactor is fabricated using a series of combination of MEMS bridges and fixed-value MEMS capacitors as shown in Figure 6.6, which offers a capacitance ratio ranges from 2 to 5, resulting a large phase shift with minimum IL [3]. A well-designed MEMS-based varactor has a series resistance less than 0.15 Ω, which offers excellent phase shifter performance at mm-wave frequencies with a phase shift of 90°/dB and 72°/dB at 60 and 75 GHz, respectively. The digital version of DMTL phase shifter is completely free from electrical circuit noise, which is generally found in an analog version of the device.

6.5.3.1   Design Parameter Requirements for Digital, 360° Phase Shifter

Distributed lump elements of the bridge and the equivalent circuit parameters of this phase shifter are shown in Figure 6.6 [3]. This phase shifter involves a periodically loaded line using MEMS bridges in series with fixed-value lumped element MIM capacitors. The overall line capacitance (C_L) of the CPW line is the series combination of the MEMS bridge capacitance (C_b), and the total lumped capacitance C_s which can be written as

$$C_L = \left[\frac{C_s C_b}{C_s + C_b} \right] \tag{6.21}$$

It is important to mention that the bridge capacitance in the upstate position of a MEMS bridge (C_{bu}) is smaller than the C_s and therefore, the effective capacitance of the line in the upstate position (C_{Lu}) will be very close to C_{bu} and the subscript b stands for bridge and u stands for upstate position. Under the above-mentioned conditions, the performance of this phase shifter remains independent of MEMS bridge capacitance ratio (i.e., the bridge capacitance in downstate position to bridge capacitance in upstate position) as long as the downstate bridge capacitance is much greater than the total lumped capacitance (C_s) and a MEMS bridge capacitance ratio is greater than 10.

This phase shifter when fabricated on a 500 μm thick quartz substrate will offer large phase shift with lower IL even at mm-wave frequencies. As stated earlier, the phase shift and the return loss are determined by the impedance change and the selection of the Bragg frequency. Low reflection coefficients are possible when the Bragg frequency is at least three times the highest operating frequency. This means to achieve low reflection coefficients a Bragg frequency of 30 and 36 GHz is needed, if the maximum operating frequency is 10 and 12 GHz, respectively.

Preliminary calculations indicate that for a 2-bit, DMTL phase shifter cascaded in 90° and 180° sections and fabricated on a quartz substrate, the impedance of the distributed line in the upstate position cannot be higher than 60 Ω to ensure a reflection coefficient of −15 dB and 70 Ω to achieve a reflection coefficient of −10 dB over the band of interest. In brief, maximum phase shift is possible with impedance values of 60/40 Ω to ensure −15 dB reflection coefficient and 70/38 Ω to ensure reflection coefficient of −10 dB over 4–12 GHz spectral region. Calculated values of ILs in the upstate and downstate positions and maximum phase shifts as a function of frequency are summarized in Table 6.9 for a 18-section bridge MEMS phase shifter designed with a Bragg frequency of 30 GHz and unloaded CPW line impedance of 102 Ω to ensure good return losses.

These calculations have assumed the number of MEMS bridge sections of 18, unloaded impedance for the CPW line of 102, periodic spacing (s) of 884 μm for the

Table 6.9 Computed Values of IL and Phase Shift for the Digital 360° Phase Shifter as a Function of Frequency

Frequency (GHz)	Insertion Loss (dB)		Phase Shift (°)	
	Upstate	Downstate	60/44 Ω	70/38 Ω
4	0.38	0.41	71	122
6	0.41	0.43	106	183
8	0.40	0.65	142	244
10	0.41	0.76	177	305
12	0.42	0.86	212	366

upstate and downstate impedances levels of 60/44 Ω and 764 μm for the impedance levels of 70/38 Ω and Bragg frequency of 30 GHz. The overall bridge length has been computed using the bridge spacing of 884 μm for upstate and downstate impedance levels of 60/40 Ω and 767 μm for impedance levels of 70/38 Ω, bridge length of 350 μm, and number of MEMS bridge sections of 18. ILs are computed assuming the CPW center conductor width of 284 μm, total center conductor width of 900 μm, bridge width of 60 μm, and bridge height of 1.5 μm above the 500 μm thick quartz substrate. The bridges are connected to two MIM capacitors with Q of 50–100 at 10 GHz. The pull-down voltage of 40 V is needed due to series division of the applied voltage between the MEMS bridge and the total lumped capacitance (C_s). An increase of total capacitance to 230 fF can significantly improve the overall performance of the DMTL-based phase shifter.

6.5.3.2 Insertion Loss Contributed by MIM Capacitors and Its Effect on CPW Line Loss

Data presented in the table below indicates that the IL contributed by the MIM capacitors is very small compared to MEMS bridge loss because the MIM series resistance is extremely small (typical value less than 0.2 Ω) compared to the series resistance of the bridge (typical value between 0.8 and 1.0 Ω). The series resistance of the MIM capacitor is written as

$$[R_s]_{\text{MIM}} = \left[\frac{1}{\omega C_{\text{MIM}} Q_{\text{MIM}}} \right] \qquad (6.22)$$

where C_{MIM} and Q_{MIM} are the MIM capacitance and unloaded Q, respectively. Computed values of MIM resistance as a function of frequency and Q are summarized in Table 6.10.

These values indicate that the MIM series resistance is significantly small compared to bridge series resistance and thus, will have negligible effect on the CPW line loss.

In summary, the digital, wideband cascaded phase shifter comprising of 90° and 180° sections shows excellent return loss of −15 dB and maximum IL of 0.76 dB over 4–10 GHz range. Computed data indicates that the upstate IL increases as $\sqrt{f}$

Table 6.10 Series Resistance of the MIM Capacitor (Ohm, Ω)

MIM Q	30 GHz	60 GHz	90 GHz
10	0.354	0.177	0.118
15	0.236	0.118	0.079
20	0.177	0.088	0.059

whereas the phase increases as f. The computed results are comparable to state-of-the art X-band phase shifter using Lange couplers with IL of 1.5 dB for 360° phase shift. IL due to series resistance of the MIM capacitor will be very small even at mm-wave frequencies (Table 6.10).

6.5.3.3 Phase Noise Contribution from DMTL Phase Shifters

A MEMS bridge and MIM capacitor associated with a DMTL-based phase shifter will generate phase noise power, which is dependent on the bridge height, phase delay, spring constant of bridge, bridge capacitance, and other related parameters. Phase noise power generated by the DMTL phase shifter can be written as

$$P_{PN} = (0.5) \left[\frac{1}{1 + \gamma} \right]^2 \left(\frac{X}{h} \right)^2 (\Phi^2)/\text{Hz} \tag{6.23}$$

where
  $\gamma = 0.8$
  $X = 1/k$
  $k = 55$ N/m
  $h = 1.4–1.6$

Φ is the phase delay which is expressed as

$$\Phi = \left[\frac{\omega C_{bu} Z_0}{2} \right] \tag{6.24}$$

where C_{bu} is the bridge capacitance in the upstate position of the bridge and Z_0 is the output impedance of the phase shifter.

Assuming a spring constant of 55 N/m, output impedance of 77 Ω, bridge capacitance of 142 fF, bridge height of 1.4 µm, design frequency of 24 GHz, and constant γ of 0.8, and inserting these parameters in above equations, one gets

$$\Phi = \left[\frac{6.283 \times 24 \times 10^9 \times 142 \times 10^{-15} \times 77}{2} \right] = [0.8247]$$

$$\Phi^2 = [0.6802]$$

$$P_{PN} = \left[0.5 \left(\frac{1}{1 + 0.8} \right)^2 \times \left(\frac{1}{55} \right)^2 \times \left(\frac{1}{1.4 \times 10^6} \right)^2 \times 0.06802 \right]$$

$$= [4.918 \times 10^{-17}]$$

$$[P_{PN}]_{dB} = [-170 + 6.92]$$

$$= -162.08 \text{ dBc/Hz at } Z_0 = 77 \ \Omega$$

$$= -165.16 \text{ dBc/Hz at } Z_0 = 50 \ \Omega$$

Table 6.11   Phase Noise Power Levels in the DMTL-Based Phase Shifters (dBc/Hz)

Frequency (GHz)	Output Impedance (Ω)	
	77	50
10	−170.7	−174.4
20	−164.7	−168.7
30	−161.2	−164.9
60	−155.2	−158.9
90	151.7	−155.3

Computed values of phase noise power levels as a function of various parameters are summarized in Table 6.11.

6.6   Multi-Bit Digital Phase Shifter Operating at *K* and *K*$_a$ Frequencies

6.6.1   Introduction

Precision missile tracking radar, tactical fighter fire control radar, and modern communications systems require low-loss, TTD phase shifters to meet stringent performance requirements. Because a precision high power, phased array missile tracking radar deploys thousands of phase shifters to provide multitarget tracking capability with high reliability, a low-loss TTD phase shifter is best suited for this application. Here low-cost, fast response, improved tracking accuracy, and high reliability are the principal requirements. MEMS switching devices and variable capacitors or varactors are used to realize switched-time phase shifters, which suffer from high cost and complexity. A TTD phase shifter configuration using the distributed MEMS phase shifter design and requiring a small analog control voltage to vary the height of the MEMS bridge offers a limited capacitance ratio not exceeding 1.5:1. This ratio is not enough to provide a large phase shift with minimum loss.

The topology of a multi-bit digital-distributed MEMS (MDDM) phase shifter deploys a synthetic transmission line whose phase velocity can be varied by snapping down the membrane or bridge of MEMS loading capacitors in the downstate position [4].

6.6.2 Design Aspects and Critical Elements of the MDDM Phase Shifter

Critical elements of the MDDM, series configuration topology for the MEMS-based capacitors and fixed capacitors, and the expression for the total capacitance are shown in Figure 6.8. Note the total capacitance is 0.01 pF for the upstate position and 0.075 pF for the downstate position of the bridge or membrane. The 1-bit phase shifter can be fabricated using a CPW transmission line on a low-cost glass substrate with relative dielectric constant of 5.7 and loss tangent of 0.001. The width of the center conductor (*W*) and the gap (*G*) can be selected as 100 and 60 μm, respectively, yielding a total width of the CPW line of 220 μm. A silicon nitride layer with thickness of 0.5 μm can be used as an insulation layer. The MEMS bridge height of 3 μm is used above the center conductor. The

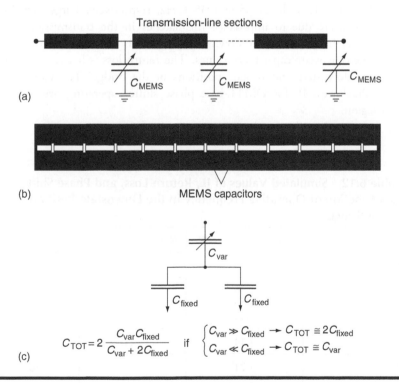

Figure 6.8 Design configuration of a multi-bit phase shifter showing (a) circuit elements such as MEMS capacitors and various CPW line sections, (b) locations of MEMS capacitances, and (c) expression for total capacitance.

membrane or bridge width and the spacing between the two bridges are 40 and 300 μm, respectively [4].

The design configuration of this phase shifter offers high reliability, reduced pull-down voltage, and very high-capacitance ratio (C_{max}/C_{min}) approaching to 7.5:1. Although the design concept described here is for a 1-bit phase shifter, which can be extended to multi-bit configuration by means of simple length scaling of the device. The MDDM phase shifter design is capable of providing a phase shift of 360° with IL not exceeding 2.0 dB and return loss close to −14 dB over 5–30 GHz bandwidth. Computer simulation values of IL, return loss, and differential phase shift as a function of frequency are shown in Table 6.12.

It is important to mention that the height of the bridge above the CPW center conductor determines the actuation voltage and the change in phase velocity and a phase shift induced at the output of the phase shifter. The phase shifter circuit under downstate-biased conditions produces the phase shift that varies linearly with the frequency as illustrated in Table 6.12. The downstate ILs are roughly 1.21 and 1.71 dB at 25 and 35 GHz, respectively. Large variations in the return loss are due to impedance mismatches in the transmission-line segments. Ripples in the IL in the downstate position is strictly due to nonuniformities of the downstate capacitance values. The return loss is better than −12 dB in both the downstate and upstate positions of the bridge. The tabulated data indicates the lowest IL for DMTL-base phase shifter operating over multiband spectral region.

Table 6.12 Simulated Values of IL, Return Loss, and Phase Shift as a Function of Operating Frequency in the Downstate Position of the Bridge

Frequency (GHz)	Insertion Loss (dB)	Return Loss (dB)	Phase Shift (°)
5	0.55	−24	35
10	0.86	−14	70
15	0.95	−16	108
20	1.14	−20	143
25	1.21	−23	184
30	1.65	−28	227
35	1.71	−13	280

6.7 Ultrawide Band Four-Bit True-Time-Delay MEMS Phase Shifter Operating over dc-40 GHz

6.7.1 Introduction

Four-bit reflection phase shifters have been designed at X-band frequencies using MEMS capacitive switches and Lance couplers. Resonant delay line phase shifters are designed for K_a-band operations. These phase shifters have demonstrated good IL, but suffers from narrow bandwidths and higher phase offset errors which are not acceptable for certain applications. Recently distributed phase shifters have been developed using CPW transmission line with MEMS bridges to vary the phase velocity of the transmission line. In this case, higher pull-down voltage (>30–60 V) is required, which may not be acceptable in applications where compatibility with microelectronics is the principal requirement. The TTD network concept for MEMS phase shifters will overcome the above-mentioned problems and limitations to achieve good phase shift performance and bandwidth that exceed those associated with MEMS capacitive switches. However, the MEMS switches will permit fabrication of high-quality switched-line TTD networks for phase shifters operating over wide microwave and mm-wave bands. Note these switches use a movable metal shunting bar to connect two RF signal lines when a dc ES actuation is applied. Furthermore, the bits can be fabricated individually for a multi-bit phase shifter with minimum cost and complexity.

Precision-phased array tracking radar incorporating ultrawide band TTD phase shifters are capable of acquiring frequency-dependent targets under multipath environments, because the TTD phase shifters offer higher phase accuracies, lower tracking errors, and minimum group delay ripples. The TTD networks can produce flat delay time over the dc to 40 GHz bandwidth and can provide full 360° phase shift capability [5]. A 16-stage, prototype phase shifter with 360° capability has demonstrated close match in delay times and IL less than 2.5 dB. The worst group delay ripple was found less than 3 ps over dc to 30 GHz range, which was well within the single-bit delay time of 5.8 ps. The phase delay can be computed using Equation 6.24.

6.7.2 Design Requirements and Parameters to Meet Specific Performance for a Wideband 4-Bit, TTD Phase Shifter

This ultra-band (dc to 40 GHz) TTD phase shifter can be fabricated on a GaAs substrate ($\varepsilon_r = 11.8$) to minimize the size of the device at lower microwave frequencies with minimum cost and complexity. Microstrip transmission lines with width 50–60 μm can be deposited on a 75 μm thick GaAs substrate to provide the required time delay. The gap between the two capacitors is 80 μm and the signal line width is kept narrow to 40 μm in the line section surrounding the switch. Smooth corners must be used in the delay line sections to minimize reflections and delay ripples. Two tee-sections, each consisting of two metal-to-metal contact MEMS switches, are used to route the RF signal through the desired delay path. Serious mismatch can prevent the TTD circuit from working particularly at higher microwave frequencies and this mismatch comes from the open-stub transmission lines connected to the open-switches at the tee-junctions. By reducing the open-stub length close to 135 μm, which corresponds to a quarter-wavelength at 162 GHz, the RF signals will be completely reflected at the tee-junction at this frequency. Even at lower frequencies, the mismatch introduced by the open-stub transmission-line capacitance can produce increased dispersion in the group delay. But, adding an inductive microstrip matching section in series with the tee-junction will reduce the impedance mismatch and phase dispersion over the band of interest. The group delay introduced by the series inductor (L) and parallel capacitor (C) can be expressed as

$$[\tau]_{group} = \left[-\frac{d\theta}{dt} \right] \tag{6.25}$$

$$\text{Tan } \theta = \left[\frac{(\omega CZ_0) + \left(\frac{\omega L}{Z_0} \right)}{2 - \omega^2 LC} \right] \tag{6.26}$$

where Z_0 is the characteristic impedance of the transmission line.

Note a reactive capacitance element degrades the delay flatness when used alone. However, a proper choice of inductive element (L) can reduce the delay ripple cause by the capacitive element (C) due to the second-order term in the denominator of the above equation. In fact for a well-matched TTD with appropriate value of the inductive element, the group delay ripple can be as low as 1 ps over the entire bandwidth.

6.7.3 Performance Parameter of the Device

The monolithic TTD network comprising of 16 MEMS switches is capable of providing a delay time from 107 to 194 ps with an interval of 5.8 ps from dc to 40 GHz frequency range. The delay time interval or step of 5.8 ps offers a phase shifts of 22.5° around 11 GHz, which can be realized with 600 μm long delay lines. Eight separate bias lines are used to control the MEMS switch positions. Good metal-to-contact for all the switches is necessary to ensure reliable performance of the MEMS switches. For a well-matched TTD network, the IL for all 16 states can vary from 2.2 to 2.7 dB at 10 GHz and from 3.5 to 4.4 dB at 30 GHz as shown in Table 6.13. The phase difference between the shortest and longest line paths is close to 345° at 11 GHz. Note the phase shift can be calculated using Equation 6.19. The return loss for a well-matched TTD network is well below −15 dB for most of

Table 6.13 Critical Performance Parameters of the Four-Bit TTD Networks Using Metal-to-Metal Contact MEMS Switches

Frequency (GHz)	Insertion Loss (dB)	Return Loss (dB)	Phase Shift (°)	Delay Difference (ps at 10 GHz)
0	0.2	−15	0	0 at 0th position
10	2.3	−19	72	22 at 4th position
20	2.7	−17	148	45 at 8th position
20	4.3	−22	212	67 at 12th position
40	4.7	−18	285	90 at 16th position

frequency range from dc to 40 GHz. The group delay ripple is generally around 2 ps for the shortest delay path and 3 ps for the longest delay path. However, the delay time ripple can be as high as 8 ps in the 30–40 GHz spectral region because of higher mismatch. The IL for all the 16 states is about 0.1 dB per switch and must be added to the conductor losses contributed by the delay lines.

The group delay ripple for a well-matched TTD network can be achieved well below 2 ps upto 30 GHz and up. For the unmatched TTD network the group delay ripple can be as high as 6 ps above 18 GHz. A single-bit TTD circuit exhibits a conductor loss of about 0.4 dB. One can expect an additional maximum loss of 0.22 dB from the transmission-line bends, tee-junctions, and contact resistances per switch at 10 GHz. These losses will be slightly higher above 15–40 GHz spectral range. Important performance parameters of the 4-bit TTD network such as IL return loss, and phase shift as a function of frequency are summarized in Table 6.13.

It is important to mention that the delay difference for the matched TTD network in picoseconds is dependent on the switch position and is estimated at 10 GHz. These values will be higher at higher frequencies. The TTD network performance can be further improved by inserting a high-impedance matching section. Nearly flat delays up to 40 GHz can be achieved for all 16 switch states incorporating high-impedance matching sections [5].

6.8 Two-Bit, V-Band Reflection-Type MEMS Phase Shifter

6.8.1 Introduction

Studies performed by the author reveal that reflection-type phase shifters using RF switches and 3 dB CPW couplers are best suited for integration in monolithic applications because of small size and compact packaging. Furthermore, advancement in MEMS technology offers reduced IL in the reflection-type phase shifters operating at mm-wave frequencies. The major loss contributor is the 3 dB CPW Lance coupler due to RF field concentration at the edges of the coupler. This problem is solved by the development and availability of a low-loss, air gap overlay (OL) CPW-MMIC 3 dB couplers with tight coupling. In addition, metal-to-metal contact MEMS series switches can be used for switching the CPW transmission-line sections. Both these techniques can reduce losses significantly in the V-band phase shifters [6].

6.8.2 Design Aspects and Performance Capabilities

The 2-bit, V-band (50–75 GHz), reflection-type phase shifter can be designed using the air gap OL CPW 3 dB couplers and metal-to-metal contact MEMS series switches. The four phase states are $0°$, $-45°$, $-90°$, and $-135°$. Three CPW line sections comprise of open-ended lines acting like open-ended stubs, each with an electrical length of $30°$ at the center frequency of 60 GHz and separated by MEMS series switches [6]. The RF signal propagates through the switchable CPW transmission lines and reflects from the open-ended stub. When the phases of the reflected signals of both coupler arms are the same, the RF signals add in phase and appear at the output of the coupler. If electrical length of each arm can be changed by $22.5°$ by actuating each pair of switches, a phase shift of $45°$ is achieved at the design frequency of 60 GHz. If the line lengths of both coupler arms are tuned for maximum phase shift, a phase shift with optimum flatness can be achieved over wide band.

The overlap 3 dB couplers and the metal-to-metal MEMS switch are the key components of this reflection-type phase shifter. The coupler is made from electroplated gold and can be fabricated on quartz substrate ($\varepsilon_r = 3.75$ and loss tangent = 0.0001). The coupler length is about 650 μm at the designed frequency of 60 GHz and the metal thickness at the top and bottom are 3 and 2 μm, respectively. The air gap between the gold-plated conductors is 1.5 μm to meet tight coupling requirements. Airbridge posts are placed with quarter-wave spacing along the length of the coupler to prevent bending and to provide appropriate support.

The 3 μm thick gold transmission lines used in the MEMS series switch are electroplated to achieve low line loss at V-band frequencies. Silicon nitride dielectric layer with thickness of 300 nm must be deposited on the part of CPW ground plate, which is used as the driving electrode. The switch contact metal must be electroplated using 2 μm thick gold to keep the contact resistance low. When a threshold dc bias between 35 and 40 V is applied between the driving electrode and the ground, the suspended driving electrode is pulled down so that the contact bar made of 2 μm thick nickel metal touches the RF signal line and the MEMS switch is on-state.

Table 6.14   Electrical Length of the Open-End Stub as a Function of VSWR

VSWR	Electrical Length (°)
1.25	6
1.50	25
1.75	32
2.00	36
2.50	43

Open-ended stub electrical length determines the reflection coefficient or the voltage standing wave ratio (VSWR) and specific electrical length is required to provide the maximum acceptable VSWR as shown in Table 6.14. The physical length of an open-ended stub as a function of frequency is shown in Table 6.15.

The tabulated data provides the empirical impedance matching techniques and parameters to the mm-wave MEMS phase shifter design engineers and research scientists.

The estimated ILs are about 0.4 and 0.7 dB and the return losses are roughly 17 and 18 dB at 40 and 60 GHz, respectively for the air gap OL 3 dB CPW coupler. The IL for the 2-bit, V-band phase shifter can be expected around 2.6, 3.6, 4.6, and 5.6 dB for the 0°, 45°, 90°, and 135° phase states, respectively. The

Table 6.15   Physical Length of the Open-Ended Stub as a Function of Frequency and Electrical Length (μm)

Frequency (GHz)	Wavelength, λ (cm)	22.5°	30°
10	3.000	1875	2500
20	1.500	938	1250
30	1.000	625	833
40	0.750	468	625
60	0.500	312	416
80	0.375	234	312
90	0.333	208	277

estimated phase error is about 0°, 3.5°, 5.7°, and 6.4° for the 0°, 45°, 90°, and 135° phase states, respectively. The return loss close to 12 dB can be expected over 55–75 GHz range. The average phase errors for all the switching states is 5.74 percent at 60 GHz and the average IL is about 4.3 dB at 60 GHz. Note large phase errors are due to the phase unbalance from the air gap 3 dB CPW couplers and the transmission lines. Specific details on phase shifter schematic, cross section of air gap 3 dB CPW coupler, MEMS switch performance, and phase shift per phase bit are shown in Figure 6.9. However, details on the physical dimensions of the air gap 3 dB coupler, critical elements of the series MEMS switch, and locations of switch pairs, 3 dB CPW couplers, and input and output ports are shown in Figure 6.10.

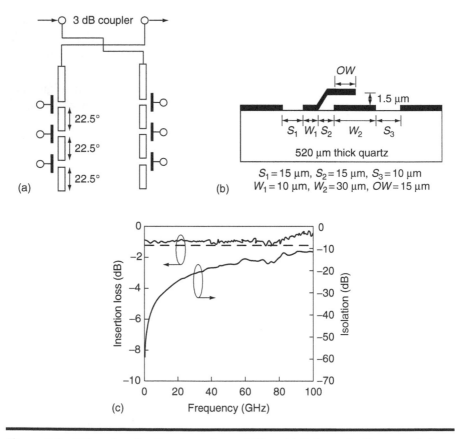

Figure 6.9 V-band, reflection-type phase shifter. (a) Schematic diagram of phase shifter, (b) cross section of CPW coupler, (c) insertion loss and isolation of the metal-to-metal contact switch,

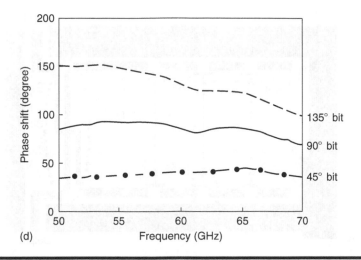

(d)

Figure 6.9 (continued)   and (d) phase shift per bit.

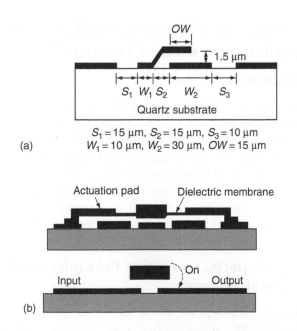

(a)

$S_1 = 15\ \mu m,\ S_2 = 15\ \mu m,\ S_3 = 10\ \mu m$
$W_1 = 10\ \mu m,\ W_2 = 30\ \mu m,\ OW = 15\ \mu m$

(b)

Figure 6.10   Specific details on a 2-bit, V-band, reflection-type phase shifter. (a) Schematic diagram of the air gap CPW coupler, (b) critical elements of the MEMS series switch,

(*continued*)

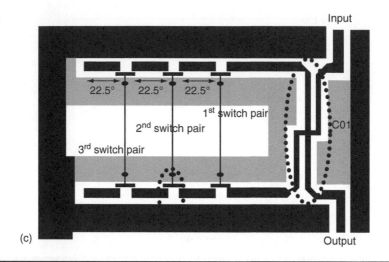

(c)

Figure 6.10 (continued) and (c) schematic diagram of a 2-bit, V-band, reflection-type phase shifter showing the locations of CPW couplers and MEMS series switches. (From Park, J.H., Kim, H.T. et al., *J. MEMS*, 11, 808, 2002. With permission.)

6.9 Three-Bit, Ultralow Loss Distributed Phase Shifter Operating over K-Band Frequencies

6.9.1 Introduction

Low-loss phase shifters are of critical importance for modern-phased array tracking radar and covert communications systems comprising of thousands such devices. Recently developed MEMS capacitive switches have demonstrated low IL, reduced parasitics, excellent linearity, and improved performance over conventional RF switches. Deployment of such switching devices in multi-bit phase shifters can significantly reduce the IL, weight, size, and power consumption [7]. Note 1-bit distributed MEMS phase shifter design can be extended to cascade several such devices to develop a multi-bit phase shifter with much lower IL.

6.9.2 Design Aspects, Operating Principle, and Description of Critical Elements

A 3-bit distributed MEMS phase shifter consists of three 1-bit phase shifters designated for 180°, 90°, and 45° phase shift, respectively. It is important to mention that each one-bit phase shifter comprises of a CPW transmission-line loaded periodically using several shunt MEMS capacitors. The estimated length of the phase shifter circuit is 11 mm and the device can be fabricated on a glass substrate with dielectric constant of 5.7 to keep the fabrication cost to minimum.

The phase shift of the synthetic CPW transmission line can be varied by switching the MEMS-based capacitors in up and down positions under appropriate bias conditions. The dc control bias for each 1-bit phase shifter is applied between the RF signal line and the ground pad of the CPW transmission line. This dc bias generates strong electric field beneath the MEMS switch membrane, which will force the membrane to snap down. The spacing between the MEMS capacitors (*s*) is 780 μm. The CPW line center conductor width is 100 μm and the ground-to-ground spacing is 190 μm [7]. With these dimensions, computed impedance of the transmission line is about 67 Ω (high impedance) in the upstates and is 37 Ω (low impedance) in the downstates. MIM capacitors with 0.6 μm thick silicon nitride layers are used as the dc blocks to prevent signal leakage. Note when the switches are snapped down, the CPW transmission line is periodically loaded with MEMS downstate capacitors and will yield a low impedance of the transmission line. The upstate capacitance is 5 fF and the downstate capacitor is 0.1 pF. The downstate capacitance, if necessary, can be reduced by tapering the center conductor width from 100 μm to an appropriate value. The up and down positions of the membrane provide the values of the impedance and capacitance parameters. The membrane width and span are 30 and 200 μm, respectively. The actuation voltage varies between 58 and 62 V depending on the membrane's spring constant and other mechanical parameters. The ES actuation voltage can be significantly reduced, if the gap between the membrane and the actuation pad is kept to minimum [8].

The 3-bit phase shifter can be designed for 360° phase shift, which will have eight phase states, namely, 0°, 45°, 90°, 135°, 180°, 225°, 270°, and 315° phase states with 45° phase step. The phase shift can be computed using Equation 6.24 using appropriate values of the parameters involved. The IL can be computed using Equation 6.20 using given values of parameters involved. Computed values of phase shift and phase errors for the 3-bit, K-band phase shifter at 26 GHz using appropriate values are summarized in Table 6.16.

In summary, the average IL is about 1.8 dB at the design frequency of 26 GHz and the return loss is better than −8 dB, which is due to impedance mismatch between the consecutive sections of the CPW transmission line. The maximum phase error for all the eight switching states is roughly 8.5° at the center frequency.

Table 6.16   Phase Shifts and Errors for the 3-Bit, K-Band MEMS Phase Shifter

Phase State	0°	45°	90°	135°	180°	225°	270°	315°
Phase shift (°)	0	49.5	85.6	143.3	183.7	219.3	263.0	321.3
Phase error (°)	0	−4.5	4.4	−8.3	−3.7	5.7	7.0	−6.3

Source: From Liu, Y., Borg, A. et al., *IEEE Microw. Guided Wave Lett.*, 10, 415, 2000.

6.10 Three-Bit, V-Band, Reflection-Type Distributed MEMS Phase Shifter

Recent research and development activities on MEMS switches and air gap 3 dB CPW couplers have demonstrated substantial reduction in IL, fabrication cost, and size of the elements, the critical elements required in the design of multi-bit phase shifters operating at mm-wave frequencies. These research activities further reveal that the phase shifters using the surface micromachined technology offer significant reduction in power consumption, larger phase shift per decibel loss (i.e., 75°/dB at 60 GHz), and minimum IM distortions compared to those using PIN-diodes or FETs. Furthermore, the reflection-type distributed MEMS phase shifter is fully compatible with monolithic technology, which yields low fabrication cost.

6.10.1 Design Aspects and Critical Performance Parameters

Design aspects and critical elements of a 2-bit, V-band, reflection-type MEMS phase shifter have been described in Section 6.7. Note that 1-bit phase shifter needs two sections capable of yielding 0° and 180° phase states, a 2-bit phase shifter needs three sections or three switch pairs as shown in Figure 6.10 capable of providing 0°, 45°, 90°, and 135° phase states. The 3-bit, V-band MEMS phase shifter will need four sections or four switch pairs using the same air gap CPW couplers and direct contact-type MEMS series switches as mentioned in the 2-bit, V-band MEMS phase shifter case involving three separate sections. In the case of a 3-bit, V-band phase shifter, four sections are needed to provide eight phase states, namely, 0°, 45°, 90°, 135°, 180°, 225°, 270°, and 315° phase states. Essentially, a 3-bit phase shifter can be realized by cascading a 1-bit and 2-bit phase shifters.

The dimensions of the 3 dB, air gap CPW coupler, critical elements of the MEMS switch, and the specific details of the 2-bit phase shifter comprising of three switch pairs are shown in Figure 6.10. As stated earlier, the 3-bit phase shifter will have four switch pairs. Note with actuation of each pair of contact switches, the electrical length of the CPW transmission-line section changes by 22.5°, resulting in a phase shift of 45° in the RF output signal at the design frequency of 60 GHz or the center frequency of V-band (55–75 GHz). All the eight states of the 3-bit phase shifter can be achieved successfully by actuating the four switching pairs.

It is evident from Figure 6.10 that the contact bar is suspended at 1.5 μm above the separated signal line gap, which is made from 2 μm thick electroplated gold structure to reduce the contact loss. The total length of the bridge-type MEMS contact switch is about 510 μm and the bridge width is 100 μm [9]. The 3 dB CPW couplers are fabricated by electroplating with gold to reduce the coupler loss. All phase shifters components can be fabricated on a 520 μm thick quartz substrate to minimize ILs at mm-wave frequencies.

The gap between the electrostatically actuated pad and the bottom ground plate is 2 μm. When dc actuation voltage ranging from 30 to 35 V is applied between the actuation pad and the ground bottom plate, the contact electrode makes metal-to-metal electrical contact with the RF signal line crossing the separated signal line gap. A silicon nitride layer of about 0.6 μm thickness is used between the contact bar and the actuation pad to provide the required isolation during the on-state position of the switch. Electroplated nickel structures are recommended as the structural material for the actuation pads of a MEMS switch to provide high mechanical strength and reliability for the contacts over extended duration. The thickness of the CPW transmission line and CPW coupler are 3 and 2 μm, respectively, to achieve optimum performance. Interdigitated metal structures are incorporated in the driving and contact parts to minimize the effect of residual stress in the dielectric membrane and to preserve the mechanical integrity of the membrane structure.

6.10.2 RF Performance of the 3 dB CPW Coupler and the 3-Bit, V-Band Phase Shifter

The estimated maximum ILs for the CPW coupler are 0.38 and 0.68 dB at 40 and 60 GHz, respectively. The return losses for the CPW coupler are −17 and −18 dB at 40 and 60 GHz, respectively. The contact-to-contact MEMS switch loss is 1.10, 1.15, and 1.20 dB at 40, 60, and 80 GHz, respectively. The total IL for four pairs of switches is about 4.40, 4.60, and 4.8 dB at 40, 60, and 80 GHz, respectively. The isolation for each contact switch is estimated to −20, −16, and −15 dB at 40, 60, and 80 GHz, respectively. The average IL for the 3-bit phase shifter including the losses in the transmission-line sections, coupler, and contact switches is about 4.9 dB at 60 GHz and at a bias voltage of 35 V. The switching time of the contact MEMS switch is close to 5 μs. The estimated phase shift is about 266° at 60 GHz with a phase error of 49° in the 315° phase state, which appears too high. The estimated values of phase shift and phase errors for various phase states for the 3-bit, V-band phase shifter are summarized in Table 6.17.

6.10.3 Maximum Phase Shift Available from a Multibridge DMTL Phase Shifter

The maximum phase shift available per decibel IL is a function of several parameters such as number of MEMS bridges, operating frequency, MEMS bridge resistance, CPW line center conductor width, loaded and unloaded impedances, bridge capacitance ratio, and effective dielectric constant of the substrate involved. Computer-simulated phase shift per decibel loss for a 32-bridge DMTL phase shifter on a quartz substrate with effective dielectric constant of 2.37 and assuming an unloaded impedance of 100 Ω, loaded impedance of 48 Ω, capacitance ratio of 1.17, a Bragg frequency of 180 GHz, and the center conductor width of 100 μm are summarized

Table 6.17   Estimated Values of Phase Shift and Phase Errors for 3-Bit Phase Shifter

Phase State (°)	Phase Shift (°)	Phase Errors (°)
0	0	0
45	39.9	5.1
90	81.1	8.9
135	121.1	14.9
180	143	37
225	180	45
270	222	48
315	266	49

in Table 6.18. One can visualize the maximum available phase shift (degrees per decibel loss) as a function of MEMS bridge resistance (R_b) and operating frequency from the data summarized in Table 6.18.

These simulated data indicates the trend for the maximum available phase shift per decibel loss as a function frequency and MEMS bridge resistance. It is interesting to mention that the phase shift per decibel loss drops from 105°/dB at zero bridge resistance to 52°/dB for a bridge resistance of 0.3 Ω.

Table 6.18   Maximum Available Phase Shift from a 32-Bridge DMTL Phase Shifter as a Function of Frequency and Bridge Resistance

Frequency (GHz)	Bridge Resistance, R_b (Ω)				
	0	0.05	0.1	0.2	0.3
20	38	38	37	37	36
40	58	56	54	51	49
60	75	70	66	62	55
80	88	81	73	64	54
100	105	92	81	63	52

6.11 Summary

RF/microwave phased shifters can be classified into two distinct categories, namely, absorption type and reflection type. Absorption-type phase shifters are widely used at RF/microwave frequencies. But reflection-type phase shifters are best suited for mm-wave applications. A phase shifter is the most critical element in electronically steerable-phased arrays for missile tracking radar, covert communications systems, reconnaissance satellites, missile seeker receivers, and unmanned aerial vehicles, where fast target tracking, high-tracking accuracy, and stability are the principal design requirements. DMTL phase shifters incorporating nanotechnology (NT) materials are discussed in great detail, because they offer fast response, minimum insertion, low-power consumption, high reliability and reduced phase errors at mm-wave frequencies. The overall performance of these phase shifters is unmatched by conventional phase shifters using PIN-diodes or FET devices. Design requirements and fabrication processes for a bulk micromachined MEMS phase shifter fabricated on a high-resistivity silicon wafer are summarized. This particular design offers significant reduction in IL, dispersion loss, and power consumption at mm-wave frequencies. Design requirements for the critical elements of a phase shifter, namely, MEMS bridges, dielectric membranes, air gap, 3 dB CPW couplers, dc bias circuits, contact pads, input/output electrodes, and structural materials are defined with emphasis on performance, reliability, and cost. Important properties of potential dielectric substrates, structural materials, and insulation layers widely used in the design of mm-wave phase shifters are summarized. Numerical examples to compute IL, Bragg frequency, periodic spacing between the MEMS bridges, actuation voltage, phase delay, phase shift, and CPW transmission-line parameters are provided for the benefit of readers. Optimum design aspects such as Bragg frequency, center conductor width, spacing between the MEMS bridges, electroplating materials for contact pads and electrodes, and dc applied voltage are identified for ultrawide band 1-bit, 2-bit, 3-bit, and 4-bit distributed MEMS phase shifters for mm-wave applications. Performance capabilities and limitations for wideband MEMS phase shifters operating in X-band, K-band, V-band, and W-band frequencies are summarized with emphasis on IL, return loss, phase shift, and phase error. Failure mechanisms in critical phase shifter elements are identified. Design aspects and performance capabilities for MDDM phase shifters are discussed with emphasis on actuation voltage and reliability. Design requirements and performance parameters for a wideband 4-bit, RF-MEMS TTD phase shifter operating over dc-40 GHz range are discussed. Critical design aspects and parameters for air gap, overlap, 3 dB CPW coupler, actuation mechanism, and metal-to-metal MEMS series switch used in a V-band reflection-type MEMS TTD phase shifter are summarized with emphasis on IL, phase error, and return loss.

References

1. N.S. Barker and G.M. Rebeiz, Optimization of distributed MEMS transmission line (DMTL) V-band and W-band phase shifter designs, *IEEE Transactions on MTT*, 48(11), November 2000, 1957–1965.
2. J.S. Hayden and G.M. Rebeiz, 2-bit MEMS distributed X-band phase shifter, *IEEE Microwave and Guided Wave Letters*, 10(12), December 2000, 540–542.
3. J.Y. Hayden and G.M. Rebeiz, Low-loss cascaded MEMS distributed X-band phase shifter, *IEEE Microwave and Guided Wave Letters*, 10(4), April 2000, 142–144.
4. A. Borgioli, Y.W. Li et al., Low-loss distributed MEMS phase shifter, *IEEE Microwave and Guided Wave Letters*, 10(1), January 2000, 7–9.
5. M. Kim, J.B. Hacker et al., A DC-40 GHz four-bit RF MEMS true-time-delay network, *IEEE Microwave and Wireless Components Letters*, 11(2), February 2001, 56–58.
6. H.T. Kim, J.H. Park et al., A compact V-band, 2-bit, reflection-type phase shifter, *IEEE Microwave and Wireless Components Letters*, 12(9), September 2002, 324–326.
7. Y. Liu, A. Borg et al., K-band, 3-bit, low-loss distributed MEMS phase shifter, *IEEE Microwave and Guided Wave Letters*, 10(10), October 2000, 415–417.
8. A.R. Jha, Potential actuation mechanisms for MEMS device applications, Technical Report, by Jha Technical Consulting Services, Cerritos, California, for Royal Institute of Technology, Stockholm, Sweden, 16 September 2006.
9. J.H. Park, H.T. Kim et al., 3-bit, V-band reflection-type phase shifter using CPW couplers and RF switches, *Journal of MEMS*, 11(6), December 2002, 808–813.

Chapter 7

Applications of Micropumps and Microfluidics

7.1   Introduction

This chapter concentrates on micropumps (MPs) and microfluidics, and their applications involving microelectromechanical systems (MEMS) as well as nanotechnology (NT). Microfluidics is an important branch of MEMS technology involving potential applications in chemical detection, microbiology, biological detection, clinical analysis, pressure transducers, microfluidic-integrated circuits, and heat transfer in microelectronic devices. An MP is the critical component of a self-contained microfluidic system. Active research and development activities have been targeted toward the design of MPs for various applications. Most MP designs, such as floating-wall check valves or cantilever-beam "flapper-valves" known as pneumatic valves, are passive. MPs using fixed valves or having no moving parts can be fabricated with minimum cost and complexity and are highly reliable due to their simplicity. Fixed valves have no ability to close and therefore, MPs with fixed valves need to operate in resonant modes to achieve flow rates and pressure compared to other pump designs. Optimum resonant response requires specific design parameters, which are possible with a low-order linear model capable of predicting the resonant behavior. In addition, an independent determination of the optimum valve shape can be achieved for maximum valve action over a specific Reynolds number range. A linear model will be able to identify all design parameters for the valves. This model shall be capable of making reliable investigation and optimization of several geometrical and material models, such as the electrical-equivalent model and the fixed-valve low-order model.

7.2 Potential Applications of Micropumps

Potential applications of various MPs can be summarized as follows:

1. An MP with electrostatic (ES) and piezoelectric actuation mechanisms
2. Self-priming and bubble-tolerant piezoelectric silicon (MP) for liquids and gases
3. Design of an ES pump for drug delivery applications
4. Design of a self-priming plastic MP
5. Valveless diffuser/nozzle-based fluid pump
6. Design, fabrication, and testing of fixed-valve pumps
7. Fixed-valve MPs for transport of particle-laden fluids
8. Electrofluidic full-system modeling of a flap-valve MP
9. Static modeling of pumps with electrical-equivalent networks
10. Dynamic modeling and simulation of MPs
11. Valveless, polysilicon-based planar MP
12. Piezoelectric valveless pump performance enhancement analysis
13. Numerical design analysis of a valveless diffuser pump using a lumped-mass model
14. Design analysis of high-performance MPs with no moving parts
15. Parametric design analysis of fixed-geometry microvalves

7.3 Design Aspects of Fixed-Valve Micropumps

MPs with fixed valves are widely deployed in various applications, where minimum cost and high reliability are the principal requirements [1]. It is important to mention that fixed valves do not open or close, but due to their geometrical shapes they provide less flow resistance in one direction compared to the flow in the opposite direction. Earlier MP designs involve converging–diverging, axis-symmetrical nozzle-diffuser valves, simple diffuser valves, and silicon-etched nozzle-diffuser valves. It is critical to mention that MPs with fixed-valve configuration offer maximum design flexibility and have greater clearances compared to other mechanical valves, and, hence, have capabilities to transport large amount of fluid in a given time. In addition, these valves offer optimum resonant frequency, because the operating frequency (f_o) is not limited to the time required for mechanical valves to open or to close.

7.3.1 Models Most Suited for Performance Optimization

Pump performance optimization requires large degree of freedom modeling such as finite element structural analysis or fluid dynamic computational modeling. Low-order modeling is best suited for a system involving few discrete elements such as mass and resistance. This type of modeling is most ideal for describing the critical design aspects of the entire MP system. Large degree of freedom modeling is not suitable for design and optimization of time-dependent MP systems involving fixed-valve concepts, which typically operate in the kilohertz frequency range. Low-order nonlinear models are used for MPs incorporating mechanical valves. Present non-linear models for diffuser pumps are very complex and require several assumptions and empirical coefficients, which cannot guarantee the modeling accuracy. Each pump can involve seven coefficients, five of which are based on steady-flow theory and two of which need adjustments for best fit to the data obtained for the turbulent flow. The velocity profiles for both the steady flow (laminar flow) and turbulent flow (unsteady flow) must be included in the modeling. Furthermore, any parameter adjustment in one area may yield errors in another parameter.

7.3.2 Reliable Modeling Approach for MPs with Fixed Valves

A modeling approach is considered reliable, which divides the overall design method-ology into two parts. The first part must consist of a low-order, linear model to maximize the resonant behavior through optimization of all the design parameters. However, such model cannot predict net pressure and flow characteristics, other than the relationship between the geometrical parameters and material properties of the pump discrete components needed to meet the resonant frequency and oscilla-tion requirements. Discrete elements for a fixed-valve MP or a straight-channel system are shown in Figure 7.1. The second part of the above model must determine the optimum shape and size to optimize the nonlinear pump performance in terms of net pressure and flow rate. Linear modeling can be used to optimize pump membrane thickness for a given valve size and to improve the resonant frequency and, therefore, the pump performance in terms of net pressure and flow rate. Regardless of the models used for MP design, experimental data involving critical performance parameters must be obtained to verify the accuracy and reliability of the model obtained through the computer simulations.

It is important to mention that the finite analysis offers exact solution for an oscillatory flow in straight rectangular channels to obtain appropriate values for the fluidic elements. However, a straight-channel fluid flow behavior is considered an approximate solution of the complex flow in actual values, because the differential behavior of fixed-valve MP is small compared to that of mechanical valve MP.

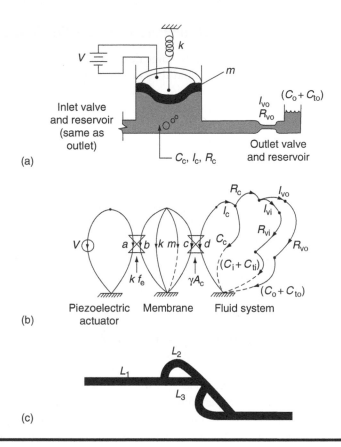

Figure 7.1 Fixed-valve MP. (a) Schematic diagram showing the critical elements, (b) linear graph representing the device elements, and (c) diagram showing a typical Tesla-type valve and two definitions of valve length. For $L \simeq L_{eq}$ segments were combined as parallel and series impedances, i.e., $L_{eq} = L_1 + (1/L_2 + 1/L_3)^{-1} + \cdots$. For $L \simeq L_{ave}$ the shortest and longest path through the valve were averaged, i.e., $L_{ave} = (L_1 + L_3 + \cdots + L_1 + L_2 + \cdots)/2$.

Both the viscous and inertial effects must be taken into account to obtain exact solution of the Navier–Stokes equation. These effects are critically important, when the pump operating frequency (f_o) is in the vicinity of the valve cutoff frequency (f_{cutoff}).

7.3.2.1 Electrical and Mechanical Parameters for Low-Order Model

The low-order model is best suited to describe the dynamic behavior of the fixed-valve MP or straight-channel device as illustrated in Figure 7.1 [1]. The inlet and

Table 7.1 Material Properties for the PZT-Actuator Layer

Properties	Pyrex	Silicon	Epoxy	PZT-5A
Young's modulus (GPa)	61	130	5.2	70
Poisson's ratio	0.20	0.22	0.30	0.28

outlet ports of a fixed-valve MP are connected to open fluid reservoirs through the tubes that are significantly larger than the *a–b* and *c–d* values shown in Figure 7.1b. The linear graph shown in the figure is dependent on 15 discrete elements between the electrical (piezoelectric actuation), mechanical (membrane), and fluid domains. Parametric values of pump electronics can be derived from at least 36 geometrical dimensions and physical properties [1]. Matlab and MathCad software programs can be used to solve the output variables in the frequency domain. The centerline velocity–amplitude of the membrane must be the primary output variables due to the dynamic response of the membrane rather than the flow rate of pumped fluid. Important information on the mechanical parameters of the MP can be obtained from the free-air resonance frequency (f_n) of the membrane or the pump containing no fluid. The electrical- and mechanical-lumped parameters for a single-large piezoelectric bimorph can be determined from finite element analysis (FEA) using ANSYS 5.5 software program, which uses the equations best suited for an analytical solution.

The bimorph model uses lead–zirconate–titanate (PZT)-5A piezoelectric material bonded to a Pyrex backing plate with conductive epoxy and the membrane support structure includes the silicon pump housing. Material properties for various materials for a single PZT-actuator layer are summarized in Table 7.1.

7.3.2.2 Mathematical Expression for Critical Pump Parameters

The spring constant (k) can be calculated using chamber pressure (P_c) acting on the internal surface of the membrane of area A_c. The expression for the parameter k can be written as

$$k = \left[\frac{(F_s)A_c P_c}{\delta_{mem}} \right] \tag{7.1}$$

where F_s is the shape factor and δ_{mem} is the centerline displacement of the membrane.

$$F_s = \left[\frac{\Delta V_{mem}}{A_c \delta_{mem}} \right] \tag{7.2}$$

where ΔV_{mem} indicates the volume displacement by the membrane, which is equal to the volume displaced by a piston having the same centerline displacement. Now the volume swept by the elastic membrane surface as a result of voltage applied across the piezoelectric actuator or the displacement per volt can be written as

$$\delta_V = \left[\frac{\delta_{\text{mem}}}{V} \right] \tag{7.3}$$

where V is the actuation voltage applied across the piezoelectric actuator.

The product of $k\delta_V$ is called the gyrator coupling, which couples the electrical to mechanical energy. But the product of $F_s A_c$ is known as the gyrator coupling coefficient, which couples the mechanical component (or membrane) to the fluidic component.

The fundamental free-air resonant frequency or the natural frequency (f_n) of the membrane system can be written as

$$f_n = \left[\frac{1}{2\pi} \left(\frac{\sqrt{k}}{m} \right) \right] \tag{7.4}$$

where m is the mass of the membrane system and k is the spring constant. Multiplying the numerator and denominator in Equation 7.1 by the chamber area A_c, one gets

$$k = \left[\frac{F_s A_c P_c}{\delta_{\text{mem}}} \right] \left[\frac{A_c}{A_c} \right]$$
$$= \left[\frac{F_s A_c^2 P_c}{\Delta V_{\text{mem}}} \right] \tag{7.5}$$

where ΔV_{mem} is the product of membrane displacement and chamber area.

7.3.2.3 Chamber Parameters

The chamber capacity is equal to the sum of the chamber capacity (Cc) with fluid and chamber housing capacity and can be written as

$$Cc = \left[(Cc)_f + (Cc)_h \right] \tag{7.6}$$

The housing capacity can be calculated using the FEA model and can be expressed as

$$[Cc]_h = \left[\frac{\Delta V_h}{P_c} \right] \tag{7.7}$$

where ΔV_h indicates the change in housing volume due to chamber pressure P_c.

From Equation 7.5, one can write as

$$\left[\frac{\Delta V_{mem}}{P_c}\right] = \left[\frac{F_s A_c^2}{k}\right] \tag{7.8}$$

The total change in the chamber volume is equal to the sum of change in membrane volume and change in housing volume and thus, it can be written as

$$\Delta V_c = [\Delta V_{mem} + \Delta V_h] \tag{7.9}$$

where the subscript c stands for chamber and h stands for housing. It is important to point out that the capacity of the pump fluid must include the compressibility effect of the liquid and gas, if any. Thus, the chamber capacity with fluid can be written as

$$[Cc]_f = \left[\frac{V_c}{G_{bm}}\right] \tag{7.10}$$

where V_c is the chamber volume and G_{bm} is the bulk modulus of the fluid. Similarly, the contribution to the chamber capacity due to adiabatic compression of the gas can be written as

$$[Cc]_g = \left[\frac{V_c V_{eff}}{\gamma P_c}\right] \tag{7.11}$$

where V_{eff} represents the effective volume of gas per unit volume of liquid or fluid and γ represents the ratio of specific heat for constant pressure to constant volume. The total chamber capacity is equal to the sum of chamber capacity due to fluid and chamber capacity due to gas, and can be written as

$$[Cc]_{total} = V_c \left[\frac{1}{G_{bm}} + \frac{V_{eff}}{\gamma P_c}\right] \tag{7.12}$$

The first term in the equation will be zero, if the pump fluid is filled with gas and the value of parameter V_{eff} is unity. The bulk modulus of water can be assumed as 2.19 GPa and the specific heat ratio can be assumed as 1.4 for air. If the fluid is other than water and the gas is other than air, then appropriate values for the parameters must be used to get the exact value for the total chamber capacity.

7.3.2.4 Fluidic Valve Parameters and Their Typical Values

The impedance of the fluid channel with rectangular configuration on each side of a circular chamber will define all the parameters of the fluid. The impedance of the valve consists of two components, namely, the resistance and the inductance of

Table 7.2 Computed Values of Cc with Water as Fluid

mm^3	$(Cc)\ mm^5/N$
1	0.009
2	0.019
3	0.027
4	0.036

Source: From Morris, C.J. and Forster, F.K., *J. MEMS*, 12, 325, 2003. With permission.

the inlet and outlet valves. The valve resistance for a steady, viscous flow is a function of absolute viscosity, channel length, channel depth, channel output ratio (width/depth), and a constant p_n, which is equal to 4.712, 7.854, 10.996, and 14.137 when n is equal to 1, 2, 3, and 4, respectively. The valve inductance is proportional to mass density of the fluid and channel length, but inversely proportional to channel width and channel output ratio. Computed values of chamber capacity with fluid obtained using Equation 7.10 are summarized in Table 7.2.

The lumped parameters for the valves are strictly determined by the length and the transverse dimensions of a straight rectangular channel also called duct, fluid density ($2.205\ lb/mm^3$, if water is used as fluid), absolute viscosity of the fluid, and frequency of oscillation. Typical geometrical parameters for various straight-channel MPs are summarized in Table 7.3.

The pressure drop across each branch is dependent on the shortest and longest path through the valve as illustrated in Figure 7.1c. The fluidic load parameters such as the inlet and outlet capacities, namely, C_i and C_o, and the open reservoir inlet and outlet capacities, namely, C_{ri} and C_{ro} are attached to the MP. The reservoir capacity with water can be computed using the following expression involving various reservoir parameters:

Table 7.3 Typical Geometrical Parameters for Straight-Channel MPs

Pump No.	Chamber Diameter (mm)	Etching Depth (mm)	Channel Volume (mm³)	Membrane Thickness (mm)	PTZ-Wafer Diameter (mm)
1	3	0.110	0.777	0.250	2.5
2	3	0.152	1.074	0.250	2.5
3	6	0.148	4.185	0.500	5.0
4	6	0.115	3.252	0.500	5.0

$$\text{Reservoir capacity}(C_\text{r}) = \left[\frac{\pi D^2}{4\rho g}\right] \tag{7.13}$$

where

D is the reservoir diameter

ρ is the density

g is the acceleration due to gravity (10.04 m/s^2)

Assuming a diameter of 1 mm, water density of 62.43 lb/ft^3 and standard value of acceleration, one gets

$$C_\text{r} = \left[\frac{0.7854 \times 10^{-6}}{62.43 \times 35.3 \times 10.04}\right]$$
$$= 1.6 \times 10^{-10} \text{m}^3$$

7.3.2.5 Description of Micropumps with Straight-Channel Configurations

MP devices with straight-channels or valves are widely used for various scientific, clinical research, and medical applications. The valves as well as the reservoirs can be etched together to reduce the fabrication costs. Four distinct MPs with Tesla-type valves or straight-channel widths are available. These MPs can be used to predict the resonance behavior under various flow conditions and to investigate the effects of fluid composition. Typically, the MP's straight channels are 120 μm wide and 1.4 mm long for a 3 mm-diameter chamber and roughly 2.25 mm long for a 6 mm-diameter chamber. The membrane or the Pyrex cover is bonded to the pump structure, while a piezoelectric actuator is attached onto the Pyrex with a conductive epoxy. After attaching the piezoelectric actuator, the thickness of the epoxy layer is reduced to 20 mm.

Various category and geometrical parameters of Tesla-type valves are shown in Figure 7.2. Critical parameters of some selected Tesla-type valves are summarized in Table 7.4 with emphasis on cell category.

Member thickness plays a key role in determining the mechanical parameters, namely, the spring constant (k) and the velocity per volt (V_{volt}). The chamber diameter affects the chamber capacity and the mechanical parameters. Resonant frequency tests can be performed using air-amplitude with inlet and outlet reservoirs, which are typically 15 cm long and 1 mm in diameter. The actuator voltages must be high enough to achieve a good velocity signal and low enough to minimize entrance effects in the straight channel. The actuation voltage can vary from 5 to 25 V depending on the spring constant and other valve parameters.

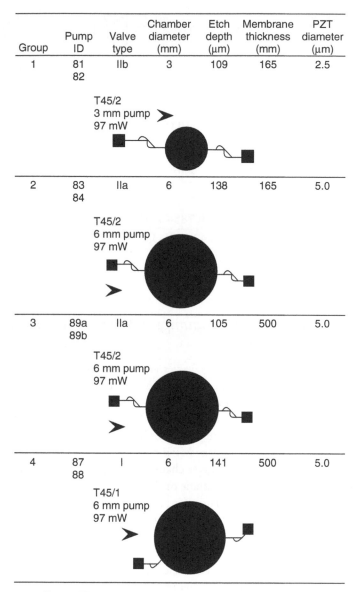

Group	Pump ID	Valve type	Chamber diameter (mm)	Etch depth (μm)	Membrane thickness (mm)	PZT diameter (μm)
1	81 82	IIb	3	109	165	2.5
2	83 84	IIa	6	138	165	5.0
3	89a 89b	IIa	6	105	500	5.0
4	87 88	I	6	141	500	5.0

Note: Type 1 means single-cell, Tesla-type valve pump
Type 2 means double-cell, Tesla-type valve pump

Figure 7.2 Geometric parameters for various Tesla-type valves. (From Morris, C.J. and Forster, F.K., *J. MEMS*, 12, 325, 2003. With permission.)

Table 7.4   Tesla-Type Valves and Their Geometrical Parameters

Tesla-Type Valve	Equivalent Path Length (mm)	Average Path Length (mm)	Cell Category
Type I	1.43	1.90	Single-cell
Type II (a)	1.91	2.64	Dual-cell
Type II (b)	2.13	3.03	Dual-cell

Source: From Morris, C.J. and Forster, F.K., *J. MEMS*, 12, 325, 2003.

Free-air resonance frequency also known as natural resonance frequency is the most important parameter to evaluate the dynamic performance of an MP, which can be computed using Equation 7.4. Computed values of resonant frequency as a function of spring constant, mass, and dimensional parameters of device and chamber are summarized in Table 7.5.

7.3.2.6   Impact of Viscosity and Membrane Parameters on Valve Performance

Membrane velocity as a function of frequency appears linear in liquid-free MPs. The membrane thickness and valve width can be optimized for maximum deflection per unit actuation voltage. The valve length to valve width ratio of 1.5 yields optimum pump performance in terms of peak velocity flow rate and peak Reynolds number. Both the peak centerline velocity and peak Reynolds number are dependent on membrane thickness and valve width. It is important to point out that linear modeling is most efficient and practical to optimize the multitude of pump design parameters. Experimental verification is needed to verify both the size

Table 7.5   Computed Values of Resonant Frequency of MPs

Pump No.	Chamber Diameter/Membrane Thickness (mm/mm)	Spring Constant k (N/m)	Mass m (N)	Natural Frequency (kHz)
1	3/0.165	4.62×10^6	0.462×10^{-5}	159.2
2	3/0.250	7.89×10^6	0.509×10^{-5}	198.2
3	6/0.165	1.37×10^6	1.53×10^{-5}	47.6
4	6/0.500	8.56×10^6	2.18×10^{-5}	99.6

and shape parameters of the valve obtained through simulations. This model provides excellent results for fixed-valve MPs, when the valves are replaced by straight channels with appropriate dimensional parameters. Computation errors are possible in straight-channel approximation due to valve fluidic characteristics. An MP can be designed using a linear model for optimum resonant behavior through adjustments of geometrical parameters and material properties. However, determination of optimum valve size and prediction of optimum pump performance in terms of net pressure and flow rates would require fluid dynamics analysis. In summary, fluid dynamics analysis is essential, if optimum performance and accurate prediction of valve shape and size are the principal requirements.

7.4 Dynamic Modeling for Piezoelectric Valve-Free Micropumps

7.4.1 Introduction

Piezoelectric MPs are most suitable for low flow rates. Such pumps are best suited for drug delivery systems, chemicals, process control systems, and cooling of small electronic circuits or MEMS devices. The fundamental problem in such small pumps is the construction of the valve that concerts the reciprocal motion of the disk into position pumping action. Deployment of valve-free pumps will eliminate the problems associated with other pumps. An MP design using diffuser-nozzle elements instead of conventional valves will eliminate the moving parts that are sensitive to the presence of particles in the fluid. A dynamic model [2] is best suited to undertake accurate simulation of the pump performance including the resonance frequency, flow rate, and pressure drop. In addition, this model is found most useful to investigate the effects of the driving frequency and optimum dimensional parameters for the inlet and outlet pipe sections. It is important to mention that most of the early models are based on kinetic consideration, which permits calculation of the flow rate based on known volume displacement produced by the piezoelectric membrane motion. The kinetic model can be modified by taking into account the acceleration of the fluid in the nozzles. This model takes into account the frictional and accelerational pressure drop effects.

7.4.2 Modeling for the Piezoelectric Valve-Free Pump

The operating principle of the piezoelectric valve-free (PEVF) pump is based on the unique diffuser nozzles, which provide preferred directions for the liquid flow as shown in Figure 7.3. This pump consists of a cylinder with piezoelectric membrane, which acts like a movable piston. The piezoelectric membrane is made of a piezoelectric element and a support disk. The movement of the piezoelectric

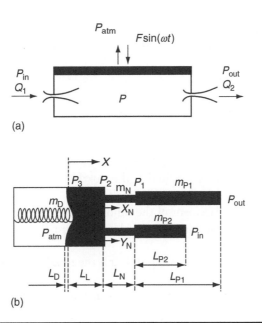

(a)

(b)

Figure 7.3 Piezoelectric MP. (a) Schematic diagram of the piezoelectric MP and (b) structural details of the piezoelectric MP. (From Ullmann, A. and Fono, I., *J. MEMS*, 11, 655, 2002. With permission.)

membrane allows the liquid to flow from a lower pressure input port (P_1) to a higher pressure output port (P_2). When an actuation voltage is applied to the piezoelectric device, a periodic force is generated acting on the center of the disk. The elasticity of the membrane is provided by the spring with a spring constant k. The fluid flow is assumed incompressible. The input and output line consists of nozzles (N) and the leading input and output pipes p1 and p2. Expressions for the balanced forces acting on the outlet pipe can be written as

$$[P_1 A_{p1}] = \left[P_{out} A_{p1} + m_{p1} \left(\frac{d^2 X_N}{dt^2} \right) \right] + \left[\left(\frac{32 \mu L_{p1}}{D p_1^2} \right) \left(\frac{dX_{p1}}{dt} \right) A_{p1} \right] \quad (7.14)$$

The balanced force acting on the nozzle N can be written as

$$[P_2 A_N] = \left[P_1 A_N + m_N \left(\frac{d^2 X_N}{dt^2} \right) \right]$$

$$+ \left\{ (0.5 A_N K_L \rho) \left[\left(\frac{dX_N}{dt} \right) \left(\frac{A_N}{A_{min}} \right) \right]^2 \right\} \quad (7.15)$$

where

P_1 and P_2 are the pressure before and after the nozzle

A_N is the nozzle area

m_N is the nozzle mass

m_{p1} is the inlet pipe mass

A_{min} is the minimum or nozzle throat area

X_N distance travel in nozzle

(dX_{p1}/dt) is the average velocity

(d^2X_{p1}/dt^2) is the acceleration in the pipe

(dX_N/dt) is the velocity of the fluid at the nozzle

(d^2X_N/dt^2) is the acceleration at the nozzle

X_N is the average displacement of the fluid in the nozzle

K_L is the friction proportional factor at low pressure

μ is the friction coefficient

ρ is the density of the fluid passing through the nozzle or pipe

Based on continuity flow relations, one can now write expressions for velocity and acceleration as

$$\left[\frac{dX_{p1}}{dt}\right] = \left[\left(\frac{A_N}{A_{p1}}\right)\left(\frac{dX_N}{dt}\right)\right] \tag{7.16a}$$

$$\left[\frac{d^2X_{p1}}{dt^2}\right] = \left[\left(\frac{A_N}{A_{p1}}\right)\left(\frac{d^2X_N}{dt^2}\right)\right] \tag{7.16b}$$

Substituting the parameters for the inlet line and nozzle section in Equations 7.14 through 7.16, one can rewrite the equation for the outlet pipe section as

$$[P_2A_N] = [P_{in}A_N] + \left\{\left[m_{p2}\left(\frac{A_N}{A_{p2}}\right)^2 + m_N\right]\frac{d^2y}{dt^2}\right\} + \left[(0.5K_NA_N\rho)\left(\frac{A_N}{A_{min}}\right)^2\left(\frac{dy}{dt}\right)^2\right]$$
$$+ \left[\left(\frac{32\mu L_{p2}}{D_{p2}^2}\right)\left(\frac{A_N}{A_{p2}}\right)\left(\frac{dy}{dt}\right)(A_N)\right] \tag{7.17}$$

Expression for a balance force of the liquid section L can be written as

$$[P_3A_L] = \left[P_2A_L + \left(\frac{m_L d^2X_L}{dt^2}\right)\right]$$
$$= [(\text{pressure} \times \text{area}) + (\text{mass} \times \text{acceleration})]_{\text{liquid}} \tag{7.18}$$

$$X_L = \left[\frac{X \, K_L A_D}{A_L} \right] \qquad (7.19a)$$

$$\frac{d^2 X_L}{dt^2} = \left[\left(\frac{d^2 \, X}{dt^2} \right) \frac{K_L A_D}{A_L} \right] \qquad (7.19b)$$

where
A_D and A_L indicates the area of disk and liquid section, respectively
X is the displacement of the membrane center
X_L is the displacement liquid
K_L is the correction factor for the line

Based on the relationship between the central force and the center disk deflection δ or X, one can quote the expression for the disk displacement at the center from the Timoshenko book as

$$\delta \text{ or } X = \left[\frac{0.55 R^2}{E L_D} \right] \qquad (7.20)$$

where
R is the disk radius
E is Young's modulus of the disk
L_D is the disk length or thickness

The value of E is 190 GPa for the steel disk, 66 GPa for aluminum nitride (AIN) piezoelectric material and 70 GPa for the PZT piezoelectric material.

7.4.3 Natural Frequency of the Micropump System

The natural frequency (f_n) of the PEVF system is one of the most critical parameters, because the best MP performance occurs only at the natural frequency. As stated earlier, the natural frequency is proportional to the square root of the ratio of spring constant of the piezoelectric membrane to equivalent mass (M_E) of the system. For a special case in which the length and the diameter of the inlet pipe and the outlet pipe are equal, the expression for the equivalent mass [2] can be written as

$M_E = $ [mass of pipe] + [mass of nozzle] + [mass of liquid + mass of disk]

$$M_E = \left[K_V K_P \left(\frac{A_D}{2 A_{p1}} \right)^2 (2 m_{p1}) \right] + \left[K_V K_P \left(\frac{A_D}{2 A_N} \right)^2 (2 m_N) \right]$$

$$+ \left[\left(\frac{A_D}{A_L} \right)^2 K_V K_P m_L + K_D m_D \right] \qquad (7.21)$$

Numerical example showing the equivalent mass for each item, is equal to the volume times the density of the material. The parameters assumed [2] are as follows:

K_V is a correction factor that concerts the virtual volume into actual volume and its assumed value is 0.4. K_D is a disk shape correction factor and has an assumed value of 0.235, and K_P is the correction factor that converts the continuous force acting on the membrane into central force and has an assumed value of 0.41 (approximately), stainless steel density $= 7850$ kg/m^3, density of water $= 1000$ kg/m^3, and $N = 4.45$ lb, volume of inlet pipe $= 254.5$ mm^3, volume of nozzle 0.277 mm^3, volume of liquid (water) $= 15.71$ mm^3, volume of the disk (steel plus piezoelectric material) $= 11.78$ mm^3. Pipes, nozzles, and disk are made of stainless steel. Inserting the volume, density, and constants in Equation 7.21, one gets the value of equivalent mass as

$$M_E = 10^{-4}[3353 + 5438 + 0.1085 + 0.4507] \text{ lb}$$

These calculations indicate that the mass of pipes and nozzles, which are the stationary components, is the major contributor to the equivalent mass. However, the average mass of the liquid and disk, which are constantly moving, is the source of natural frequency. Therefore, average equivalent mass that contributes to natural frequency can be written as N

$$[M_E]_{\text{ave}} = 10^{-4}\left[\frac{0.1085 + 0.4507}{2}\right] = [10^{-4} \times 0.2796]\text{lb} = [10^{-4} \times 0.0628]N$$

Because the membrane consists of piezoelectric element ($E = 66$ GPa) and stainless steel supporting disk ($E = 190$ GPa), the value of the E will be used in computation of the spring constant k, which is defined as

$$k = \left[\frac{12\,EI}{L_D^3}\right] \tag{7.22}$$

where I is the moment of inertia of the disk, which is defined as

$$I = \left[\frac{bd^3}{12}\right] \tag{7.23}$$

Assuming 190 GPa for E, 0.15 mm for the disk length (L_D), 0.25 mm for dimension b and 8 mm for dimension d and inserting these values in above equations, one gets

$$I = 10.66 \text{ mm}^4 \text{ and } k = 78.72$$

Inserting these parameters in the expression for the natural frequency Equation 7.4, one gets the magnitude of the natural frequency as

$$f_n = \left(\frac{1}{2\pi}\right)\sqrt{(k/M_{ave})}$$

$$= (0.159)\sqrt{(78.72/0.0628 \times 10^{-4})}$$

$$= 562 \text{ Hz}.$$

The corresponding natural frequency was reported as 550 Hz in Ref. [2].

7.4.4 Pump Performance in Terms of Critical Parameters

Flow rate, efficiency, and natural frequency are considered the most critical pump performance parameters. The liquid flow rate through the nozzles can be obtained by multiplying the fluid velocity to the area of nozzle. The expressions for the input port and the output port can be written as

$$Q_{in} = \left[\left(\frac{dX_N}{dt}\right)A_N\right] \tag{7.24a}$$

$$Q_{out} = \left[\left(\frac{dY_N}{dt}\right)A_N\right] \tag{7.24b}$$

where
  Q is the flow rate
  A_N indicates the nozzle area
  X_N and Y_N are the fluid movements in the nozzle along X and Y directions

MP flow rates as a function of natural frequency and pressure drop are shown in Figure 7.4. The maximum flow rate occurs at a natural frequency of 550 Hz, regardless of pressure drop (ΔP) as illustrated in Figure 7.5. The curves shown in Figure 7.4 indicate that the flow rate decreases with pressure differences [2]. At high-pressure difference, the liquid flow may be reversed, which can allow the liquid to flow from high pressure at the outlet to the low pressure at the inlet, and this could be opposite to the pump's normal action. Furthermore, these curves indicate that the pump performance in terms of flow rate versus pressure is very close to a linear change. The maximum pressure drop occurs at zero flow rate.

The pump efficiency (η_p) can be defined as the ratio output work to input work and for most MPs the efficiency seldom exceeds 1.2 percent, which occurs only at the maximum natural frequency. For this type of pump flow rate is more important than the efficiency. Preliminary studies performed by the author indicate that the natural frequency is very sensitive to the inlet and outlet pipe length. The natural frequency

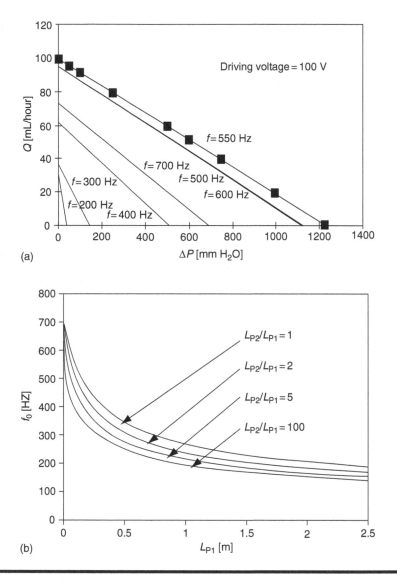

(a)

(b)

Figure 7.4 Performance of MP (a) Pump flow rate as a function of frequency and differential pressure and (b) impact of inlet/outlet length ratio on natural frequency. (From Ullmann, A. and Fono, I., J. MEMS, 11, 655, 2002.)

drops rapidly as the lengths of the pipes increase due to the increased liquid mass in the line. If the inlet pipe length is larger than the outlet pipe length, the natural frequency further decreases as the ratio of the inlet pipe length to outlet pipe length increases. However, when the length of both pipe lines is equal, neither the flow rate nor the maximum pressure at the natural frequency is sensitive to the pipe lengths. It is interesting to note that the natural frequency decreases with the increase in pipe

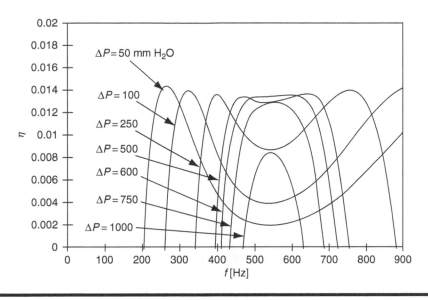

Figure 7.5 Pump efficiency and natural frequency as a function of maximum differential pressure from water. (From Ullmann, A. and Fono, I., *J. MEMS*, 11, 655, 2002.)

lengths, but the flow rate at lower and higher natural frequencies remains the same. The amplitude of the membrane increases as the length of pipe lines increases.

The studies further indicate that negligible frictional losses occur in the inlet/outlet pipes for the zero pressure difference. Note viscosity of the liquid or fluid can introduce additional frictional losses in the pipe lines. However, such losses are negligible for pipe lengths not exceeding 1 m. But frictional losses in longer pipe lengths can decrease the pump performance. In addition, decrease in pipe diameter can cause a drastic increase in frictional related pressure drop. However, such an effect is negligible for pipe diameters greater than 2 mm or so. Any change in pipe internal diameter can affect the natural frequency due to change in the effective mass. In summary, the best pump performance which involves maximum flow rate and pressure drop can be obtained at the natural frequency. Dynamic model calculations are considered ideal for investigating the pump properties and overall performance of a PEVFMP.

7.5 Design Aspects and Performance Capabilities of an Electrohydrodynamic Ion-Drag Micropump

7.5.1 Introduction

This particular pump uses the interaction of an electric field with the electric charges, i.e., electrons or the particles embedded in a dielectric fluid to move the liquid in a pipe. The electric charges can be injected into a fluid by sharp points or by adding a small amount of liquid containing high-density ions [3]. The major driving force in an electrohydrodynamic (EHD) ion-drag pump is due to the movement of the ions across the applied electric field, which is established by a charged electrode (emitter) and an electrode called collector. The electrons present in the liquid from ionized molecules can be accelerated to ionize under strong electric field. Note the electrons at high speed can act as ion producers. The Coulomb force (F_c) produced by an excited electric field will affect all the charges in the fluid and the net resulting force will remain the same. Because the electron mass is negligible compared to an ion mass, the majority of the fluid motion is produced by the motion of the ions. The friction between the moving ions and the working fluid drags the fluid toward the collector, thereby leading the fluid into motion.

7.5.2 Design Concepts and Critical Parameters of an EHD Pump

Electrode configuration, electrode dimensional parameters, and formation of sharp points on the emitter electrodes at which the charge injection occurs are the most important design aspects of this pump. Critical elements of the pump are shown in Figure 7.6. The dielectric liquids can be pumped by the injection of ions in an

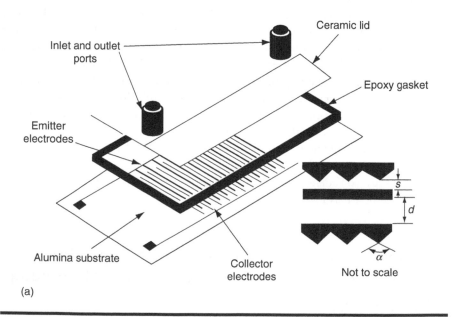

(a)

Figure 7.6   EHD ion-drag MP. (a) Schematic of the EHD-ion-drag MP,

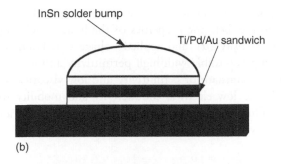

(b)

Figure 7.6 (continued)   (b) 3-D bump structure details. (From Darabi, J., Rada, M. et al., *J. MEMS*, 11, 684, 2002. With permission.)

applied electric field of appropriate strength [3]. The maximum static pressure that can be produced in an EHD ion-drag pump from a dielectric fluid with a plane electrode can be written as [3]

$$P_{\text{max}} = \left[1.125(C\varepsilon) \left(\frac{V}{d} \right)^2 \right] \qquad (7.25)$$

where C is a constant whose value can vary between 0 and 1 depending on the charge emission laws at the electrodes and electrode dimensional parameters and ε is the permittivity of the liquid to be pumped.

The expression for the force density inside the ion-drag pump can be written as

$$F_{\text{d}} = \left[\frac{\rho V_0}{d} \right] + \left[\left(\frac{\rho^2}{\varepsilon} \right) (x - 0.5d) \right] \qquad (7.26)$$

where
  ρ is the density of the liquid to be pumped
  V_0 is the threshold voltage below which no pressure is obtainable
  d is the spacing between the electrodes

It is important to mention that both the maximum pressure and force density assume the presence of uniform electric field. In actual practice, the electric field gets distorted that makes accurate prediction of this pump comprising of electrodes with sharp features extremely difficult. Furthermore, nonuniform pressure could exist within the fluid because of EHD force, which varies with the fluid's distance x from the emitter. Note a force gradient can generate an internal stirring within the fluid that can form electroconvection leading to reduction in the output pressure. It is important to mention that the EHD pump performance relies strictly on the electric and the dielectric properties of the liquids such as permittivity (ε), conductivity (σ), viscosity (μ), and applied voltage (V). High performance with EHD ion-free pump is possible with high permittivity and low viscosity. Furthermore, high dielectric constant or permittivity and low viscosity will generate high flow velocity, whereas low electrical conductivity and mobility would yield high pump efficiency. Quadratic increases are possible in flow rate and pressure as a function of applied voltage.

7.5.3   Benefits of EHD Ion-Drag Pumps

EHD pumps designed with microfabrication technology are absolutely vibration free because they have no moving parts. These pumps are electronically controllable and require very little power consumption, require no or little maintenance, and are extremely light. These pumps are best suited for microsystems such as microfluidic systems. Because of above-mentioned benefits, EHD have potential applications including pumping of minute quantities of a large variety of liquids, implantable medicine dosage control, precision fuel injection, microcooling devices, and microscopic fluid or gas sampling.

Several design configurations of this pump have been investigated by various scientists. An EHD pump employing the concept of a traveling electrical wave charge imposed between the electrodes positioned in a nonconducting fluid was designed to move transverse to the electrodes by applying a sinusoidally applied voltage. The pump consisted of ten parallel electrodes spaced 10 μm apart and 0.75 μm thick. But no experimental data is available on this pump.

A micromachined ion-drag EHD pump comprising of pairs of facing perforated planar grids was developed. The fluid can move through the planar with an applied voltage of 500 V. The pump with a gap of 350 μm between the grid electrodes demonstrated a maximum static pressure head of 2500 Pa (or 0.362 psi) and a pumping rate pf 14 mL/minute for the ethanol fluid at an applied voltage of about 500 V. This flow rate is about three orders of magnitude higher than that of piezoelectric or thermally excited MPs when the pumping fluid was ethanol. The pressure head increases with the increase of the electric field intensity regardless of the electrode configuration as illustrated in Figure 7.7. Impact of inlet and outlet pipe length ratio on natural frequency of pump is shown in Figure 7.7. The pump flow rate increases with the increase in both the natural frequency and the pressure head.

In 2004, scientists developed several versions of this pump including the ion-drag MP using traveling wave design concept for drug delivery applications. The working fluid was ethanol. The pump deployed planar electrodes in a 3 μm wide channel with a channel height of 200 μm and demonstrated a flow rate of 55 mL/minute at an applied voltage of 60 V. A polarization MP was developed recently for

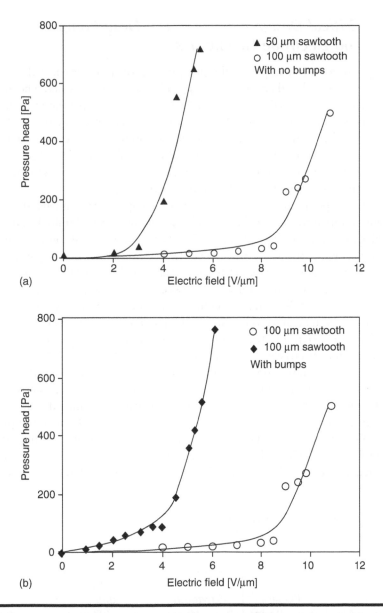

Figure 7.7   Pump pressure head as a function of electric field and (a) sawtooth electrode with no bump and (b) electrode with bump structure.

microelectronics cooling applications. An active evaporation cooling surface, a polarization MP, and temperature sensors were incorporated into a single chip. The prototype MP demonstrated a maximum cooling capacity of 65 W/cm^2 with a pumping head of 250 Pa (0.036 psi).

7.5.4 Critical Design Aspects of Ion-Drag Pump and Electrode Geometries

In the case of an ion-drag pump, the pumping action is achieved, if the electrical shear stresses are higher than shear stresses produced by the viscosity. The locations at which ions are generated and to which they are accelerated play an important role in determining the overall pump efficiency. The design of electrode pair is of critical importance. The pump uses a series of planar comb finger electrodes known as emitters and collectors and the electrodes allow the fluid to flow through them. Two-dimensional (2-D) and three-dimensional (3-D) electrode configurations must be investigated to maximize the EHD pump performance.

The EHD pump is made of three micro components, namely, the bottom substrate (alumina) containing the EHD electrodes, top cover (alumina), and sidewall made of epoxy gasket as shown in Figure 7.6. Note the electrode consists of emitters and collectors with specific geometry. The electrodes can have planar geometry with specific electrode gap or separation (*s*) and distance (*d*) between stages or sawtooth geometry with specific gas and distance between stages or sawtooth geometry with hemispherical-shaped 3-D bump structures. The bumps are placed on every other sawtooth with bump diameters less than 1 mm and bump height less than 15 percent of the base diameter. Electrical power consumption versus generated pumping pressure data for various electrode designs is summarized in Table 7.6.

The estimated power consumption data indicates that decreasing the electrode spacing or gap and increasing the number of pump stages can significantly improve the EHD pump performance in terms of pumping capability and efficiency. Note a high-pressure closed-loop system is not required for a fluid with low vapor pressure, low permittivity, low viscosity, and high thermal conductivity. It is important to mention that thermophysical properties of heat transfer fluids determine the applied voltage requirements and overall EHD pump performance. Pumping pressure is a

Table 7.6 Estimated Power Consumption versus Pumping Pressure for Various Electrode Geometries (mW)

		Electrode Geometry		
Pumping Pressure (Pa)	Planar (s = 50 μm)	Sawtooth (s = 100 μm)	Sawtooth with Bumps (s = 100 μm)	Sawtooth (s = 50 μm)
200	38	36	10	2
400	64	38	12	3
600	85	42	16	4
800	110	45	20	5

function of applied voltage, electrode gap, and electrode geometry. The data summarized in Table 7.6 indicates that higher pumping performance is possible from an electrode with sawtooth geometry incorporating bump structures. Furthermore, an applied voltage close to 900 V is required to produce a pumping head of 600 Pa with a planar electrode, whereas only 200 V is needed for an electrode with sawtooth geometry. Pressure head as a function of electric field intensity for various electrode configurations is shown in Figure 7.7. The increase in pressure head is due to the higher electric field gradient possible with sawtooth geometry, which accelerates the injection process leading to higher pumping efficiency. Operating voltage less than 200 V is possible, if the space between the emitter electrode and the collector electrode is reduced between 15 and 50 μm. Pumping effects are improved with presence of 3-D bumping structures with optimized shape and size.

7.6 Capabilities of a Ferrofluidic Magnetic Micropump

7.6.1 Introduction

This section will focus on the design aspects and performance capabilities of a ferrofluidic magnetic (FM) MP, which uses a magnetic actuation mechanism to push fluid through a microchannel [4]. This particular pump design could lead to development of "lab-on-chip" technology that can offer a miniaturized and integrated system capable of providing chemical analysis with minimum error, lower cost, and higher throughput. Currently available MPs suffer from cost, complexity, and power consumption. MPs equipped with diaphragms have to deploy either electrothermal or piezoelectric or ES or pneumatic actuation mechanism. Furthermore, these pumps suffer from limited pulsatile flow and generation of air bubbles. Both piezoelectric and ES actuation methods require higher actuation voltage and complex structures. Electrothermal actuation can degrade biological fluids by denaturing proteins. In addition, diaphragms-based MPs require valves with moving parts that can cause clogging. EHD ion-drag pumps require very high actuation voltages anywhere from 200 to 1,000 V. The ferromagnetic pumps overcome most of the above problems and at same time offer maximum design flexibility and other benefits such as flow reversal, uniform flow rate, and nanosize ferromagnetic particles.

7.6.2 Design Aspects and Critical Performance Parameters

The pumping mechanism strictly depends on the magnetic actuation to move a ferromagnetic plug capable of pushing the fluid of interest. A ferrofluid contains nanosize ferromagnetic particles suspended in an oil-based solvent. For the pumping applications, the ferrofluid must be immiscible with the fluid being pumped. Ferrofluids have been applied as an actuation mechanism. Critical elements of an

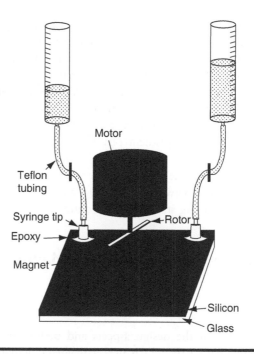

Figure 7.8 Critical elements of the FM MP (From Hatch, A., Holman, G. et al., *J. MEMS*, 10, 215, 2001.)

FM MP are shown in Figure 7.8. Locations of the stationary ferrofluid plug and stationary permanent magnets are clearly depicted in Figure 7.9.

The closed-loop pump configuration must be selected for continuous pumping capability. A circular geometry permits easy implementation of the moving magnetic actuator as illustrated in Figure 7.9. A ferrofluidic valve is held in the desired place by a stationary magnet M and is always present in the channel between the inlet and outlet of the pumping loop. A ferrofluidic plug acting as a piston is drawn inside the channel by an external permanent magnet attached to a rotor of the electric motor. As the mobile plug is moved in the pumping loop, the fluid is drawn into the pumping loop through the inlet port and forced out through the outlet port. The force that holds the ferrofluid in place is proportional to the gradient external field and the saturation magnetization of the ferrofluid, which is strictly dependent on the volume ranging from 100 to 1000 G. More leakage is possible, if the cross-sectional area of the pumping loop increases. However, if the cross-sectional area of the loop is reduced, one can expect more resistance to fluid flow, and therefore higher plug velocity will be needed to maintain an equivalent volumetric flow rate. It is important to mention that the maximum flow rate and pressure that can be generated are strictly dependent on the fluidic resistance of the channel.

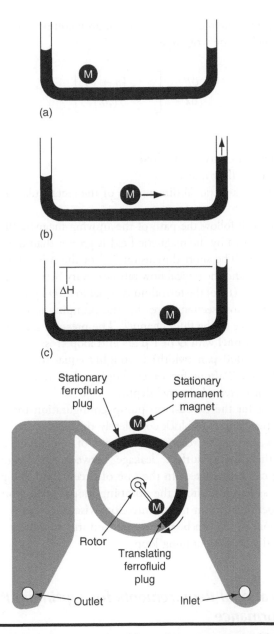

Figure 7.9 Details of the FM MP with various magnet positions. (a) Magnet stationary, (b) magnet moving, and (c) magnet stationary with fluid column.

The expression [4] for the pressure gradient for an incompressible laminar flow in a rectangular channel can be written as

$$\frac{dP}{dx} = \left[8\mu Q \frac{(a+b)^2}{(ab)^3} \right]$$

(7.27)

where
 μ is the dynamic viscosity of the fluid
 Q is the volumetric flow rate
 a and b are the cross-sectional dimensions of the rectangular channel

Note the plug will follow the path of the moving magnetic field as long as the resultant force generated by the magnetic field is greater than the force provided by the pressure gradient. The short dimension "a" of the microfluidic channel varies from 10 to 1000 μm and the typical flow rates can vary between 0.001 and 10 μL/s. The dynamic viscosity (μ) of the ferrofluid is equal to 0.09 kg/m, which is 100 times that of water. Large cross-sectional area for the pumping loop is recommended to minimize the resistance of the fluid channel. The pressure gradient using Equation 7.27 for a straight channel with Q of 1 μL/s and loop cross-sectional dimensions of 250 μm (depth) × 2000 μm (width) is roughly equal to 29.12 Pa/mm for the ferrofluid and 0.2912 Pa/mm for water. Higher flow rates require greater plug translational rate or increased channel depth.

It is important for the closed-loop pump configuration that adequate fluid is introduced into the channel to block completely the inlet or outlet of the pumping loop at all times during the regeneration phase of operation. When the pumping fluid wets the channel surface, leakage between the channel wall and the ferrofluidic plug can be expected. In the case of excess fluid, the pump is designed to expel the excess ferrofluid from the pumping loop. It is important to point out that the pump performance in the counterclockwise direction is as good as in the clockwise direction, which can be due to nonuniform alignment of the translating magnet associated with the pumping loop.

7.6.3 *Operational Requirements for Optimum Pump Performance*

Dynamic leakage, which usually occurs at high pressures, must be eliminated to achieve satisfactory pump performance over extended durations. To minimize dynamic leakage, operation with higher flow rates must be avoided. Note leakage is directly proportional to the hydrostatic pressure and is more pronounced in the nonlinear portions of the flow rate curves, because the ferrofluid is displaced for a longer duration. Optimum as well as steady-flow rate is the principal requirement for this kind of pump, which is based on the volume of the pumping loop and the revolution of the electric motor. Typical motor revolution rate ranges from 4 to 8 rpm (revolutions per minute). Assuming an outer loop diameter of 11 mm, inner diameter of 9 mm, and channel depth of 0.25 mm, the computed volume of the pumping loop comes to about 7.92 μL (1 L is equivalent to 1000 cc). About 25 percent of this volume is filled with ferrofluid, which effectively reduces the pumping loop volume to 75 percent. At a motor speed of 4 rpm, the volumetric flow rate comes to 23.76 μL/minute. If the speed increases to 8 rpm, the flow rate will increase to 48 μL/minute. High efficiency of the ferrofluidic plug is only possible, when there is no leakage of the fluid, the device surfaces are completely wetted by the ferrofluid and contamination through the contact with the ferrofluid can be entirely eliminated.

The flow rate gradually increases as the plug separates from the ferrofluid pool and the ferrofluid near the permanent magnet remains in place. The flow is maximum when it passes through the inlet port and then rapidly drops as the moving plug reaches the outlet port of the pump loop.

Studies performed by the author on various actuation methods indicate that the ferrofluid appears to be a robust technique for actuation, because it readily conforms to the geometry of the microchannel and the ferrofluidic plugs can be separated or merged without any problem. Acting as a valve, the plugs can be maintained at pressures greater than 1800 Pa. Optimum plug performance can be maintained even at linear velocities close to 3.9 mm/s using a translating permanent magnet, while generating fluid pressures exceeding 1200 Pa. Continuous flow rate can be achieved through one pump cycle, provided the channel is uniform in size and the translating plug is moving at an uniform velocity. If continuous flow rate over longer duration is desired, the pump must be designed with multiple pumping loops with one loop in regeneration mode at all times.

The feasibility of ferrofluidic plugs as a fluid actuator opens the door for potential applications for the FM MP. Because of low manufacturing cost and ease of fabrication, this particular pump technology can be used in many formats. This pump technology is most attractive for various other device geometries and materials. The major advantage of this pump is that it can block the entire cross section of the fluid channel and it can adapt to changes in the channel shape and size. This pump is best suited for applications where short duration pumping is needed. This pump is capable of pumping water at flow rates of 23 and 46 μL/minute with minimum back pressure for motor speed of 4 and 8 rpm, respectively. The maximum pressure head generated are roughly 118 and 136 mm of water at motor speed of 4 and 8 rpm, respectively. Precise control of flow rates and much lower flow rates are possible by changing the geometry of the device and speed of the electric motor. Note the highest power consumption at the maximum motor speed of 8 rpm is typically 12 mW.

7.7 Summary

This chapter described several types of MPs using MEMS- and NT-based devices and materials with particular emphasis on PEVFMP, EHD ion-drag MP, and FM MP. Design aspects and performance capabilities of FM and EHD MPs are discussed in detail. Self-priming and bubble-tolerant piezoelectric silicon-based MPs for liquids and gases, valveless nozzle-based MPs for fluid transfer, and fixed-valve MPs for transport of particle-laden fluids are described with emphasis on volumetric

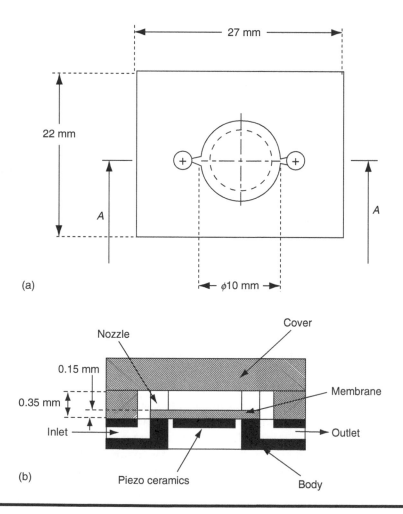

Figure 7.10 Piezoelectric valveless MP. (a) Plan view of the MP and (b) cross-sectional view of the critical elements of the pump with physical dimension in the vertical plane only. (From Darabi, J., Rada, M. et al., *J. MEMS*, 11, 684, 2002.)

flow rate, differential pressure, efficiency, and natural frequency. Design aspects and performance capabilities of MPs with no moving parts are described with emphasis on reliability and fabrication costs. Benefits and device parameter requirements for low-order linear and dynamic nonlinear modeling of MPs are identified. Low-order modeling is best suited for an MP system involving small number of discrete elements. However, determination of correct valve size and prediction of optimum pump performance in terms of net pressure and flow rates require additional fluid dynamics analysis. Numerical examples to compute natural frequency, differential pressure, equivalent mass, chamber volume, pressure gradient, dimensional parameters for straight channels, and flow rate are provided for the benefits of readers. Natural frequency, spring constant, and equivalent mass calculations for a PEVFMP are provided assuming appropriate values of dimensional and mechanical parameters for various elements of the pump. Curves showing the flow rates as a function of natural frequency and differential pressure are provided for the convenience of the readers. Performance capabilities and potential applications of EHD MPs are summarized. Design features and performance capabilities of FM MPs are briefly discussed with emphasis on flow rate and benefits of magnetically actuated ferrofluidic plugs as a means of pumping. Note the dynamic modeling of a piezoelectric valveless pump shown in Figure 7.10 is straightforward and the simulation data obtained can be used to evaluate the performance of parameters such as flow rate and pressure head for other MPs.

References

1. C.J. Morris and F.K. Forster, Low-order modeling of resonance for fixed-valve micropumps based on first principle, *Journal of MEMS*, 12(3), June 2003, 325–334.
2. A. Ullmann and I. Fono, The piezoelectric valve-less pump dynamic model, *Journal of MEMS*, 11(6), December 2002, 655–663.
3. J. Darabi, M. Rada et al., Design, fabrication and testing of an electrohydrodynamic ion-drag micropump, *Journal of MEMS*, 11(6), December 2002, 684–689.
4. A. Hatch, G. Holman et al., A ferrofluidic magnetic micropump, *Journal of MEMS*, 10(2), June 2001, 215–220.

Chapter 8

Miscellaneous MEMS/ Nanotechnology Devices and Sensors for Commercial and Military Applications

8.1 Introduction

This chapter will describe the performance capabilities of microelectromechanical systems (MEMS)-based and nanotechnology (NT)-based devices and sensors best suited for commercial, medical, satellite, and military applications. Discussions will be limited to devices and sensors, which have not been described previously. Performance capabilities and unique design aspects will be summarized for devices and sensors including MEMS varactors or tunable capacitors, tunable filters, accelerometers, NT-based tower actuators using multiwall carbon nanotubes (MWCNTs), strain sensors, micromechanical (MM) resonators, biosensors using carbon nanotube (CNT) arrays, ultrasonic transducers, and MEMS sensors using smart materials to monitor health of structures, weapon systems, and battlefield environments. In addition, topics of great interest such as micro-heat pipes, photovoltaic cells, radar absorbers, photonic detectors, Li-ion microbatteries (MBs), and microminiaturized deformable mirrors best suited for laser beam corrections will be discussed.

Smart materials, which have potential applications in fabrication of MEMS-based and NT-based sensors or devices, are discussed with emphasis on their thermal, mechanical, electrical, and dielectric properties. Smart materials best suited for commercial, medical, aerospace, military, and satellite applications will be identified. Applications of nanowires (NWs), nanotubes, nanocrystals, and nanorods are briefly described.

8.2 MEMS Varactors or Tunable Capacitors

MEMS varactors are also known as tunable capacitors because of their ability to provide wideband tuning capability in radio frequency (RF) and microwave regions. MEMS varactors are best suited for microwave and MM-wave phase shifters, wideband tuning voltage-controlled oscillators (VCOs), RF amplifiers, and reconfigurable RF filters. Critical elements of a MEMS varactor are shown in Figure 8.1. It is important to mention that the MEMS varactor provides ultra-wideband tuning capability because of its high-capacitance ratio under appropriate bias conditions. This ratio is strictly dependent on the MEMS bridge position. Critical elements of a MEMS varactor are illustrated in Figure 8.2. Note the MEMS bridge provides two values of the MEMS varactor depending on the position of the bridge, namely, the capacitance due to bridge-up position (C_{bu}) and the capacitance due to bridge-down position (C_{bd}). The value of these capacitances is dependent on the position of the MEMS bridge, actuation voltage, and the distributed lumped parameters of the

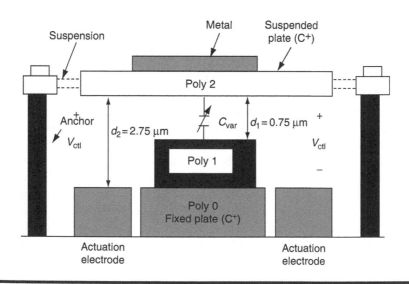

Figure 8.1   Cross-sectional view of a MEMS varactor showing its critical components. (From Rebeiz, G.M., *IEEE Trans. MTT*, 50, 1316, 2002.)

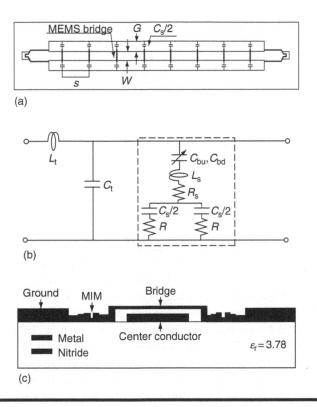

Figure 8.2 **CPW-based phase shifter showing (a) bridge and CPW line parameters, (b) distributed lumped parameters of the bridge and CPW line, and (c) cross section of the device showing the center conductor, bridge, and MIM capacitance on the quartz substrate ($e_r = 3.78$).**

bridge and coplanar waveguide (CPW) transmission line as shown in Figure 8.2. Note the bridge impedance, phase velocity of the CPW transmission line, and the insertion loss (IL) in the distributed transmission line are function of bridge capacitance in up-position. The zero-bias capacitance of the bridge (C_{b0}) plays a key role in the design of MEMS phase shifters. Its value is dependent on the periodic spacing (s), unloaded transmission impedance (Z_{ul}), and the transmission-line capacitance per unit length (C_t). The parameter can be defined as

$$C_t = \sqrt{\frac{\varepsilon_{\text{eff}}}{cZ_0}} \tag{8.1}$$

where
  ε_{eff} is the effective dielectric constant of the substrate material
  c is the velocity of light
  Z_0 is the characteristic impedance of the CPW transmission line

The zero-bias capacitance of the bridge can be written as

$$C_{b0} = s \left[\left(\frac{L_t}{Z_{ul}^2} \right) - C_t \right] \tag{8.2}$$

where

s is the periodic spacing

Z_{ul} is the loaded CPW line impedance when the bridge is in up-position

L_t is the inductance of the CPW transmission line per unit length, which can be written as

$$L_t = \left[C_t Z_0^2 \right] \tag{8.3}$$

It is important to point out that the bridge capacitance plays a key role in achieving high Bragg frequency needed for the optimum performance of MEMS phase shifter operating at mm-wave frequencies. Note a distributed MEMS phase shifter using CPW transmission-line technology may contain 8 or 16 or 24 MEMS bridge sections depending on the phase shift requirements. Typical values of transmission-line inductance and capacitance per unit length are roughly 10 pH and 40 fF, respectively, assuming a CPW transmission-line characteristic impedance of 100 Ω and effective dielectric constant of 2.5 for the quartz substrate ($e_r = 3.78$). The computed value for the zero-bias bridge capacitance comes between 34 and 36 fF assuming a periodic spacing (s) of 200 μm, unloaded or characteristic CPW line impedance of 100 Ω and CPW center conductor width of 100 μm. The zero-bias capacitance decreases to 21 fF with a periodic spacing of 110 μm, while all other parameters remain the same. The computed values of these parameters have been obtained by inserting the assumed values in Equations 8.1 through 8.3. Computed values of zero-bias capacitance of the MEMS bridge as a function of spacing and loaded line capacitance in upstate position are summarized in Table 8.1.

It is important to mention that optimum values of loaded line impedance must be selected to achieve maximum change in MEMS bridge capacitance.

Table 8.1   Computed Values of Zero-Bias Bridge Capacitance, C_{b0} (fF)

Periodic Spacing (μm)	Loaded Line Impedance (Ω)			
	48	47	46	45
100	17.3	18.2	19.3	20.1
200	34.5	36.4	38.5	40.6
300	51.8	54.6	49.9	60.8

8.2.1   Benefits and Shortcomings of MEMS Varactors

MEMS varactors play a key role in the design and development of numerous RF and microwave devices and components such as tunable filters, diplexers, duplexers, VCOs, and distributed RF filter approach. This device relaxes the stringent requirements for the high-performance duplex filter by introducing several tunable filters of moderate performance with minimum cost. Duplex filters with base communication station deploy coaxial resonators with a Q factor in the order of several thousands, but at the expense of a big form factor, excessive weight, and high cost. However, multiple small filters using the distributed filter approach look very promising for a MEMS technology implementation. Note that MEMS varactors for tunable impedance matching structures to realize good selectivity suffer from circulatory currents and power-handling capability. Currently, the Q of MEMS varactors range from 20 to 80, but the efforts are under way to achieve significant improvement of the Q factor. Research and development activities are directed to eliminate these problems soon.

Another potential application of MEMS varactors is in wide-tuning VCOs, which are widely used in the design of microwave synthesizers. Note conventional varactors used by the VCOs generate flicker or shot noise that can degrade the synthesizer performance. On the other hand, MEMS varactors are passive devices, which do not exhibit flicker noise. Furthermore, the separation of control terminal and RF port is an attractive design feature in the case of MEMS varactors. Note the RF-MEMS devices have certain mechanical inertia properties, which could be exploited with the regular electrical loop filter in a phase-locked loop (PLL) to further improve the Q factor of the MEMS varactor. An active filter topology using MEMS varactors would result in a Q-factor enhancement.

A reconfigurable filter combines RF-MEMS varactor technology with other technologies such as RFIC, logic, and digital-to-analog (D/A) converter technology for driving the various MEMS elements. In addition, RF-MEMS varactors offer compact packaging, minimum size, zero or negligible power consumption, and numerous reconfigurable RF functions.

8.2.2   MEMS Varactor Design Aspects and Fabrication Requirements

MEMS technology has the potential of realizing RF variable capacitors with a performance that is superior to conventional solid-state aviator diodes in terms of nonlinearity, flicker noise, and losses. An electrostatically actuated MEMS bridge offers a MEMS parallel-plate (PP) variable capacitor that can be achieved with simple fabrication, higher yield, and minimum cost. It has built-in capability for tuning over bandwidths exceeding 100 percent at microwave and mm-wave frequencies.

A MEMS varactor or variable capacitor can be designed with wide-tuning capability by making the actuation electrodes spaced differently for the capacitor plates as shown in Figure 8.1. The MEMS variable capacitor as illustrated in this figure consists of two movable plates with an insulation dielectric layer on the top of the bottom plate. This tunable capacitor has demonstrated extended tuning range when two plates touch each other. An improved design of the variable capacitor can be achieved using two structural layers, three sacrificial layers, and two insulation layers of silicon nitride. The top plate of the device can be fabricated from a 24 μm thick nickel, which is covered by a 2 μm thick silicon nitride. The bottom plate can be made of polysilicon covered with a 350 nm silicon nitride. Note the nitride oxide outlines the area, which will be used to etch a trench in the silicon substrate. This nitride layer forms the bottom cover of the polysilicon layer, which is also a part of the bottom plate of the variable capacitor. The second nitride layer is deposited on the top of the polysilicon layer to form an isolation area, which prevents the electrical contact between the two capacitor plates, thereby eliminating the sticking problem. A dc actuation voltage from 0 to 40 V can be applied in appropriate steps to the variable capacitor to investigate the tuning range of capacitance. Preliminary studies performed by the author indicate that a capacitance greater than 2 pF can be achieved with a plate area of 200×300 μm^2. This plate area has demonstrated a tuning range exceeding 280 percent at a microwave frequency of 1 GHz.

8.2.3 Effects of Nonlinearity Generated by MEMS Capacitors

Nonlinearity effects are generated by the movable MEMS membrane of the MEMS variable capacitor, which can introduce very low-level phase noise close to the carrier. The electrostatic (ES) force produced by the applied voltage between the MEMS bridge and the bottom electrode provides the tuning capability of the MEMS varactor. Note the intermodulation (IM) effects are extremely low as compared to other switching elements. The studies further reveal that a two-tone third-order (TTTO) intermodulation intercept point (IIP3) is greater than +40 dBm for a bridge spring constant (k) greater than 10 N/m and over a resonance frequency bandwidth of 3 to $5f_0$, where f_0 is the mechanical resonance frequency. The IIP3 level can increase to +80 dBm for a difference signal of 5 MHz. Note the IM product is inversely proportional to eight power of the height of the membrane or gap (g_m). If the bridge or membrane height is reduced by 30 percent, the IM level will increase by a factor of 10 log (1/0.7) or 12.4 dB. This means lower the membrane, higher will be the IM product level, regardless of operating frequency. Estimated IIP3 levels for a single MEMS varactor as a function of bridge (or membrane) spring constant and membrane height and the PP capacitance as a function of gap are summarized in Table 8.2.

Computed values of PP capacitance (femtofarad) as a function of MEMS bridge height or gap and impact of membrane height reduction on IM at a fixed bias level of 14 V are shown in Table 8.3.

Table 8.2   Intermodulation Intercept Point (IIP3) (dBm) Level as a Function of Spring Constant (k) and Bridge Gap (g_m)

IIP3 Level (dBm)				
	Gap (μm)			
Spring Constant k (N/m)	1	2	3	4
1	29	41	48	54
5	36	48	55	61
10	39	51	58	63
50	48	58	66	70
100	49	61	68	74

As stated earlier the MEMS varactor generates very low IM products compared to conventional varactors, which make them most attractive for MEMS switches, RF synthesizers, reconfigurable RF amplifiers, MEMS phase shifters best suited for electronically steering phase array antennas, matching networks, covert communications systems, and tunable filters [1]. Note the additional phase noise generated by the MEMS varactor known as Brownian noise must be taken into account in some communications systems. This phase noise power [1] be written as

Table 8.3   Computed Values of PP Capacitance and Impact of MEMS Bridge Reduction on IIP3 Level Generated by MEMS Capacitor

Gap (μm)	PP Capacitance (fF)	Reduction of Gap (Percent)	Intermodulation (IM) (dBm)
1	174	10	3.66
2	87	20	7.75
3	58	30	12.39
4	44	40	17.75

Note: The PP capacitance computations assume plate area of 140 × 140 μm^2 and a fixed bias level of 15 V.

$$[P_{PN}] = \left[\frac{1}{(1+\gamma)^2}\right]\left[2\left(\frac{X_{ave}}{Z_0}\right)^2\right]\left[\frac{1}{(gk)^2}\right] \bigg/ \text{Hz} \qquad (8.4)$$

where

X_{ave} is the average capacitive reactance $(1/2\pi fC)$ in the upstate position of the MEMS bridge and its value is 50 Ω assuming the RF frequency of 64 GHz and capacitance of 50 fF

γ is a constant with typical value of 0.35

Z_0 is the characteristic impedance of the CPW line with a typical value of 100 Ω

g is the MEMS bridge or membrane height and a typical value of 3 μm can be assumed

k is the spring constant (K/m) with an assumed value of 10 N/m

Inserting these values in Equation 8.4, one gets the phase noise value at 64 GHz as

$$[P_{PN}] = \left[\frac{1}{(1+0.35)^2}\right]\left[2\left(\frac{50}{100}\right)^2\right]\left[\frac{1}{(3\times 10\times 10^6)^2}\right]$$
$$= [0.5487][0.5]\left[1.11\times 10^{-15}\right]$$
$$= \left[3.048\times 10^{-16}\right]$$
$$[P_{PN}]_{dB} = [160-4.8] = 155.2 \text{ dBc/Hz}$$

The author has performed parametric analysis on the phase noise generated by the MEMS capacitor using the previous parameters and as a function of bridge spring constant (k) and membrane height or gap (g) and the results have been presented in Table 8.4.

Table 8.4 Phase Noise Level at 64 GHz as a Function of Spring Constant (k) and Membrane Height or Gap (G)

Spring Constant k (N/m)	Phase Noise Power P_{PN} (dBm)	Gap G (μm)	Phase Noise Power P_{PN} (dBm)	
			k = 10	k = 50
10	−155	1	−146	−160
20	−161	2	−152	−166
50	−169	3	−155	−169
100	−175	4	−158	−172

These parametric computations indicate that the phase noise levels generated by the MEMS capacitors are significantly less than those produced by the conventional varactors. It is important to mention MEMS capacitors with such low phase noise levels are best suited for applications in mm-wave synthesizers, stable local oscillators (LOs), tunable VCOs, and covert communications systems as illustrated in Figure 8.3.

8.3 Micromechanical Resonators

8.3.1 Introduction

Vibrating mechanical tank components such as crystal and surface acoustic wave (SAW) resonators are widely used for frequency selection in telecommunications and communications systems. The high-Q SAW and bulk acoustic mechanical resonators are used to realize ultralow phase noise and high-frequency stability for the LOs. Despite their high quality factors (Q's) (2000–5000), ultralow noise levels, and exceptionally high-frequency stability against thermal vibrations, mechanical vibrations and aging, these devices suffer from large packaging, excessive weight, and high power consumption [2]. Furthermore, it is impossible to achieve miniaturization and portability, which are needed for aerospace sensors and satellite communications systems. MM resonators are best suited for applications, where microminiaturization, portability, and minimum power consumption are the critical design requirements. Schematics of potential MM resonators are depicted in Figure 8.2.

Deployment of high-Q on-chip MM resonators, which may be referred later on as micro-resonators, will not only eliminate the previously mentioned shortcomings but also will provide a quality factor Q exceeding well over 50,000 under vacuum and center frequency temperature coefficients better than -10 ppm/°C. A two-stage band-pass filter using micro-resonators has demonstrated IL well below 2 dB at 35 GHz and over a bandwidth close to 2 percent. Three-stage micro-resonator filters and 20 kHz, high-Q oscillators have demonstrated much lower IL and 64 dB of rejection only at lower frequencies. Because many portable communication equipments operate at very high frequency (VHF) and ultrahigh frequency (UHF), aggressive research and development activities must be undertaken to develop MEMS devices that could provide both miniaturization and portability at higher frequencies.

8.3.2 Types of Micro-Resonators and Their Potential Applications

Substantial size difference between the MM resonators or micro-resonators and the macroscopic counterparts is quite evident from Figure 8.4. This figure compares a typical 100 MHz SAW oscillator with a clamped beam MM resonator fabricated using the polysilicon surface micromachining technology. Note the surface

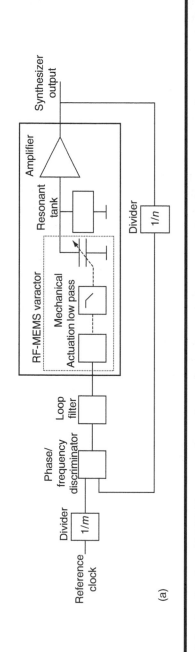

Figure 8.3 Applications of MEMS varactor in various microwave components. (a) Use of MEMS varactor in an RF synthesizer and

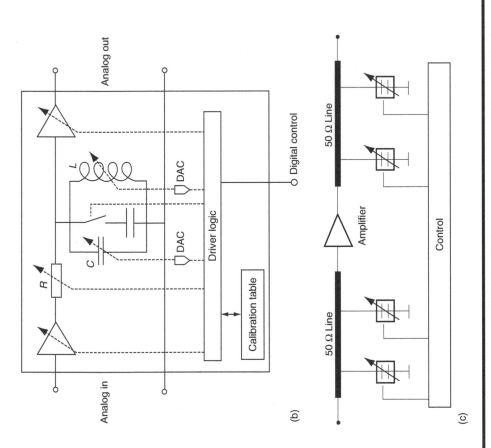

(b)

(c)

Figure 8.3 (continued) (b) reconfigurable RF filter using MEMS varactor, and (c) tuning control elements.

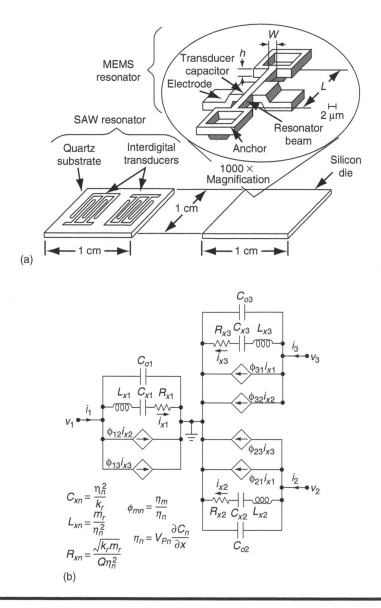

Figure 8.4 MM resonator showing (a) size comparison between a current SAW technology and high-Q MM resonators and (b) small-signal equivalent circuit for a 3-port MM resonator. (From Nguyen, C.T.C., *IEEE Trans. MTT*, 47, 1486, 1999.)

micromachining technology is fully compatible with complementary metal-oxide semiconductor (CMOS) processing technology. These micro-resonators are best suited for VHF/UHF communications systems, intermediate frequency

(IF)/radio frequency (RF) filters, and high-Q oscillators due to their outstanding benefits such as compact packaging, virtually zero-power consumption, large stiffness-to-mass ratios, and high reliability under harsh operating environments.

Three distinct types of micro-resonators have been developed, namely, free-free beam (FFB) micro-resonators, clamped-clamped beam (CCB) micro-resonators, and folded-beam, comb-transducer lateral (FBCTL) micro-resonators. High stiffness is necessary to achieve large dynamic range. High power-handling capability is considered most ideal for communications applications. The highest Q close to 20,000 and above has been reported for VHF micro-resonators using submicrometer technology. The required stiffness is too small to provide the needed large dynamic range. This means one must undertake trade-off studies between the stiffness and the quality factor Q to meet specific performance requirements. Limited studies performed by the author indicate that the FFB micro-resonators have demonstrated remarkable performance with center frequency from 30 to 90 MHz, stiffness from 30,000 to 80,000 N/m, and quality factor Q as high as 8400. Larger stiffness is possible from CCB micro-resonators, but at the expense of increased anchor dissipation and lower resonator Q. Because the FFB micro-resonators incorporate the flexural-mode beam design with strategically located supports, the FFB micro-resonators will offer lower anchor losses, high Q's, and high stiffness [2].

8.3.3   Free-Free Beam High-Q Micro-Resonators

As stated earlier that the FFB micro-resonators incorporating the basic flexural-mode beam design offer several benefits such as high stiffness, lower anchor losses, large dynamic range, and high power-handling capability [3]. These performance parameters are most desirable for covert communications applications, where large suppression of interference in adjacent channels is of critical importance. Critical components of an FFB micro-resonator are shown in Figure 8.5a. FFB micro-resonators with high Q's are best suited for image rejection filters with center frequency anywhere from 800 to 2500 MHz, IF filters with center frequency anywhere from 455 kHz to 255 MHz, and low phase noise LOs operating in the 10–2500 MHz range, where minimum size, weight, power consumption, and adjacent channel interference are of paramount importance.

8.3.3.1   Structural Design Aspects and Requirements of FFB Micro-Resonators

The flexural-mode beam design uses strategically located supports and anchors, which provide stiffness from 30,000 to 80,000 N/m and quality factors (Q's) in excess of 8500. The FFB micro-resonator as shown in Figure 8.5 consists of a beam supported at its flexural node points by four torsional beams, each of which is anchored to the substrate material by rigid anchors. ES actuation voltage is applied

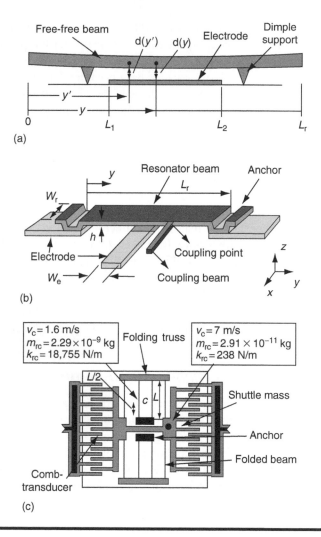

Figure 8.5 **Schematics of various micro-resonators: (a) of a FFB micro-resonator, (b) of a CCB micro-resonator, and (c) of a folded-beam micro-resonator showing critical elements. (From Wang, K., Wong, A.C. et al., *J. MEMS*, 9, 351, 2000.)**

to an electrode located underneath the FFB. This particular device is a one-port device and therefore, its output current must be taken directly from the micro-resonator structure. The bias voltage is supplied through a bias-tee comprising of an inductor and the coupling capacitor. The torsional support beams are designed with quarter-wavelength long to achieve an impedance transformation needed to isolate the FFB from the rigid anchors. Under these conditions, the FFB sees zero impedance into its supports and the anchor dissipation mechanisms, if any, are suppressed leading to much higher quality factor for the device. The transducer

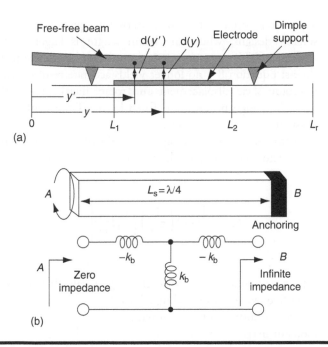

Figure 8.6 **FFB resonator. (a) Cross-sectional view of FFB resonator showing critical components and physical parameters and (b) quarter-wavelength torsional beam parameters and equivalent acoustic network elements.**

capacitor gap, which is dependent on the height of the dimple, enhances the Q of the device. When a dc bias (V_P) is applied between the electrode and the micro-resonator, the entire device structure comes down and rests upon the dimples (Figure 8.6). Deployment of dimples to set the capacitor gap spacing permits the use of thicker sacrificial oxide spacers, which will provide high mechanical rigidity and improved reliability under harsh operating environments.

It is important to point out that optimum design of the FFB micro-resonator requires proper selection of geometries, which will provide optimum resonant frequency, improved support isolation, and suppression of spurious modes. For a narrow FFB or CCB with uniform cross section, the resonant frequency for an uncoupled beam can be written as

$$f_{res} = 1.03\sqrt{\left(\frac{E}{\rho}\right)\left(\frac{h}{L_r^2}\right)} \tag{8.5}$$

where
E is Young's modulus of the structural material
ρ is the density of the material
h is the beam thickness
L_r is the resonator length

Note this equation is based on the Euler–Bernoulli equation and is valid for low frequency resonator designs, where the beam length is larger than the beam thickness. In the case of VHF resonator design, the beam length begins to approach the beam thickness; Equation 8.5 no longer yields accurate results, because it does not take into account shear displacement and rotary inertia. For high-frequency resonator designs, the design procedure defined by the Timoshenko equation will yield most consistent and accurate results.

When the electromechanical coupling, electrical stiffness, and mechanical stiffness of the beam factors are considered, the net resonant frequency can now be expressed as

$$f_{net} = \left[(F_P) \left(\frac{1}{2\pi} \right) \right] \left[\sqrt{\frac{k_r}{m_r}} \right] \tag{8.6}$$

where
F_m is the frequency modification factor with a typical value between 0.87 and 1.0 at operating frequency not exceeding 70 MHz
k_r is the mechanical stiffness of the resonator
m_r is the resonator mass

8.3.3.2 Operational Requirements and Parameters FFB Micro-Resonator

The resonant frequency is the most important performance parameter. When factors such as electromechanical coupling, electrical stiffness, and mechanical stiffness are taken into account, the resonant frequency is different from the one available in absence of these factors. Furthermore, acoustic network analysis modeling plays a key role in determining the resonant frequency and dimensional parameters of a torsional beam as shown in Figure 8.6. The length of a torsional beam support can be obtained through appropriate acoustic network analysis in terms of quarter-wavelength at the operating frequency. The expression for the support-beam length shown in Figure 8.6 can be written as

$$L_s = \left[\frac{1}{4f_{net}} \left(\frac{\sqrt{GC_t}}{J_s \rho} \right) \right] \tag{8.7}$$

where
L_s is the support-beam length
G is the torsional modulus of the torsional beam
C_t is the torsional constant
J_s is the polar moment of inertia of the beam
ρ is the density

The torsional constant for a resonator with rectangular cross section and when the thickness-to-width ratio is equal to 2, can be written as

$$C_t = \left[0.229 h W_s^3\right] \tag{8.8}$$

where h is the thickness of the beam and W_s is the width of the support. Typical value of this constant is about 0.47 μm^4.

The polar moment of inertia for the support can be defined as

$$J_s = \left[\frac{(h W_s)(h^2 + W_s^2)}{12}\right]$$

Under normal operating conditions, the FFB micro-resonator must be pulled down onto its supporting dimples under the influence of dc bias voltage (V_p). Furthermore, when the dimples are down, the electrode-to-resonator gap spacing (d) is small enough to provide the necessary electromechanical coupling for most applications. To avoid further pull-down of the FFB into the electrode after the dimples are down, the dc bias voltage (V_P) must satisfy the following voltage relationship:

$$V_C > V_P > V_D \tag{8.9}$$

where V_C is the catastrophic voltage and V_D is the voltage when the dimples are down. Typical values of these voltages for the FFB micro-resonator as a function of frequency are shown in Table 8.5.

Table 8.5 Calculated Values of Frequency and Typical Values of Dimple Down Voltage, dc Bias Voltage, and Catastrophic Pull-In Voltage for the FFB Resonator as a Function of Frequency

Voltage (V) and Frequency (MHz)	Frequency (MHz)			
	30	*50*	*70*	*90*
V_D	9.2	25.3	57.4	98.2
V_P	22	86	126	76
V_C	232	521	1024	1262
f (Timoshenko)	30.63	50.83	71.39	90.99
f (Euler–Bernoulli)	31.62	53.51	76.57	99.29

Source: From Wang, K., Wong, A.C. et al., *J. MEMS*, 9, 351, 2000.

It is important to mention that the dc bias requirement increases with the target frequency. However, in the case of a 90 MHz resonator, it is possible that the resonator beam may not have been fully down on its dimples at the dc bias level of 76 V, which is less than the predicted pull-down voltage of 98.2 V.

Numerical samples

1. Using Timoshenko equation (Equation 8.6) and assuming resonator stiffness of 27,425 N/m and resonator mass of 74×10^{-14} kg, one gets

$$
\begin{aligned}
f_0 &= \left[\left(\frac{1.0}{2\pi} \right) \sqrt{\frac{27,425}{74 \times 10^{-14}}} \right] \\
&= \left[0.1594 \times 10^8 \times \sqrt{3.703} \right] \\
&= \left[10^8 \times 0.3067 \right] \\
&= 30.67 \text{ MHz}
\end{aligned}
$$

2. Using Euler–Bernoulli equation (Equation 8.5) and assuming $E = 150$ GPa, $h = 2.05$ μm, $L_r = 23.3$ μm, $\rho = 2800$ kg/m^3 and using appropriate units, one gets the net frequency as

$$
\begin{aligned}
f_{res} &= 1.03 \left[\sqrt{\frac{10^{12} \times 354.3 \times 2.05 \times 10^6 \times 4.45}{2800 \times 2.207}} \right] \\
&= \left(1.03 \times 10^6 \right) \left[\sqrt{\frac{1350 \times 1000}{1388}} \right] \\
&= \left[10^6 \times 1.03 \right] \left[\sqrt{972.6} \right] \\
&= \left[10^6 \times 1.03 \times 31.186 \right] \\
&= 32.12 \text{ MHz}
\end{aligned}
$$

It is important to mention that the resonant frequency calculation using Timoshenko equation (Equation 8.6) provided more accurate results than those obtained using Euler–Bernoulli equation as evident from the data summarized in Table 8.5.

8.3.4   Folded-Beam Comb-Transducer Micro-Resonator

Critical elements, design aspects, and dimensional parameters of this micro-resonator are shown in Figure 8.5. Specific details on the micro-resonator structure and shuttle mass and locations of the anchors and folding trusses are clearly identified in this figure. This particular micro-resonator is best suited for lower VHFs. Quality factors for this micro-resonator as high as 50,000 have been demonstrated at 416 kHz. This Q is reduced to as low as 15,000, if a suboptimal polysilicon material is used in the fabrication of the device. The peak gain of the folded-beam, comb-transducer (FBCT) resonator can be as high as 35 dB at an optimum dc bias voltage of 50 V. Note the resonator gain is dependent on the bias level. Because the Q of the device and the gain are contingent on various conditions, the design requirements and fabrication procedures will not be discussed further.

8.3.5 Clamped-Clamped Beam Micro-Resonator

Important components of this micro-resonator such as the electrode, resonator beam, coupling beam, and the anchors are shown in Figure 8.5b. Critical device parameters include the resonator beam dimensions (length, width, and thickness), electrode width, initial gap, Young's modulus (E), resonator stiffness and mass, bias voltage, and catastrophic voltage. This particular CCB device uses PP capacitive transducers with small gap spacing, which make these resonators to retain high mechanical integrity even under severe operating environments. There are several advantages of PP capacitive transducers with thin gaps. The electrode-to-gap spacing greater than 100 nm is strongly recommended to ensure higher loaded Q with manageable magnitudes of applied voltage. High unloaded Q for this resonator is possible at 70 MHz or so. Note the anchor dissipation is considered a dominant loss mechanism for the CCB resonator with high stiffness at VHFs. In the case of FFB resonators, the use of nonintrusive supports will greatly reduce the loss mechanism problem. The Q of CCB resonator decreases as the operating frequency increases from 50 to 70 MHz, whereas the Q of the FFB remains fairly constant over this frequency range. This suggests that smaller resonator length-to-thickness ratio and larger axial stiffness will lead to larger deformations or the displacements at the anchor supports leading to degradation of overall Q of the resonator. Note the Q of a CCB flexural-mode mechanical resonator is inversely proportional to the square of the displacement at the anchor. Because the bias voltage is less than the pull-down voltage, it violates the validity of the Equation 8.9. In essence, the resonator beam was not down on the dimples and therefore, there is an error in the magnitude of bias at 90 MHz case.

Note the electrode-to-resonator gap spacing (d) determines the effect of the electromechanical coupling on the resonant frequency. The actuation voltage or the bias level generates both the mechanical and electrical spring stiffness factors. Note the overall resonator spring stiffness (k_r) is comprised of two spring stiffness factors, namely, the electrical spring stiffness factor and mechanical spring stiffness factor. However, the bias-induced electrical spring stiffness factor (k_e) generates minor shift not exceeding 0.3 percent in the resonant frequency of VHF flexural-mode MM resonators operating at 50–70 MHz frequencies. The large mechanical spring stiffness factor (k_m) of the MM resonator contributes larger shifts in the resonant frequencies rather than electrode-to-resonator gap spacing.

The impact of frequency modification factor on the resonant frequency of the resonator is very small or negligible, when the modification factor is equal to unity. However, if the modification factor is less than unity, large errors could appear in resonator's resonant frequency, regardless of the electrode-to-resonator gap spacing. At lower modification factors, the resonator surface topography and the finite anchor elasticity greatly affect the resonant frequencies of the CCB resonators. However, in the case of FFB resonators, both the surface topography and the anchor elasticity have no effect on resonant frequencies.

8.3.5.1 Effects of Environmental Factors on Micro-Resonator Performance

Because of involvement of several materials in the fabrication of the MM resonators, one must investigate the impact of temperature on the resonant frequency of the resonator. The structure-to-substrate thermal expansion mismatches can plague the performance of the CCB resonators. Preliminary studies performed by the author indicate that thermal performance of an FFB (Figure 8.6) resonator (13 ppm/°C) is better than CCB (Figure 8.7) resonator (17 ppm/°C) in terms of temperature coefficients [3]. Deployment of integrated owen-control technique has reduced the temperature coefficient of a capacitive-comb transduced micro-resonator less

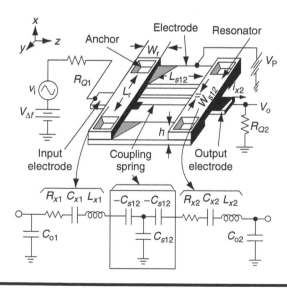

Figure 8.7 Schematic of a 2-resonator MM filter showing the equivalent circuit, bias network, and sensing circuit.

than 2 ppm/°C, but at the cost of a more complex micromachining process. Note the degree of improvement in thermal performance of FFB resonator is not large. This could be because that the mechanical spring stiffness of high-frequency resonators is so large close to 60,000 and the resonator length is so small. These facts indicate that the thermal stability of these MM resonators is very impressive even at operating temperatures close to 400 K. In other words, the changes in spring stiffness factor due to thermal expansion stresses are insignificant, thereby indicating very little effect on the thermal stability of the resonant frequency (f_0).

The effects of contaminants are unknown at the present. However, contaminants that absorb onto the micro-resonator surfaces at low temperatures are either burned-off or evaporated-off the resonator surfaces as the operating temperature increases. This could remove the excess mass of the resonator, which would raise the resonant frequency of the resonator in the beginning. As soon as the contaminants are removed, one can see that the resonant frequency decreases with the temperature due to negative temperature coefficient in Young's modulus of the resonator structure. The studies suggest that the vacuum encapsulation at the wafer or package level must be seriously considered to alleviate the adverse effects of environmental contaminants including the outgassing from the inserted electronic circuit boards.

8.3.5.2 Performance Summary of Various Micromechanical Resonators

Generation of spurious modes can degrade the performance of the MM resonators. In the case of an FFB micro-resonator (Figure 8.6), any attempt to maximize the resonator Q would require complex design, which may lead to generation of spurious modes. The entire resonator and the support-beam structure can vibrate in a direction perpendicular to the substrate, if the dimples are not held rigidly to the substrate by the application of the dc bias voltage V_P as shown in Figure 8.6. Spurious mode generated under these conditions, if not suppressed, can interfere with the performance of a filter or oscillator using this particular micro-resonator design. Sometimes, support material modifications can alleviate the mode generation problem. Measurements made on a 55 MHz FFB MM resonator revealed no spurious responses, except a 75 dB down spurious mode at 1.7 MHZ, which is sufficiently far from the operational frequency of 55 MHz. Shielding measures at both the board and substrate levels must be planned during the design phase to alleviate the parasitic feedthrough problem.

Finite widths of the dimples can rub the substrate during the resonance vibration. The Q of the resonator can be affected depending on how the dimples are held down on the substrate. As stated earlier, any change in resonator's Q will definitely affect the resonator performance. It is critical to mention that the Q of the capacitive transduced MM resonators strictly depends on the applied dc bias voltage (V_P) regardless of the presence or the several dimples deployed. Deployment of quarter-wavelength torsional supports attached at node points and optimum selection of electrode-to-resonator gap in the FFBMM resonator design will eliminate the anchor dissipation and processing problems, which have plagued the CCB resonator design.

The present MM resonator designs yield Q's in excess of 8500, which is more than sufficient for most IF filters widely used in the cellular and cordless communications transceivers. The high stiffness capability of these MM resonators is necessary to maintain adequate dynamic range required in both the oscillator and filtering applications.

8.4   Micromechanical Filters

Pilot tone filtering using second-order single-pole band-pass filter centered at 16.5 kHz is used in mobile phones. However, the second-order filter characteristics are not adequate for majority of communications applications. Note band-pass filters with flat passband response, sharp skirt selectivity, and high stopband rejections are best suited for above applications and such performance capability is possible from MM filters.

8.4.1   Micromechanical Filter Design Aspects

Many MM resonators must be linked together by soft-coupling springs as illustrated in Figure 8.7 to achieve the above-mentioned performance characteristics of the filters. The soft-coupling spring can be designed using ideal mass-spring-damper elements. Linking resonators through mechanical springs can lead to development of a coupled resonator capable of displaying several modes of vibration. It is important to point out that in the lowest frequency mode, all micro-resonators vibrate in phase.

Simultaneous increase in resonator spring constant or stiffness factor (k_r) and resonator mass (m_r) are required to maintain the desired center frequency of the filter. However, simultaneously scaling of both the resonator stiffness and mass are needed to maximize the filter bandwidth without drastically altering the resonator dimensions. The effective dynamic stiffness and mass of a given resonator are strictly dependent on the locations on the resonator. Note that different locations on a vibrating resonator move with different velocities, which will make the dynamic mass and stiffness of the resonator as strong functions of the velocity. Note the magnitude of the dynamic resonator stiffness and mass as seen by coupling beam is strictly dependent on the coupling location. The fundamental-mode folded-beam resonators coupled at their shuttle masses present the minimum stiffness to the coupling beam, because of the highest velocity at the shuttle masses. However, the same resonators coupled at locations to their anchors as shown in Figure 8.5, where the velocities are much smaller, present very large dynamic stiffness to their respective coupling beams. But this will allow filters with much smaller bandwidths for the same coupling beam stiffness.

Deployment of folded-beam micro-resonators in filters provides low-velocity coupling without major changes in the design of resonators, while retaining coupling at the resonator folding trusses. This particular filter configuration permits variations in the resonance velocity magnitude of the folding truss as a function of ratio of the outer beam length to inner beam length. The resonator velocity with folded truss can be written in terms of filter center frequency (f_0), displacement of the shuttle mass and parameter β, which is defined as the ratio of the outer beam length to the inner beam length. The effective dynamic stiffness (k_{rt}) and effective dynamic mass (m_{rt}) at the resonator folding trusses [3] can be written as

$$k_{rt} = (k_{rs})\left[1 + \beta^3\right]^2 \tag{8.10}$$

$$m_{rt} = (m_{rs})\left[1 + \beta^3\right]^2 \tag{8.11}$$

where k_{rs} and m_{rs} are the effective dynamic stiffness and mass, respectively, at the resonator shuttle, where the maximum resonance velocity occurs. The filter center frequency can be expressed as

$$f_0 = \left[(0.1594)\sqrt{\frac{k_{rs}}{m_{rs}}}\right] \tag{8.12}$$

Assuming a value of 1258 N/m for the effective dynamic stiffness and 2.48×10^{-10} kg for the effective dynamic mass, one gets the filter center frequency as

$$f_0 = \left[(0.1594)\sqrt{\frac{1258}{2.47 \times 10^{-10}}}\right]$$

$$= \left[0.1594 \times 10^5 \times 22.568\right] = 359.7 \text{ kHz}$$

$$= 360 \text{ kHz (approximately)}$$

The plot showing the normalized effective stiffness at the folding truss versus a function of parameter β (Figure 8.8) indicates a six orders of magnitude variation, seven orders of magnitude variation, and eight orders of magnitude variation in stiffness, when the parameter β has a value of 10, 15, and 20, respectively. With a coupling beam width of 2 μm in a 360 kHz filter, these stiffness variations correspond to range of percentage bandwidths from 0.7 to 3×10^{-6} percent to 1.2×10^{-7} to 7.5×10^{-7} percent. To meet the filter passband ripple requirement, the Q of the end resonators must be controlled to specific values dictated by filter synthesis techniques.

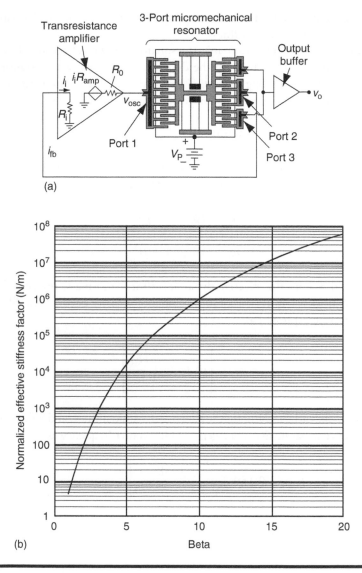

(a)

(b)

Figure 8.8 **(a) Schematic of a 16.5 kHz oscillator and (b) normalized effective stiffness factor as a function of folded-beam length ratio (Beta).**

8.4.2 Critical Elements and Performance Parameters of Micromechanical Filters

The critical elements of an MM resonator filter are the comb-transducer, anchor, folded-beam, coupling beam, and folding truss. The resonators for the filter must be designed such that their folding truss resonance velocities are the velocity at the shuttle when the parameter β has a value of 1.53. It is important to mention that the shuttle moves faster than any other location on the resonator during the resonance. The maximum velocity occurs at the shuttle location point and the coupling at the folding trusses in this particular filter at 7/32 (or 0.22 percent) of the maximum coupling velocity and which occurs when β is equal to 1.53. Note 50 percent of the maximum coupling velocity occurs when β is equal to unity. Note the filter with 50 percent of the maximum coupling velocity requires coupling beam widths of 1 μm, which will yield larger band width of 760 Hz with filter overall Q of 474. If the coupling beam width is increased to 2 μm, the band width is reduced to 401 Hz with a Q of 898 using the formula, Q_{filter} = filter center frequency/bandwidth. This indicates that the wider coupling beams of the lower velocity-coupled filter are less susceptible to fabrication process variations than are the thinner beams of higher velocity-coupled ones. In addition, low-velocity designs are best suited in achieving smaller percentage bandwidths, improved accuracy, and overall better filter performance. A three-stage MM band-pass filter with a Q of 898 demonstrated an IL less than 0.6 dB and stopband rejection better than 66 dB. Such performance is only possible with macroscopic high-Q crystal filters.

8.4.3 Performance Summary of a Two-Resonator High-Frequency Filter

High-frequency filters require resonators with much smaller mass. As a result, filters using folded-beam resonators are not appropriate for high-frequency filter applications. Rather CCB resonators are best suited for high-frequency filters. Furthermore, higher actuation voltages are required to achieve needed electromechanical coupling via comb-transducers. Efficient transducers are needed to achieve more practical operating voltages and higher resonator Q's. This high-frequency two-resonator filter consists of two MM CCB resonators as illustrated in Figure 8.7. The resonators are coupled mechanically by a soft spring and are suspended at 100–200 nm above the substrate. Conductive strips are attached under each resonator, which serves as a capacitive transducer electrodes to induce resonator vibrations in the direction perpendicular to the substrate surface. The MM resonator-to-electrode gaps are determined by the thickness of the sacrificial layer, which must be quite small (e.g., 100 nm or so) to maximize the electromechanical coupling.

Under normal operation, the filter is excited capacitively by a signal voltage applied to the input electrode. The output signal is taken at the other end of the filter structure. As soon as the input is supplied with a suitable frequency, the resonators begin to vibrate in one or more flexural modes in a direction perpendicular to the substrate. If the excitation voltage has a frequency within the passband of the filter, both resonators will vibrate. The vibration of the output resonator couples to the output electrode, providing an output current. The output current and the resistor in the output circuit provide the termination impedance for the MM filter. The resonance frequency of the filter can be expressed as

$$f_0 = \left[(0.1594) \sqrt{\frac{k_r}{m_r}} \right] \tag{8.13}$$

where k_r and m_r are the resonator stiffness and mass, respectively. Assuming the resonator stiffness of 82,000 N/m and resonator mass of 2.5×10^{-13} kg, the computed value of resonance frequency of the filter comes to

$$f_0 = \left[(0.1594) \sqrt{\frac{82,000}{2.5 \times 10^{-13}}} \right]$$
$$= \left[0.1594 \times 10^8 \times 5.727 \right]$$
$$= 91.29 \text{ MHz}$$

Note the loss to the substrate through anchors becomes more important with the increasing frequency of the filter. This is because that the stiffness of the resonator beam increases with the frequency, leading to larger forces exerted by the beam on its anchors during the vibration of the resonator. The anchor-related loss mechanisms can be significantly reduced by using anchorless resonator design. Such filter design employs an FFB resonator suspended by four torsional supports attached at the flexural nodal points. Deployment of quarter-wave support dimensions will null out the impedance presented to the beam by the supports, thereby leaving the beam virtually levitated and free to vibrate. This resonator can use an electrode-to-resonator gap spacing of 0.2 μm to limit the large applied bias voltage needed for sufficient output voltage. This type of device can provide an unloaded Q of the resonator close to 8000 for a micro-resonator operating at the center frequency around 71 MHz.

The dynamic range in the passband region of an MM filter is strictly dependent on the nonlinearity of the electromechanical transducers and the noise generated by the terminal resistors. The dynamic range of the filter is defined by the ratio of the maximum input power to the minimum detectable signal level, which depends on the stiffness location on the beam. Note in the case of a CCB filter, PP capacitively transduced resonators are used. For this filter, the dynamic range is proportional to the square of the electrode-to-resonator gap spacing (d), but proportional to the resonator stiffness. Pull-in must be avoided at all costs, if the gap spacing is extremely small. Trade-off studies must be undertaken to select optimum value of gap spacing to avoid pull-in, while retaining high dynamic range. Use of more linear and efficient transducers is strongly recommended to enhance the dynamic range of the filters.

In summary, high-Q filters and oscillators using MM resonators have demonstrated several advantages such as compact and small, high skirt selectivity, fast switching capability, high stiffness needed for wide dynamic range, and virtually zero-power consumptions. These benefits will significantly relax the power requirements of the associated for the components such as low-noise amplifiers (LNAs), mixers, and A/D converters, which are widely used in the transceiver stages. In addition, the IC-compatible MM resonators would be able to achieve vibrational frequencies well into gigahertz range soon. More research and development activities must be directed toward improvement in frequency-dependent loss mechanisms, electromechanical coupling, and matching tolerances, which will ultimately affect the overall performance of the filters and oscillators.

8.5 Transceivers

8.5.1 Introduction

Transceiver is the most critical component of the communications systems. The front end of a communications transceiver or wireless transceiver is comprised of several off-chip high-Q components and circuits that are potentially replaceable by MM versions. The components recommended for replacement include RF bandpass filters, image rejection filters with center frequencies ranging from 800 to 2500 MHz, IF filters with center frequencies ranging from 450 kHz to 255 MHz and high-Q, low phase noise LOs operating from 10 to 2500 MHz spectral region. Limited studies performed by the author indicate that the superheterodyne receiver will benefit the most from the components using MM resonator technology.

8.5.2 Transceiver Performance Improvement from Integration of Micromechanical Resonator Technology

Miniaturization is the biggest advantage for the transceiver form integration of MM resonator technology. Integration of MEMS technology into resonator will make a significant impact at the system level, by offering potential transceiver architectures capable of providing substantial reduction in power consumption and remarkable improvement in the overall performance of the transceiver. Higher level integration transistors or latest solid-state devices and alternative architectures will not only reduce cost, size, and power consumption but also eliminate the need for the off-chip high-Q passive devices currently used by superheterodyne transceivers.

Spurious-free dynamic range (SFDR) of the transceiver can be affected by the presence of nonlinearity present in various components such as LNA, mixer, and A/D converter. In addition, the phase noise in the LO and electronic interference signals can desensitize the receiver by generating third-order intermodulation (IM3) distortion most likely to appear at the outputs of these components. These two problems created as a result of integration of micro-resonator technology must be carefully addressed before developing device using these resonators.

8.6 Oscillator Using Micromechanical Resonator Technology

Deployment of a folded-beam comb-transduced MM resonator fully integrated with CMOS electronics would maximize the frequency stability of the oscillator against supply voltage fluctuations. System-level schematic diagram of a 16.5 kHz oscillator using folded-beam comb-transduced MM resonator technology is illustrated in Figure 8.8. Critical components of this oscillator including the transresistance amplifier, 3-port MM resonator, and output buffer amplifier are clearly identified in this figure. The MM resonator consists of a finger-supporting shuttle mass, which is suspended around 2 μm above the polysilicon substrate by folded flexures. The folded flexures are anchored to the substrate at two central points to provide needed mechanical support. The shuttle mass is free to move around in the direction parallel to the plane of the polysilicon substrate. The fundamental resonance frequency is largely determined by the material properties and geometrical parameters. The resonance frequency is proportional to the square root of the ratio of effective stiffness (k_{rs}) of the resonator at the shutter location to effective mass of the shuttle (m_{rs}).

Assuming a value of 0.65 N/m for the effective stiffness of the resonator at the shutter and a value of 5.73×10^{-11} kg for the effective mass resonator at the shuttle [3], one gets the computed value of the resonance frequency as

$$f_0 = \left[(0.159)\sqrt{\frac{0.65}{5.73 \times 10^{-11}}} \right]$$

$$= \left[(0.159) \times 10^5 \sqrt{\frac{6.50}{5.73}} \right]$$

$$= [(0.159) \times 1.06507]$$

$$= 16.92 \text{ kHz}.$$

This computed value is not too far from the design value of 16.50 kHz. The difference is due to the assumed values of the parameters involved.

8.6.1 Design Concepts and Performance Parameters of the 16.5 kHz Oscillator

As stated earlier, the design resonance frequency is 16.5 kHz. A folded-beam comb-transduced MM resonator is selected for the design of this oscillator due to its high-frequency stability against supply voltage fluctuations. It is important to mention that the MM resonator can be switched in and out just by application and removal of the dc bias voltage. Such switching capability can be important in transceiver architectures. The two ports are embedded in a positive feedback loop in series with a sustaining transresistance amplifier, while the third port is directed to an output buffer amplifier. Note the use of the third port essentially isolates the sustaining feedback loop from the variations in the output impedance loading. The folded-beam length and width are roughly 185 and 2 μm, respectively. The unloaded Q of the resonator is close to 50,000 in vacuum at a dc bias voltage of 20 V. The total area occupied by the 16.5 kHz prototype oscillator [3] is 420×335 μm^2, approximately. However, a dc bias voltage about 30–35 V may be required to obtain the equivalent-series resistance between the ports 1 and 2. Phase noise measurements on this 16.5 kHz oscillator are difficult to obtain and thus, are not readily available. But the white noise of the oscillator can be expected to be around in the order of -70 dBc at large offsets from the carrier frequency. One can be sure that the high Q of these resonators will significantly reduce the close-to-carrier phase noise of such an oscillator. On the basis of these parameters, higher frequency oscillators will require micro-resonators with much less mass and will occupy even much smaller area, particularly at frequencies exceeding 100 MHz or so. No meaningful thermal stability data is available on such oscillators.

The polysilicon micromachining technology can play a key role in fabrication of such oscillators. Integration of CMOS technology with micromachining technology will significantly reduce the weight, size, and the power consumption in these oscillators.

8.7 V-Band MEMS-Based Tunable Band-Pass Filters

8.7.1 Introduction

Low-loss, compact tunable filters using MEMS varactors or micromachined varactors are best suited for highly integrated multifrequency/multiband communications systems, where secured and clear communication is of paramount importance [4]. MEMS-based band-pass filters have demonstrated wideband tuning capability at microwave and mm-wave frequencies including both the digital and analog versions. Note only discrete center frequencies are possible with digital type filters, whereas continuous frequency tuning is possible with analog version. It is important to point out that large frequency range can be achieved from the digital versions of MEMS filters, which deploy MEMS switches to turn on or off the resonant components to change the center frequency. A center frequency variation exceeding 50 percent is possible by switching the resonating components operating over 15–30 GHz microwave region.

In the case of analog tunable filters, the center frequency of the filter is varied by changing the gap of the cantilever-based metal–air–metal (MAM) capacitors using ES force provided by the application of appropriate actuation voltage. However, partial deflection of the cantilever beam and nonlinearity generated at pull-in can limit the frequency tuning range of the filter to less than 5 percent with high IL close to 4 dB at 25 GHz. This high IL in the analog filter is contributed by the radiation loss due to weak magnetic coupling between the microwave spiral inductors, dielectric loss in the substrate, and the conductor loss. Note the deployment of quartz substrate and low-loss capacitive coupling using MIM capacitors analog with lumped elements will reduce the radiation losses at high microwave and mm-wave frequencies.

8.7.2 Design Parameters and Fabrications Techniques for a V-Band MEMS-Filter

The tuning of the mm-wave MEMS-based tunable filters can be achieved using the MIM tuning capacitors and by changing the gap between the MIM elements. Two types of topologies are widely used, namely, the distributed-element type using CPW transmission-line sections and lumped-element type using elements as resonators. Variable MIM capacitors attached at the end of CPW transmission-line segments and fixed spiral type inductors as lumped elements can be used in mm-wave tunable filters. Note distributed type design offers smaller tuning range compared with the lumped-type filter architecture. Maximum tuning capability for the lumped-type tunable filters is estimated close to 14 percent because of maximum range control associated with MIM capacitors. Specific details on MIM capacitors and CPW transmission-line parameters and characteristics are discussed in great detail in Chapter 5.

In the case of V-band tunable filter design and fabrication, both the lumped-type resonators and MIM capacitors using bridge-type beams are recommended to achieve wider tuning range for the filter. Note bridge-type beams provide more stability, less sensitivity to bending, and reduced IL and therefore, are best suited for mm-wave tunable filter applications. The lumped inductors and capacitors can be implemented using high-impedance CPW transmission-line sections and MIM capacitors, respectively. The dimensions of the lumped elements in the resonators and the coupling capacitors must be optimized using commercial electromagnetic software programs to meet the specific filter characteristics.

The mm-wave filters must be fabricated with electroplated gold structures on a quartz substrate ($e_r = 3.78$) of optimum thickness to reduce radiation and ILs at V-band frequencies. The high-impedance (100 Ω or so) CPW transmission-line sections of 3 μm thickness and bridge thickness of 2 μm have been selected to keep the IL low and the bridge mechanical integrity high. A 300 nm thick silicon nitride insulation layer between the CPW ground plane and the bridges is used to avoid dc short circuit. The air gap must be kept around 1 μm to keep the actuation voltage to minimum. The top plates size of MEMS capacitors for the 60 GHz, two-stage MEMS tunable filter are 60×100 μm^2. The inductance of the MEMS bridge can be provided by two narrow parallel beams and the dc bias is applied between the movable bridge capacitance and the top CPW ground plate. The bias should be increased to accommodate the pull-in voltage effects of the bridge. The total inductance of the two-cascaded spiral inductor is 0.7 nH, which provides a resonance frequency greater than 88 GHz.

8.7.3 Performance Parameters of a V-Band, Two-Stage MEMS Tunable Filter

As stated earlier the resonance frequency of this filter in excess of 88 GHz, which will be able to provide satisfactory performance over the entire V-band frequencies (50–75 GHz). This particular V-band filter is comprised of two stages: one with center frequency of 52 GHz and other with a center frequency of 66 GHz. The dc bias between 22 and 25 V is adequate to obtain continuous tuning of the filter over the 50–75 GHz range. The total IL is well below 3.4 dB and the return loss is in excess of −12 dB over the entire V-band. It is important to mention that the combined effects of the lumped-element topology using MIM capacitors and the capacitive coupling between the resonators provided improved frequency tuning range with minimum IL. This low-loss and compact MEMS tunable filter can be used in highly integrated multiband/multifrequency mm-wave communications systems.

8.8 MEMS-Based Strain Sensors

8.8.1 Introduction

MEMS-based strain sensors can be used to monitor the structural health of the bridges, steam-turboalternators, jet engines, gas turbines, large commercial structures, and underground tunnels. Typically, the strain sensors are bonded to the specimen with a polymetric adhesive at appropriate locations on the structure, where maximum strain levels are expected. The strain coupling between the stiff wafer and the surrounding material is extremely poor. However, the strain coupling for a polymer-backed strain gage is excellent [5]. The stiffness of the polymer backing and the adhesive are nearly identical, which make the near-field and far-field strain gage reading identical. The sensor performance is estimated from the gauge factor of the strain sensor.

8.8.2 Design Aspects and Requirements for Strain Sensor Installation and Calibration

Proper installation, calibration, and sensor geometry requirements must be satisfied to obtain accurate and reliable data from the strain sensors. These requirements may change depending on the materials used for the strain-sensing elements and epoxy layers. Performance of the sensor is depended on the mismatch between the sensor stiffness and surrounding "g" material, because these factors affect the transfer of forces and displacements in various directions of the structure. Two distinct strain sensor designs are available, namely, a monofilament sensor in which a narrow *n*-polysilicon element is patterned on a silicon wafer and a membrane sensor in which the *n*-polysilicon element is patterned on a silicon nitride/silicon oxide membrane. Specific details on the cross section of the strain sensor and geometrical parameters on the sensor and epoxy are shown in Figure 8.9. Similar to a metallic foil strain gauge, the strain generated is transferred to the MEMS-based strain sensor through an adhesive bonding layer. Because the silicon wafer has high modulus of elasticity compared to the surrounding adhesive layer, the strain transfer is very poor. Typically, the silicon wafer is about 500 μm thick and has a modulus of elasticity of 165 GPa, which is bonded with a 200 μm thick epoxy layer with a modulus of

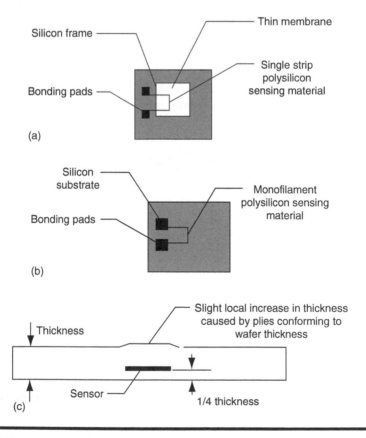

Figure 8.9 MEMS strain sensor views. (a) Top view of a single strain sensor, (b) top view of a monofilament strain sensor, and (c) location of a MEMS strain sensor. (From Hautamaki, C., Cao, L. et al., *J. MEMS*, 12, 720, 2003.)

elasticity of 4 GPa. When a strain level of 800 μm is applied to an aluminum specimen, the strain at the surface of the silicon wafer (where the polysilicon piezoresistive material is patterned) is only −16 μm. It is critical to mention that high shear occurs at the edges of the epoxy. Shear strain is zero at the middle of the glue line.

In composite materials, strain is transferred between the fiber and the surrounding polymer medium or matrix. When the strain in the stiff fiber is equal to the strain in the composite material, the load transfer length (L_{trans}) for the cylindrical fibers can be expressed as

$$L_{trans} = \left[\frac{dE_f\sigma_f}{2E_c\sigma_s}\right] \tag{8.14}$$

where
E_f is the fiber modulus
E_c is the composite modulus
σ_f is the stress in the fiber of diameter d
σ_s is the shear stress in the longitudinal direction

For a fiber with rectangular cross section, the transfer length can be written as

$$L_{trans} = \left[\frac{wt}{(w+t)}\right]\left[\left(\frac{E_{sen}}{E_c}\right)\left(\frac{\sigma_{sen}}{\sigma_s}\right)\right] \tag{8.15}$$

where
w is the width of the sensor
t is the thickness of the sensor
E_{sen} is the sensor modulus

The stress in the polysilicon wafer is dependent on the sensor geometry and combined moduli of the wafer and the polymer matrix, which can be written as

$$\sigma_{wafer} = \left[(G_f\sigma_s)\left(\frac{E_c}{E_{sen}}\right)\right] \tag{8.16}$$

where G_f is the geometry factor of the sensor. Using above equations, the geometry factor can be rewritten as

$$G_f = \left[\left(\frac{w+t}{wt}\right)L_{trans}\right] \tag{8.17}$$

Close examination of these equations will indicate that as the geometric factor increases, the longitudinal stress and the corresponding strain in the wafer increase.

These equations further indicate that the modulus of the surrounding material will impact the load transfer to the strain sensor.

The monofilament strain sensors can be mounted on the aluminum specimen and embedded in the composite. Typical dimensions of the aluminum specimen are $250 \times 13 \times 6.4$ mm³ (length, width, and thickness). Typical dimensions for the composite specimen consisting of woven glass fiber reinforced with plastic are $300 \times 64 \times 6$ mm³. The membrane sensors are embedded in the vinyl material. Typical dimensions of the vinyl specimen or sample are $160 \times 30 \times 6$ mm³.

8.8.3 Gauge Factor Computation

As stated earlier that the strain sensors are fabricated from the piezoresistive materials such as silicon or zinc oxide or aluminum nitride. However, silicon is widely used because of its unique properties and availability at minimum cost. There is a fundamental relationship between the resistance change and the mechanical strain developed in the piezoresistive materials. It is important to mention that the resistance change is contingent on the changes in the geometry of the strain gauge. The gauge factor for a metal foil strain gauge can be expressed as

$$K_{\text{foil}} = \left[1 + 2\mu + \left(\frac{d\rho}{\rho} \right)(\varepsilon_1) \right] \tag{8.18}$$

where
  μ is Poisson's ratio
  $d\rho$ is the change in the resistivity
  ε_1 is the strain in the longitudinal direction

Typically, the gauge factor for the metallic foil gauges is 2, but for the semiconductor gauges such as silicon gauges, the gauge factor is in excess of 160 due to change in resistivity under piezoelectric effect.

The piezoresistive coefficients of n-type piezoresistive materials in the (110) direction have significant effects on the gauge factor of the strain sensor. The longitudinal piezoresistive coefficient (C_l) and the transverse piezoresistive coefficient (C_t) in the (110) direction play a key role in the optimization of the strain sensor gauge factor. These coefficients can be written as

$$C_l = [d_{11} + 0.4(d_{44} + d_{12} - d_{11})] \tag{8.19a}$$

$$C_t = [d_{12} + 0.2(d_{11} + d_{12} - d_{44})] \tag{8.19b}$$

where d_{ij} and d_{ij} are the directional coefficients for the n-type polysilicon material in both the longitudinal and transverse directions. Assuming that the stresses are constant over the resistor R, the resistance change can be written as

$$\frac{\Delta R}{R} = [\sigma_t C_t + \sigma_l C_l] \tag{8.20}$$

Note the transverse stress can be written as

$$\sigma_t = [\mu \sigma_l] \tag{8.21a}$$

For linear elastic materials, one can write as

$$\sigma_l = [E\varepsilon_l] \tag{8.21b}$$

Using above equations, the longitudinal strain due to resistance change can now be written as

$$\varepsilon_l = \left[\frac{\Delta R}{R}\right]\left[\frac{1}{E(C_l + \mu C_t)}\right] \tag{8.22}$$

Using above equations, one can now write the expression for the gauge factor in the polysilicon medium as

$$K_{poly} = [E(C_l + \mu C_t)] \tag{8.23}$$

Because the term μC_t is extremely small, it can be assumed to be zero. The calculated value of C_l for the polysilicon is equal to 45.5×10^{-11} Pa^{-1} and using Young's modulus (E) value of 120 GPa, the gauge factor for the polysilicon material comes to

$$
\begin{aligned}
K_{poly} &= \left[120 \times 10^9 \times 45.5 \times 10^{-11}\right] \\
&= [1.2 \times 45.5] \\
&= [54.6] \ (\text{assuming 120 GPa for parameter } E) \\
&= [56.8] \ (\text{assuming 125 GPa for parameter } E) \\
&= [59.2] \ (\text{assuming 130 GPa for parameter } E)
\end{aligned}
$$

Note the modulus of elasticity (E) for polysilicon varies from 120 to 130 GPa. As mentioned earlier, the transfer rate will increase linearly based on the shear lag theory. The monofilament strain sensors with varying surface areas and thicknesses must be fabricated to test the validity of this theory. Typical thickness ranges from 0.25 to 0.50 mm and the surface area ranges from 2×2 to 10×10 mm^2. The highest geometry factor of 40 is available with large cross-sectional area. The gauge factor of the MEMS strain sensor can be defined as

$$K_{MEMS} = \left[\left(\frac{\Delta R}{R}\right)\left(\frac{\varepsilon_f}{\varepsilon_n}\right)\right] \tag{8.24}$$

where ($\varepsilon_f/\varepsilon_n$) is the ratio of far-field to near-field strain.

8.9 MEMS Interferometric Accelerometers

8.9.1 Introduction

Accelerometers with resolving capability in the nanogram range are best suited for specific applications such as for measuring seismic disturbances and gravitational waves. For a nanogram accelerometer, a low resolution frequency, proof mass, and a displacement sensor with subangstrom resolutions are the principal requirements. Conventional high-resolution tunneling with proof mass ranging from 10 to 50 mg can provide a resolution within 10 Å of the reference tunneling dip. Such strict tolerances will lead to crashing of the proof mass into the dip during large accelerations. It is critical to point out that the tunneling accelerometers operating in force-feedback mode will control a deflection voltage needed to keep the proof mass stationary, thereby eliminating the crashing problem [6]. Researchers have recently fabricated and tested an accelerometer using electron-tunneling transducers with closed loop configuration, which offers a resolution of 30 ng between 7 and 1100 Hz and over 90 dB dynamic range.

8.9.2 Design Aspects and Requirements for an Interferometric Accelerometer

Reliable processes are required for fabrication of proof masses capable of providing resonant frequencies as low as 60 Hz. An open loop accelerometer design demonstrated excellent performance in terms of sensitivity, noise, and linearity. Linearity of motion normal to the plane of the device is of critical importance, if one wants to operate an accelerometer without the use of linearization circuit. This requires the suspension of the proof mass with the diagonal folded pinwheel springs, which provides linear motion normal to the plane of the accelerometer as shown in Figure 8.10. Note the symmetry of the pinwheel springs allows the mass to rotate slightly, making it less sensitive to external strains.

Summary of critical elements of an accelerometer

- Laser source
- Light detector
- Proof mass
- PIN-wheel spring
- Silicon die
- Interdigitated finger
- Reference tunneling dip

(a)

Figure 8.10 Folded-pinwheel accelerator. (a) Critical elements of a folded-pinwheel interferometric accelerator and

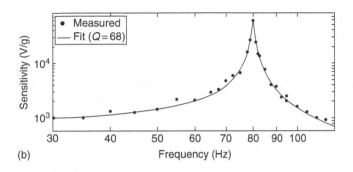

(b)

Figure 8.10 (continued) **(b) sensitivity of the device as a function of resonance frequency. (From Anon., *J. MEMS*, 11, 182, 2002.)**

The accelerometer device package consists [6] of four acrylic segments. The bottom section holds the proof mass die. The top segment holds a silicon die incorporating a p–n junction diode. A laser diode with an output of 5 mW is attached to the top segment with a lens that focuses the light onto the fingers. The proof mass of this accelerometer can contain 50 or more $175 \times 20 \times 6 \ \mu m^3$ (length $\times$ width $\times$ thickness) interdigitated fingers with a gap of 3 μm between the two fingers. The dimensions must be selected to eliminate breakage and stiction problems. The accelerometer package requires five electrical leads to maintain continuous operation of the device: three for the laser diode and two for the photodiode. A constant power driver is required to drive the solid-state laser current amplifier to the output current from the photodiode.

The sensitivity of such device for a given Q of 68 is proportional to the resonant frequency as illustrated in Figure 8.10. This means higher device sensitivity is only possible at lower resonant frequencies. Typical sensitivity of the accelerometer sensor varies from 250 times less along the fingers to 50 times less across the fingers than the sensitivity to the accelerations normal to the plane of the sensor. Note the sensor sensitivity (V/g) can be improved at lower resonance frequencies, if an optimum Q is selected for the sensor. Accurate and reliable test data on accelerometer resolution is not readily available. The accelerometer sensor resolution is strictly limited by the thermal noise of the proof mass, because the position noise of the accelerometer is less than the thermal displacement noise. Thus, the expression of the thermal noise of the proof mass can be written as

$$N_{TM} = \left[\frac{8\pi k_B T f_0}{mQ} \right]^{0.5} \tag{8.25}$$

where
  f_0 is the resonance frequency
  T is the operating temperature of the sensor (K)

k_B is the Boltzmann constant (1.38×10^{-23} J/K)
m is the proof mass
Q is the quality factor the device

Assuming a Q of 68, resonance frequency of 80 Hz, temperature of 300 K, proof mass of 30 mg and inserting these values in Equation 8.25, one gets

$$N_{TM} = \left[\frac{8 \times \pi \times 1.38 \times 10^{-23} \times 300 \times 80 \times 1000}{30 \times 68} \right]^{0.5}$$

$$= \left[25.13 \times 13.8 \times 80 \times 14.706 \times 10^{-23} \right]^{0.5}$$

$$= \left[51.00 \times 8 \times 10^{-20} \right]^{0.5}$$

$$= \left[408 \times 10^{-20} \right]^{0.5}$$

$$= 20.19 \times 10^{-10}$$

$$= 2.019 \times 10^{-9}$$

$$= 2.02 \text{ ng/Hz}$$

Computed values of thermal noise of the accelerometer sensor as a function of resonance frequency and proof mass are summarized in Table 8.6.

Note that to resolve the thermal noise, the proof mass displacement must be grater than the noise of the position sensor. Assuming the displacement of the proof mass as the quotient of the acceleration and the square of the resonance frequency, the thermal displacement noise of the MEMS sensor at an assumed resonance frequency of 80 Hz comes to about 3×10^{-7} μm/Hz. The noise of the interdigital position sensor will be dominated by the laser wavelength variations, which will have

Table 8.6   Sensor's Thermal Noise as a Function of Frequency and Mass (ng/Hz)

Resonance Frequency (Hz)	Proof Mass, m (mg)			
	20	30	40	50
100	2.55	2.08	1.80	1.61
80	2.28	1.86	1.61	1.44
60	1.97	1.61	1.39	1.24
40	1.61	1.32	1.14	1.02

Note: These calculations assume a Q of 80 and a temperature T of 300 K.

a ($1/f$) characteristic. However, this noise will typically be ten times greater than the photodetector shot noise. Because position noise of the accelerometer is less than the thermal displacement noise, one can conclude that the resolution of the sensor will be limited by the thermal noise.

8.10 MEMS-Based Micro-Heat Pipes

8.10.1 Introduction

The heat generated in the modern integrated circuits and semiconductor chips will result in high temperatures in these circuits and chips. These high temperatures can degrade the performance ultimately leading to device failure, caused by the excessive thermal stresses and fatigue, if not removed immediately from the circuits and chip surfaces. MEMS-based micro-heat pipes (MMHPs) will provide effective and immediate heat removal, leading to rapid reduction in the operating temperature. It is important to mention that the micro-heat pipes are best suited for achieving high local heat removal rates and uniform temperatures in computer chips and microcircuits. As the heat enters the heat pipe, a fraction of the working fluid is vaporized resulting in a higher vapor pressure. The pressure gradient allows the vapor to flow through the lower-temperature condenser, which condenses into liquid phase by giving up the latent heat of vaporization, and as a result the heat is transferred from the evaporator to the condenser through continuous cycles of working fluids with thermal gradient [7].

8.10.2 Design Aspects and Critical Parameters of Micro-Heat Pipes

The micro-heat pipes can be designed using p-type (100) silicon wafers to meet specific cross-sectional configurations, namely, circular, rectangular, or triangular cross sections. Specific details on the critical elements, lightly and heavily doped regions, bonding pads, substrates, and various materials used in the design and fabrication of micro-heat pipes are shown in Figure 8.11. The MEMS device to be cooled can be designed with several heat pipes ranging from 4 to 12, depending on the heat removal requirements. Each pipe can be 100 μm wide and 20 mm in length. It is important to make sure that the ratio between the pipe length and its hydraulic diameter is large enough for the working fluid to meet adequate evaporation and condensation requirements under adiabatic conditions.

A microminiaturized heater is integrated in the micro-heat pipe device as a heat source. The heater is located at one edge of the device to be cooled, while the other edge dissipates heat generated into a cold sink. The heater is incorporated in the silicon substrate to minimize heat loss to the surrounding structures. Temperature and capacitive microsensor arrays are installed along the heat pipes to monitor the temperature levels. A thin nitride membrane as shown in Figure 8.11 is provided to cover the heat pipes to visualize the two-phase flow patterns inside the pipes. The central sensing region of the device with an area of $6 \times 4 \ \mu m^2$ is lightly doped, while the remaining portions of the dumbbell are heavily doped. The electronic signals are transmitted from the metal lines installed on the silicon substrate to metal lines on the glass as illustrated in Figure 8.11. Both the nitride film or layer and the glass substrate provide good insulation between the top electrodes of the distributed capacitors. The length of the top electrode must be as small as possible. The capacitive sensor fabricated on the glass wafer provides much lower parasitic capacitance leading to improved sensor sensitivity.

The sensor can be fabricated using two distinct fabrication processes, namely, a CMOS-compatible fabrication process and a glass-based fabrication process.

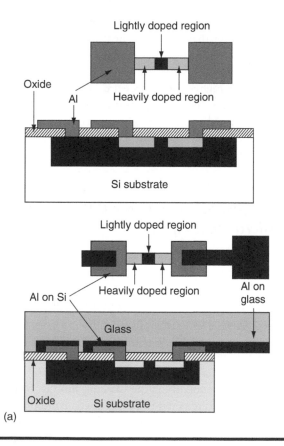

(a)

Figure 8.11 Micro-heat pipe systems. (a) Views of micro-heat pipe with temperature sensor using CMOS and glass technologies,

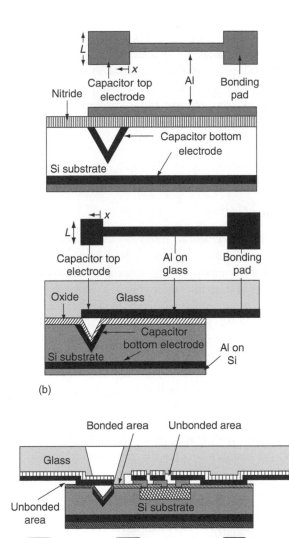

(b)

(c)

Figure 8.11 (continued) **(b) views of micro-heat pipe with capacitive sensor using CMOS and glass technologies, and (c) critical elements and materials used by micro-heat pipe system. (From Lee, M., Wong, M. et al., *J. MEMS*, 12, 138, 2002.)**

Note the glass-based process offers enhanced quality of flow visualization and minimizes the parasitic capacitance, which leads to improved sensor sensitivity. It is beyond the scope of this book to provide specific details on fabrication processes.

The fabrication of the sensor must ensure good quality of bonding, no water leakage from the heat pipes, lower parasitic capacitances, thicker metal lines buried within the substrates, and good quality of flow visualization to monitor the two-phase flow patterns. In summary, the integrated micro-heat pipe system consists of a local miniaturized heater, an array of heat pipes, temperature, and capacitive micro-sensors to monitor critical operating parameters.

8.11   MEMS-Based Thin-Film Microbatteries

8.11.1   Introduction

MEMS-based electrical, mechanical, infrared, and optical systems occupy a total volume about few tens of cubic millimeters. Such miniaturized sensors and devices need miniaturized power sources, namely, three-dimensional (3-D), thin-film MBs. Potential architectures and technologies must be investigated to meet the operational and fabrication requirements for the MBs. Earlier research and development activities [8] recommended the deployment of conformal thin-film structures because of distinct advantages over the conventional bulk MBs such as miniaturization of the geometrical dimensions. The planar two-dimensional (2-D) thin-film batteries cannot be considered as MBs, because they require large footprints of few square centimeters to achieve a reasonable battery capacity. The maximum energy density available from a thin-film battery is about 2 J/cm^3. Commercial thin-film batteries with a footprint of 3 cm^2 have a capacity close to 0.4 mA hour, which comes to 0.133 mA hour/cm^2. It appears that the 3-D, thin-film Li-ion MBs can be expected to meet the requirements for the miniaturized power sources needed by MEMS sensors.

8.11.2   Critical Design Aspects and Requirements for the 3-D, Thin-Film Microbatteries

It is evident from the above statements that the 3-D, thin-film, Li-ion MBs will able to meet the miniaturized power source requirements. Critical elements of the 3-D, thin-film MB including the current collector, graphite anode, cathode, and hybrid polymer electrolyte (HPE), and the cross section of the MB are shown in Figure 8.12. Note cathode thickness and the volume determine the maximum energy density and the battery capacity. Studies performed by research scientists indicate a significant gain in geometrical area, and therefore, cathode volume can be achieved with a perforated substrate instead of a full substrate. The area gain (AG) is strictly a function of holes or microchannels in a multichannel plate (MCP) substrate, substrate thickness, and the aspect ratio (height/diameter) of the holes. Computed values of AG as a function of microchannel diameter (d), interchannel spacing, and

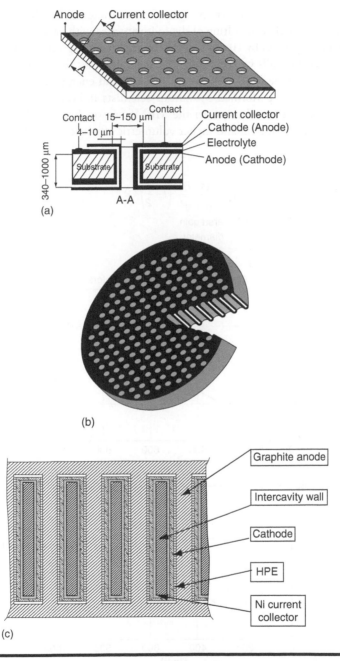

Figure 8.12 **Three-dimensional (3-D) thin-film MB showing (a) critical elements of thin-film MB, (b) MCP substrate with perforated holes, and (c) cross-sectional view of a 3-D MB showing critical elements of the device. (From Nathan, M., Golodmitsky, D. et al., *J. MEMS*, 14, 879, 2003. With permission.)**

substrate thickness (*t*) are shown in Figure 8.13a through c. Results of a mathematical model indicate that the tilting of the footprint area with various hole geometries offers slightly larger AG by 10–15 percent. In other words, geometry with hexagonal holes separated by walls with constant thickness provides the optimum AG with a perforated substrate leading to optimum capacity and energy density.

Research studies performed by various scientists indicate that the 3-D-MCP-based MBs are best suited to power the MEMS sensors because of minimum size, lowest power consumption, and higher volumetric efficiency. The 3-D-MCP-based

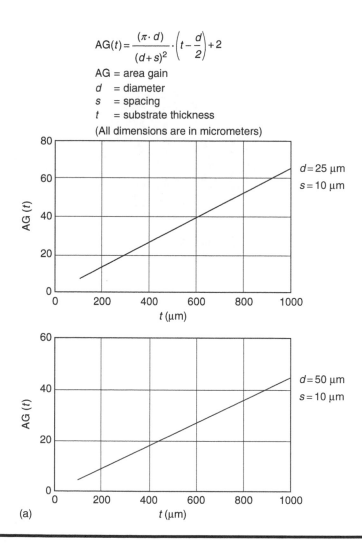

$$AG(t) = \frac{(\pi \cdot d)}{(d+s)^2} \cdot \left(t - \frac{d}{2}\right) + 2$$

AG = area gain
d = diameter
s = spacing
t = substrate thickness
(All dimensions are in micrometers)

Figure 8.13   (a) AG as a function of hole diameter, spacing, and substrate thickness,

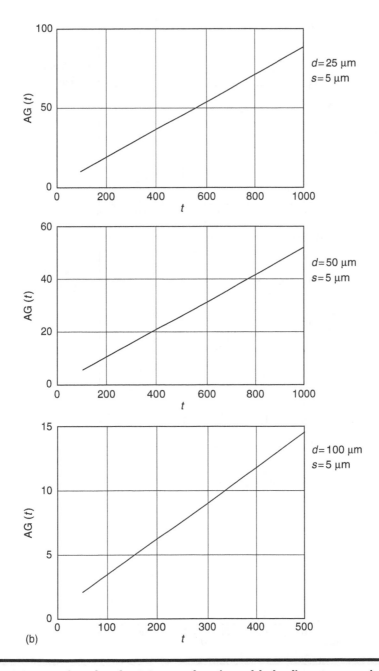

(b)

Figure 8.13 (continued) **(b) AG as a function of hole diameter, spacing, and substrate thickness,**

(*continued*)

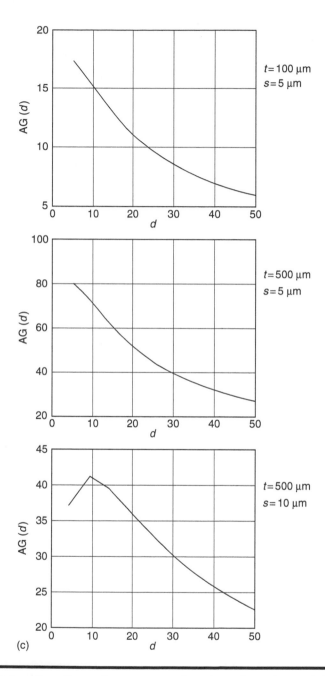

Figure 8.13 (continued) **and (c) AG as a function of spacing and thickness of multichannel plate (MCP).**

MB as shown in Figure 8.12 is comprised of a nickel cathode current collector, a low-cost, low-toxicity molybdenum sulfide cathode, a lathinated graphite anode acting as a second current collector, a highly-perforated emitter (HPE) and a MCP substrate. The MCP substrate illustrated in Figure 8.12 has a diameter of 12 mm and thickness of 0.5 mm with several 50 μm diameter hexagonal holes or micro-channels separated and is best suited to integrate in the design of a 3-D MB.

8.11.3 Projected Performance Parameters of a 3-D, Thin-Film Microbattery

The cutoff voltage based on room temperature test results on 3-D-MCP, thin-film MBs [8] varies from 1.3 to 2.2 V with current densities ranging from 100 to 1000 μA/cm². The slope characteristics of charge/discharge curves of these MBs shown in Figure 8.14 are similar to those of planar cells or lithium batteries. The discharge profile of these MBs remains practically unchanged over extended cycling period and at current densities ranging from 0.2 to 1.0 mA/cm². The test results further indicate that a 3-D-MCP device with a cathode thickness of 1 μm demonstrated 20–30 times higher capacity over a 2-D-MCP MB with the same footprint. The capacity of a 3-D MB capacity is projected near 2.0 mA hour/cm², which is roughly 15 times higher than that of a 2-D-MCP device. Computer simulations indicate that with a 3 μm thick cathode, the energy density in a 3-D-MCP device can be increased to greater than 10 mW hour/cm² with same footprint.

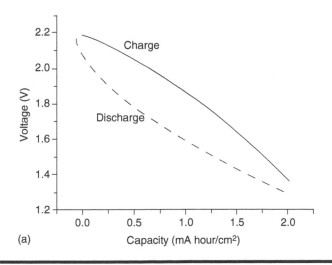

(a)

Figure 8.14 **Characteristics of 3-D MB. (a) Typical charge/discharge curves for a 3-D MB and**

(continued)

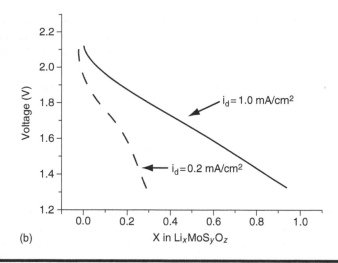

(b)

Figure 8.14 (continued)   (b) typical voltage rating for 3-D MB as a function of current density.

8.11.4   Unique Features and Potential Applications of Microbatteries

It is important to note that the 3-D-MCP-based, thin-film MBs with optimum cathode design parameters will outperform the 2-D-MCP-based, thin-film devices in terms of power capacity, energy density, and rate characteristics. The computed capacity gain varies from 25 to 35 and is more than adequate for the MBs for possible applications in MEMS sensors that require a footprint not exceeding 3×3 mm². Preliminary calculations indicate that a 3-D-MCP-based MB with a footprint of 3×3 mm² and a substrate thickness of 500 μm (or 0.5 mm) will provide enough power for various MEMS devices deployed for drug delivery and various implantable devices best suited for health-monitoring applications.

8.12   Summary

This chapter describes various MEMS- and NT-based devices and sensors best suited for scientific and medical research, aerospace and satellite sensors, health-monitoring sensors for structures, and low-cost, compact microwave components needed to improve the offensive and defensive weapon systems. Some of the sensors and devices, which were not previously discussed, will be briefly described with emphasis on their unique performance capabilities. Performance capabilities and limitations of tunable filters, tunable capacitors, MEMS varactors, ultrasonic transducers, VLF, HF, and VHF oscillators and transceivers incorporating micro-resonators, MEMS-based sensors to monitor the health of structures, weapons and battlefield environments, and accelerometers with nanogram resolution capability are summarized. Performance capabilities and applications of FFB, CCB, and folded beam-type micro-resonators are summarized with emphasis on resonance frequency. Timoshenko and Euler–Bernoulli equations are used to compare the computational accuracy of the resonance frequency. Sample calculations indicate that Timoshenko equation (Equation 8.6) offers most accurate values of the resonance frequency, which is the most critical parameter of a resonator. Performance parameters and design aspects of MBs are summarized for possible use as compact power sources for MEMS sensors. Computer simulation plots on AG for MBs as a function of micro-channel diameter, spacing, and substrate thickness are provided for the benefits of readers. Applications of NWs, nanocrystals, CNTs, and nanoparticles are identified. Performance parameters and design concepts of MM resonators are summarized with emphasis on dynamic range, resonance frequency, IL, stopband rejection and thermal stability of the oscillators incorporating MM resonators in their design configurations. Potential applications of MEMS varactors in wideband tunable filters, matching networks, phase shifters, RF synthesizers, and reconfigurable RF amplifiers are identified because of low IM products. Computed values of phase noise generated by the MEMS varactors as a function of impedance, MEMS bridge height, spring constant, capacitive reactance of the MEMS varactor, and operating frequency are summarized for the benefits of MEMS sensor design engineers.

References

1. G.M. Rebeiz, Phase noise analysis of MEMS based circuits and phase shifters, *IEEE Transactions on MTT*, 50(5), May 2002, 1316–1322.
2. C.T.C. Nguyen, Frequency selective MEMS for miniaturization low-power communications devices, *IEEE Transactions on MTT*, 47(8), August 1999, 1486–1502.
3. K. Wang, A.C. Wong et al., VHF free-free beam high-Q micro-mechanical resonators, *Journal of MEMS*, 9(3), September 2000, 351–363.
4. Hong-Teuk Kim, J.H. Park et al., Low-loss and compact V-band MEMS based analog tunable band pass filter, *IEEE Microwave and Wireless Components Letters*, 12(11), November 2002, 432–438.
5. C. Hautamaki, L. Cao et al., Calibration of MEMS strain sensors fabricated on silicon: Theory and experiments, *Journal of MEMS*, 12(5), October 2003, 720–724.
6. Anon., Sub-10 cm^3 Interferometric accelerometer with nano-g capability, *Journal of MEMS*, 11(3), June 2002, 182–185.
7. M. Lee, M. Wong et al., Integrated micro-heat-pipe fabrication technology, *Journal of MEMS*, 12(2), April 2002, 138–145.
8. M. Nathan, D. Golodmitsky et al., Three-dimensional, thin-film Li-ion micro-batteries for autonomous MEMS, *Journal of MEMS*, 14(5), October 2003, 879–885.

Chapter 9

Materials for MEMS and Nanotechnology-Based Sensors and Devices

9.1 Introduction

This chapter identifies potential materials and their important properties best suited for the design and development of nanotechnology (NT) sensors and devices, photonic components, and for a new generation of microelectromechanical systems (MEMS) devices. In addition, properties and applications of smart materials are summarized with emphasis on their unique capabilities, which will accelerate the design and development of macro-NT-based sensors critical in monitoring weapon health and battlefield environment. Carbon nanotube (CNT) arrays represent a new class of NT material with multifunctional capabilities and their potential applications are identified. The CNT arrays are best suited for a variety of military, aerospace, and commercial applications in high-efficiency photovoltaic cells. Potential applications of CNT/polymer arrays are investigated in the design of high-current density field emitters, flat displays with high resolution capability, and carbon MEMS electrodes for lithium (Li)-ion microbatteries (MBs). Electrodes made from CNT technology offer high aspect ratio leading to significant increase in surface areas. In addition, the CNTs are best suited for electrode applications in MBs. Potential applications of photonic crystals are also identified. Applications of quantum dots dye-doped silica nanoparticles are discussed with emphasis on commercial and military sensors. Cadmium selenium (CdSe) and zinc selenium (ZnSe) nanocrystals are investigated for potential applications in LEDs and nanolasers. Fluorescent nanoparticles capable of yielding high-resolution images superior to those obtained using current magnetic resonance imaging (MRI) and computerized tomography (CT) technologies are evaluated for cancer imaging [1]. These nanoparticles are most suitable for conducting clinical studies of molecular dynamics and interactions of living cells using maximum-likelihood fitting algorithm, leading to computations of the axial and radial positions of images.

9.2 Photonic Crystals

A photonic crystal is created by fabricating periodic variations in the refractive index (n) of a material, which is the critical element in the development of photonic crystal fiber (PCF), leading to hybrid materials with low effective index and improved dispersion performance. Hybrid materials are produced as a result of introduction of many air holes in silica substrates and are referred to as air–silica matrix structures. There is a special class of PCF, which is known as a photonic bandgap (PBG) fiber. This class of fiber finds applications where minimum insertion loss (IL) and dispersion over wide spectral region are of critical importance. Two distinct applications of photonic crystal are PBG devices and PCF optic cables. The solid core of a PCF (SC-PCF) yields minimum phase noise, and hence is most ideal for optoelectronic oscillators (OEO) operating at 10 GHz and beyond with exceptionally low phase noise levels. It is important to mention that the PCFs offer both the high-temperature stability and frequency stability for the OEOs. An OEO using PCFs has demonstrated a phase noise better than -143 dBc/Hz at an offset of 10 kHz from the 10 GHz carrier. Furthermore, the temperature sensitivity of the same oscillator is reduced by a factor of 3 compared to a conventional single-mode fiber (SMF)-28-based oscillator. The thermal stabilization of an OEO can improve the long-term frequency stability from -83 to -0.1 ppm/°C while maintaining a short-term frequency of 0.2 ppm/°C due to the elimination of mode-hopping effects. Performance comparison between the SMF and single-core PCF is evident from the data summarized in Table 9.1.

9.2.1 Photonic Bandgap Fiber

It is important to point out that the PBG fiber is guided by a PBG, analogous to the electronic bandgap (EBG), in semiconductor materials. As stated earlier, the PCF is created by fabrication of periodic variations in the refractive index of the material. In the case of military and space applications, the core of the PBG fiber [2] has demonstrated high mechanical integrity, enhanced reliability under harsh mechanical and thermal environments, low dispersion, and improved optical stability. Furthermore, the PBG optical fibers fabricated with high tensile strength, improved fatigue resistance, high stiffness capability, and excellent thermal shock resistance are

Table 9.1   Performance Comparison between SMF PCF and Core Diameters

Performance Parameter	SMF-28 (12.5 μm)	Solid Core of PCF (SC-PCF) (12.5 μm)	PCF Better Times
Frequency stability per °C (Δf/f) (1/ΔT)	−2.581	−0.159	4.32
Effective index change per °C (Δn/n) (1/ΔT)	11.37	4.73	2.41

best suited for optical systems operating under harsh thermal and mechanical conditions. Their performance capabilities are superior to glass-based core materials, summarized in Table 9.2 [3].

9.2.2   Core Material Requirements for PCF

Regardless of the fiber type or category, the core material must have a low dielectric constant and loss tangent to keep the IL minimal in the fiber. Studies performed by the author on various fibers [3] indicate that optimum properties of the core material can be altered by introducing certain ions in the silica. For example,

Table 9.2   Thermal and Mechanical Properties of Glass-Based Cores

Properties	E-Glass	Fused Silica	Cordierite Glass	Oxynitride Glass
Tensile modulus ($\times 10^6$ psi)	10	10.8	12.6	25.4
Tensile strength (psi)	600,000	750,000	860,000	985,000
Knoop hardness	460	685	715	835
Fatigue resistance	Fair	Good	Very good	Excellent
Coefficient of thermal expansion (CTE) (10^{-6} in./in. at 0°C)	9.92	0.57	4.50	2.82
Estimated cost ($ per meter)	0.60	3.12	8.25	14.55

introduction of lead oxide (PbO) ion in the core leads to significant improvement in optical quality, and introduction of silicon nitride (Si_3N_4) enhances thermal shock resistance. Furthermore, higher numerical apertures of the core yield higher light-coupling efficiency or lower-coupling loss. The studies [3] further indicate that the PCF offers a large numerical aperture for an SMF, higher core design flexibility, mode-free performance, more uniform index profiles both in longitudinal and cross-sectional directions, robust and stable structure with high optical power capability, higher thermal damage level, and improved optical stability.

9.2.3 Unique Properties of PCFs and Their Potential Applications

PBG technology can be applied to several devices operating in microwave and optical regions. This theory is based on the principle of light localization: the light wave whose frequency is within the bandgap is trapped inside the material and is not allowed to propagate. Because of the crystal structure, the dimensions of the periodic lattice and the properties of the component materials with enhanced propagation of electromagnetic waves in certain frequency bands (the PBGs) may be forbidden within the crystal. The formation of PCF is based on PBG technology. It is important to point out that the all-fiber laser design based on PCF technology is possible with continuous wave (CW) power output exceeding 25 W at a wavelength of 1075 nm, leading to a slope efficiency better than 75 percent, optical-to-optical efficiency greater than 65 percent, and 3 dB-line width less than 0.2 nm. Integration of PCF technology will significantly improve the performance of a vertical-cavity surface-emitting laser (VCSEL). Integration of PCF technology with VCSEL will offer several advantages over the traditional Fabry–Perot egde-emitting semiconductor (FPEES) laser. A VCSEL using PCF technology offers lower threshold level, higher SMF-output optical power, higher device efficiency, smaller device size, and lower power consumption. Cross section of a VCSEL photonic crystal showing its critical elements is shown in Figure 9.1.

It is important to point out that PCFs are single-material optical fibers with a periodic array of air holes running along their entire length. The large and controllable periodic variations of transverse refractive index properties open up exciting new opportunities for the control and guidance of light, thereby promising development of optical fibers with unique transmission characteristics. In other words, PCFs have extraordinary properties such as high dispersion control capability, reduced nonlinear effects, and single mode of operation over the entire infrared (IR) spectrum. Such unique properties are unavailable in other optical fibers. The dispersion control capability of a PCF optical fiber as a function of radius and spacing is of critical importance, because it determines the transmission capacity of information in the fiber. The dispersion in the PCF increases with an increase in

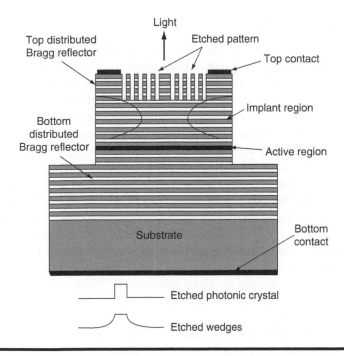

Figure 9.1 **Cross-sectional view of a photonic crystal VCSEL showing the critical elements of the device.**

core diameter, but its dispersion performance is much better than in conventional optical fibers.

PCF technology can allow fabrication of a large array on-wafer with unique performance capabilities such as improved beam quality, round and low-divergence optical beam, and low-cost beam formation scheme. These features of PCF are most ideal for near-IR applications. Potential applications of VCSEL using PCF technology include high speed optical data processing best suited for coherent side looking radar, coherent beam-steering capability, covert optical communications, and two-dimensional (2-D) optical arrays with high spectral purity.

9.3 Nanotechnology-Based Materials and Applications

This section is dedicated to NT-based materials, their unique properties, and potential applications in nanosensors and nanodevices. NT-based materials include nanotubes, nanoparticles, nanocrystals, nanodots, nanobubbles, CNT arrays, nanostructures, and nanomaterials known as smart materials. Properties and applications of each NT-based material are summarized in the following sections.

9.3.1 Nanocrystals

Nanocrystals are important in the design and development of NT-based exotic lasers, emitting specific wavelengths, best suited for biomedical research and in fabricating IR sensors that measure toxicity levels with great accuracy. Preliminary studies performed by the author on nanocrystals indicate that ZnSe- and CdSe-based nanocrystals are receiving considerable attention by scientists involved in the development of NT-based sensors. ZnSe nanocrystals are thermally very stable even at operating temperatures exceeding 200°C.

Manganese selenium (MgSe) can be produced with minimum cost and complexity when ZnSe nanocrystals are doped with copper (Cu) and manganese (Mg). Such nanocrystals offer better performance over CdSe nanocrystals. Because ZnSe nanocrystals offer minimum toxicity and higher thermal stability, they are considered most appropriate in the development of biomedical labels, light-emitting diodes (LEDs), and microlasers. CdSe nanocrystals are relatively easy to produce at minimum cost and complexity. Furthermore, the emission wavelength of a CdSe nanocrystal can be continuously tuned over visible IR spectral region merely by controlling its physical parameters. As far as toxicity is concerned, ZnSe nanocrystals are relatively less toxic.

It is interesting to note that selenium belongs to the sulfur family because it resembles sulfur both in various forms and in composition. Selenium nanocrystals exhibit both photovoltaic action—where light is directly converted into electricity—and photoconduction action—where the electrical resistance decreases with increased illumination. These unique properties make selenium most attractive for both photocells and solar cells. Elemental selenium is practically nontoxic, but some of its compounds are extremely toxic and require careful handling and storage.

Elemental cadmium often occurs in small quantities associated with zinc ores. Most cadmium is obtained as a by-product during the treatment of zinc, copper, and lead ores. It has a low coefficient of friction and offers great resistance to fatigue. Handling of cadmium requires great care because of its high toxicity.

9.3.2 Photonic Nanocrystals

If the size of a photonic crystal is reduced significantly, it is then referred to as a photonic nanocrystal. These nanocrystals have a unique ability to mold and control light in three dimensions by opening a frequency region (bandgap) in which light is forbidden to propagate. When a photonic band structure is correctly designed and fabricated, such a structure will not allow photons with a bandgap energy to penetrate the lattice, regardless of their angle of incidence. Note that a PBG is intrinsically dependent on the underlying photonic lattice topology, periodicity, and integrity. Photonic crystal-based sensors, when attached to a composite substrate, will experience a significant change in their bandgap profile due to submicron damage induced in the composite substrate. This indicates that a photonic crystal sensor is best suited for submicron damage detection, quantification, and structural diagnosis. In other words, this type of sensor is most appropriate for monitoring the health of structures and semiconductor wafers or chips.

9.3.3 Nanowires and Rods and Their Applications

NT scientists have identified various types of nanowires (NWs) and nanorods for possible applications in NT-based sensors and devices. Zinc oxide (ZnO) NWs and silicon nanorods have been developed for NT-based biomedical sensor applications.

9.3.3.1 Zinc Oxide Nanowires

Results published by Hogan [4] indicate that n-type ZnO nanostructures could lead to the development of p-type NWs, ideal for high-gain, low-power photodiodes. These photo detectors have potential applications in sensing, imaging, storage, and in covert communications. These photo detectors have demonstrated a photoconducting gain exceeding 2×10^8 and a photocurrent decay within seconds. Scientists have predicted that ZnO NWs are best suited for development of optically pumped lasers, chemical and biological sensors, and field-effect transistors (FETs) with high gain. These NWs have various optoelectronic applications including LEDs, displays, photon detectors, and lighting sources. Studies performed by the author indicate that ZnO NWs could form a viable and less expensive substitute for GaN LEDs. ZnO NWs have demonstrated high photosensitivity due to the large surface-to-volume ratio and the presence of deep-level surface trap states that prolong the lifetime of the photo carrier. In addition, one finds reduced dimensions of the active area in the NW device, which shortens the carrier transit time. Photodetectors made using ZnO NWs have unique characteristics, because they use low-dimensional semiconductors with high-density surface trap states. According to publications on ZnO NWs, p-type ZnO NWs have significantly thinner and longer structures than their n-type counterparts. Research also indicates that these p-type ZnO NWs could allow economical mass production of LEDs, photon detectors, and other optoelectronic devices widely used in industrial and military display equipment.

9.3.3.2 Silicon Nanowires

NT scientists have claimed that using p-type silicon NWs flexible NT devices can be designed and developed for various applications. Silicon NWs several micrometers long but less than 100 nm diameter conduct electricity and act as devices, if appropriate dopants are implanted and activated. Silicon NWs when deposited onto a glass substrate can form an array, which can be selectively doped to either n- or p-type. Optical inspection predicts that the NWs can be melted with laser energy density ranging between 60 and 70 mJ/cm^2, which is less than one-fourth the energy required for a silicon thin film and even lesser energy density needed for bulk silicon material. In brief, silicon NWs melt and anneal at much lower light intensity or laser energy density than bulk silicon does.

Scientists at Harvard University have claimed that silicon NWs with single-crystal structures and uniform diameters are best suited for avalanche photodetectors in integrated photonic sensors. The length of the intrinsic region, where the avalanche breakdown occurs, is dependent on the diameter of the NW and the amount of chemical doping introduced. The maximum photocurrent intensity occurs at the avalanche breakdown region. Characterization of the NW structure as a function of diameter and length can be accomplished by transmission electron microscopy. However, the growth of the NW is dependent on the chemical dopant used and the NW dimensions. An average growth from a uniform single-crystal structure is about 6.5 μm in the (112) direction along the axis. The variations among the three NW regions can be measured using force microscopy and scanning gate microscopy. Note an ES force microscopy signal can reveal a voltage drop across an intrinsic region, while any decrease in the conductance over the same region can be measured in the scanning gate microscopy operational mode. The current versus voltage characteristics can be measured at forward bias greater than 1 V and at a reverse bias level close to 35 V. An increase in breakdown voltage can cause an avalanche breakdown in NT devices and scientist or designers should make a note of this. This breakdown is a function of temperature, which is strictly based on band-to-band impact ionization. Silicon NWs can be used as avalanche photo detectors because of the avalanche multiplication factor. In summary, because of high spatial resolution and sensitivity, silicon NWs have potent applications in integrated nanophotonic sensors.

9.3.3.3 Zinc Selenium Nanowires

ZnSe NWs have demonstrated unique optical and electrical properties, which are best suited for nanoelectronic and optoelectronic device applications. It is important to point out that the fundamental optical and electronic properties of NWs depend on the size, material, and growth technique employed to produce them by controlling length and diameter at the root and at the tip. Furthermore, the NW dimensions can be varied by adjusting the laser parameters such as laser pulse rate (100–1000 Hz), pulse width (2–5 fs), energy level (1–2 mJ), and emission wavelength (400–800 nm). A wavelength of 800 nm is considered optimum for growing an array of ZnSe NWs on crystal surfaces. The 800 nm source can grow NWs on the designated position without contaminating surrounding areas, which is important for application in optoelectronic devices.

NWs can be grown with lengths ranging from 1 to 3 μm with diameters ranging from 35 to 80 nm at the root and tip, respectively, with a growth rate of 4–6 μm/s.

To obtain an array of NWs with uniform performance, one must use ZnSe crystals with a surface orientation other than 111. These NWs may contain longitudinal and transverse optical modes. A pulsed laser deposition technique will yield uniform ZnSe NWs in localized positions, generating NWs with desired optical and electronic properties best suited for optoelectronic applications.

9.3.3.4   Zinc Phosphide Nanowires

Zinc phosphide (Zn_3P_2) is cheap and is widely available without any restrictions. It has excellent optical efficiency and offers great resistance against the buildup of oxide on its surface. These properties of Zn_3P_2 make it most ideal for optoelectronic applications. When a mixture of zinc oxide and graphite is placed on silicon wafers and is exposed to excimer laser, dark yellow zinc phosphide nanostructures are formed. One can form nanostructures with sixfold symmetry, nanobelts of uniform thickness, and NWs with diameters up to 100 nm.

Zinc phosphide NWs have a bandgap between 1.4 and 1.6 eV, which make them most attractive for solar energy conversion devices. The on/off ratio (the ratio of current under light to dark conditions) is a function of time and impinging light wavelength. The NW resistance is the lowest and the on/off ratio is the highest for the high-energy light wavelength of 523 nm [5]. A quick response time less than 1 s and sensitivity when exposed to light combined with low cost and resistance to oxidation make zinc phosphide NWs most ideal for light-sensing applications including solar cells. An array of photodiodes using zinc phosphide NWs could serve as nanooptoelectronic components in high-resolution cameras, high-efficiency solar cells, and various scientific instruments needed for conducting scientific research in remote areas.

9.3.3.5   Cadmium Sulfide Nanowires

Current research indicates that combining low-temperature micro photoluminescence imaging and time-resolved photoluminescence will provide useful information on the electronic states and recombination modes of single cadmium sulfide (CdS) NWs. The scientists further state that when ten CdS NWs with diameters ranging from 50 to 200 nm and lengths from 10 to 15 μm are excited, information on spectral resolution of 70 μeV and spatial resolution of 1.2 μm are possible. CdS NWs have demonstrated great potential in optically driven applications such as lasers, photo detectors, and optical waveguides with minimum IL and dispersion. A variety of scientific observations can be undertaken using time-resolved photoluminescence and micro photoluminescence measurements on CdS NWs with diameters ranging from 10 to 20 nm. A very uniform and straight CdS NW exhibited a single, broad spectral peak at 2.525 eV. A 2-D photoluminescence image of the CdS NW showed no evidence of localized states, indicating the dominance of the near-band emission peak. The intensity of this emission can vary at different positions along the length of the uniform wire. However, in the case of nonuniform NWs, a series of sharp peaks, which are referred as secondary peaks, with lower energy levels and longer time decays will be observed. Passivating the surface states of the NWs could increase the quantum efficiency at least by an order of magnitude, which will be of significant importance in determining the quality of CdS single wires and their applications.

9.3.3.6 Iron–Gallium Nanowires

A technical paper presented at an SPIE conference on smart materials briefly describes a unique application of iron–gallium NWs. However, no specific details are provided on the composition or dimensions of the iron–gallium NWs used in the design and development of an acoustic sensor incorporating iron–gallium NW technology. The principal objective of this research program was to demonstrate the feasibility of an underwater acoustic sensor package capable of allowing sound transmission to the sensor, while keeping out moisture and salt ions. A bio-inspired sensor using these NWs to study hearing mechanisms in fish and other aquatic animals was developed. This acoustic sensor design employs the magnetostrictive property of iron–gallium NWs. Arrays of these NWs mechanically respond to incoming sound waves underwater, which create magnetic fields that are sensed by the magnetic sensor attached to the base of NWs. The NWs are free to move in the fluid medium. The acoustic properties of this NW are of critical importance. This NW has a tensile strength exceeding 500 MPa and is highly ductile, which is highly desirable in bending applications. As stated earlier, the magnetic fields created by the acoustic pressure-induced bending of the iron–gallium NWs can be picked up by the sensitive magnetic sensor attached to the base of the NWs. By having NWs of different lengths and diameters positioned over on giant magnetoresistive (GMR) devices the magnetic sensor could be made sensitive over a wide range of acoustic frequencies, which are responsive to human ears. Underwater reconnaissance and surveillance are the potential applications of an acoustic sensor using iron–gallium NWs.

9.4   Nanoparticles

This section deals with nanoparticles, which are quite different from NWs in terms of structure and properties. It is interesting to mention that optical imaging techniques that use nanometric-scale fluorescent nanoparticles as probes offer highly sensitive cancer diagnostic methods. NT-based sensors using fluorescent nanoparticles yield much higher sensitivity than existing MRI or CT imaging techniques. Precision diagnosis is absolutely necessary for achieving long-term patient survival [6] and imaging must be done at a microscopic level to ensure high sensitivity and resolution. It is important to mention that the number of nanoparticles implemented per volume of the target tissue determines the signal quality and resolution. In other words, a higher number of particles are preferable to achieve high-resolution images. Note the probes comprising nanoparticles must operate near-IR wavelengths for excitation and emission within the tissue samples to ensure high extinction coefficients and quantum yields. The probes can be designed to operate with both fluorescence and magnetic resonance imaging techniques. Quantum dots, gold nanoparticles, and dye-doped silica nanoparticles with their unique optical and magnetic properties have potential applications in cancer imaging.

Nanoparticle coatings with antireflection, antifogging, and self-cleaning properties have been observed by Massachusetts Institute of Technology (MIT) scientists. Thin films of nanoparticles are deposited on glass or silicon substrates using an aqueous-based layer-by-layer deposition technique. The multilayer coatings comprised positively charged particles of titanium oxide and negatively charged particles of silicon oxide. The average diameter of each nanoparticle is about 7 nm. Automobile windshields made from such nanoparticles offer better visibility for motorists, because nanoparticle-based coatings on glass significantly reduce the reflection losses in the visible region, leading to transmission efficiency exceeding 90 percent. The self-cleaning and anti-fogging properties of the coating have potential applications in photovoltaic cells, windows of high rise-buildings, and auto- and aircraft windshields.

9.5   Quantum Dots

Quantum dots fall under the nanostructure category. Quantum dots have extremely small width diameters ranging from 2 to 10 nm. Essentially, a quantum dot is considered a semiconductor nanocrystal—a tiny chunk of material with unique optical and electrical characteristics. The material can be made from any semiconductor, such as silicon, germanium, gallium arsenide, cadmium selenide, and zinc sulfide, to name a few. A high-quality semiconductor is considered a perfect, single crystal with almost zero defects. Studies performed by the author indicate that quantum dots with nanophosphor coatings could offer several advantages and potential applications such as a high quantum yield in the 85–95 percent range and a narrow emission spectral bandwidth with broad absorption and minimum temperature dependency.

It is interesting to mention that in the case of core-shell quantum dots, the core crystal is coated with another crystalline semiconductor to modify dot material properties, to improve emission efficiency and to enhance optical stability. Because of extremely small quantum dots, their electrical and optical properties are determined by quantum mechanics rather than classical physics. The confined electron in the quantum dot has an excess energy than the electron in a bulk semiconductor and the excess energy changes the electrical and optical properties of the quantum dots. This unique property of quantum dot materials will have a tunable bandgap region. Recent advancements in quantum dot materials have developed high-quality core materials with low polydispersity, high optical emission efficiency, and enhanced optical stability. Note that the performance of quantum dots is strictly dependent on the availability of advanced semiconductor materials, the size of the dot, and the tweaking of its chemical composition.

9.5.1 Applications of Quantum Dots

Because quantum dot technology is still nascent, potential applications have been not yet identified, except in biological sensors and light-based devices. Quantum dots are being fully evaluated and analyzed for various applications including LEDs, lasers, flat screen displays, solar sensors, and memories. Intensive research and development activities are being pursued to explore the possibility of using quantum dots in the development of next generation of flay panel displays, which will outperform liquid crystal, cathode ray tube, and plasma displays soon. Progress in research on quantum dot LEDs is currently discernible. For example, according to articles published in the May 2007 *Photonic Spectra*, the green quantum dot LED demonstrated the quantum dot efficiency better than 82 percent at a brightness of 140 cd/m^2. The red quantum dot LED can provide a quantum dot efficiency of about 3.2 percent at a brightness of 210 cd/m^2. No such data on blue quantum dot LEDs is currently available because they are still being tested in laboratories.

It is important to mention that a quantum dot-based display does not require background light and color filters, resulting in power consumption ten times less than the current array of display technologies. The display using quantum dot technology could be packaged into screen sizes ranging from a cell phone display to a 60 in. HD display. The finished display product will be thinner and lighter with enhanced color capability and improved quantum efficiency. Because nanostructures are tunable, highly emissive, and durable, these unique characteristics will bring this cutting-edge display technology into commercial markets soon.

9.5.2 Unique Security Aspects of Quantum Dots

Research scientists claim that the unique optical properties of quantum dots offer protection against almost any manner of document theft, fraud, or counterfeiting. This is because quantum dots can be tuned to emit optical energy over wavelengths ranging from 400 to 2400 nm, and therefore it will be impossible for the counterfeiter to mimic the quantum dot code programmed on the authentic product or item, even one with most sophisticated NT skills and state-of-the-art equipment. Because of combining different wavelengths, changing energy concentration levels, and using different excitation wavelengths, the encryption is extremely difficult to unravel. The material can be integrated into different inks, various fabrics, and unique paper categories, which can be read with appropriate detection systems. Emissive capability of the dots can be made to last indefinitely or disappear within a short duration. Quantum dot is a security technology that is best suited for highly sensitive applications such as currency protection. This technology is best suited for security and defense applications such as tracking and surveillance of drug suspects, base perimeters, and weapon storage areas. Scientists are working on a quantum dot laser with low threshold level and high thermal stability, which could be deployed in specific military and defense applications. The next generation of quantum dot-based systems can play key roles in detection of chemical agents and in missile defense.

9.5.3 Lead Sulfide Quantum Dots with Nonlinear Properties

South Korean scientists have revealed that successful incorporation of PbS quantum dots into zeolite-Y films will lead to the development of an intrazeolite lead sulfide quantum dot material with unique nonlinear properties. This material could be used in the design and development of IR detectors for medical and military applications, optical switches with high isolation, limiters, fast computers, and a host of laser components. High-level of third-order nonlinear optical properties of these dots make them most attractive for exotic sensor applications such as security and defense. These dots are found identical in size and are available in crystalline structures smaller than 1.5 nm. Scientists claim that the nonlinear refractive and absorption coefficients of the intrazeollite lead sulfide quantum dots are 20–300 times higher than the highest values available from other quantum dots. Scientists further claim that the third-order nonlinear properties of lead sulfide quantum dots will offer much higher sensitivities and can operate simultaneously with various wavelengths. In summary, lead sulfide quantum dots are best suited for military, defense, and security applications.

9.6 Nanobubbles

Current NT research reveals that nanobubbles may have implications for cancer therapy and advancements in ink-jet printer technology. NT scientists believe that the lifetimes of these nanobubbles are estimated to be about 100 μs [7], which can improve optical heat transfer design in nanostructures. When water is heated rapidly with 5 μs voltage pulse, its resistance increases and a nanobubble is nucleated at around 300°C. The bubble grows to tens of micrometers in diameter. When the pulse ends, the bubble (first) collapses as the temperature falls. However, if a second pulse is applied closely enough, another (second) bubble forms at lower temperature, say at 250°C or so. On the basis of this theory, scientists believe that the nanobubbles formed by the collapse of the first bubble become the nucleation sites for the growth of the later bubbles. According to scientists, the lifetimes of the nanobubbles can be estimated by changing the timing between the voltage pulses and observing the morphology and repeatability of the second pulse. The voltage pulses of known duration can be generated by a Q-switched Nd:YAG, 532 nm laser beam, which provides the electrical pulses to a platinum micro-heater element to increase the temperature of water that is needed for the formation of these bubbles.

9.6.1 Applications of Nanobubbles

Thermal cancer therapy will have advantages because it works by sending nanoscale objects into the tumors and heating them with IR radiation or with alternating magnetic fields. Each nanoscale object acts like a micro-heater, and the nanobubbles can be created if the nano-objects are heated to appropriate temperature. Scientists believe that nanobubbles can have a therapeutic effect if additional IR heat is applied and the nanobubbles produce mechanical stresses to the surrounding tissues.

These nanobubbles can offer significant improvements in ink-jet printer technology. The size and duration of the nanobubbles can be optimized to design ink-jet printers with high resolution and minimum usage of ink, thereby leading to significant reduction in printing costs. Note the temperature is the variable that optimizes the laser pulse energy and pulse duration. Studies performed by scientists indicate that pulse widths smaller than 5 μs will yield maximum economy and high resolution.

9.7 MEMS Deformable Micro-Mirrors

Astronomy scientists believe that deployment of MEMS deformable mirrors (MDMs) in adaptive optics will compensate for atmospheric aberrations in the optical path of laser beams or telescopes [8]. Airborne laser beams suffer from adverse effects of scattering, reflection, and absorption while traveling in higher atmospheric regions. These scientists further believe that incorporating adaptive optics using MDMs will lead to dramatic improvement in image quality, which has potential application in vision science. The overall manufacturing process for the MDMs is very expensive now, because it is very expensive to produce deformable mirrors from wafer-thin glass sheets with bonded ES actuators. Deformable mirror manufacturing costs can be significantly reduced by producing an ES deformable mirror using the micromachined silicon membrane technology developed at Stanford University.

MDM standard diameters come in 10, 16, and 25 mm. However, these mirrors can be fabricated as large as 125 mm in diameter using a 633 nm laser beam. For example, a 25 mm diameter mirror comprising of 37 ES actuators can be fabricated using an applied voltage of 110 V. Note the actuation voltage increases with an increase in the actuators deployed. It is important to mention that higher-order corrections in laser beams require driving the ES actuators in different patterns. A high-fidelity modeling software is needed to design an MDM. Optimum design requires appropriate values of design parameters such as mirror thickness, number of actuators, spacer size, and mirror deflection as a function of actuation voltage. A vertical comb array microactuator (VCAM) with continuous membrane mirror plays a key role in the design of an MDM. A three-dimensional (3-D) VCAM can be fabricated using a series of parallel vertical plates involving a horizontal substrate. Note a VCAM with continuous membrane mirror can move up to 1 mm in height, which can open up applications in other regions of the electromagnetic spectrum. Standard deformable mirrors are available with membrane diameters ranging from 10 to 50 nm and ES actuators ranging from 37 to 79 nm. Regarding the cost of deformable mirrors, companies involved in the manufacturing of these devices predict cost for conventional deformable mirrors as $1000 per actuator. Mirror costs for MEMS versions are in the order of $100 per actuator. However, advancement of MEMS technology and deployment of thermal actuators can reduce the cost to about $25 per actuator soon. Although the response time of thermal actuators is relatively slow, the low cost of this mirror makes it most attractive for applications where compensation for slow varying optical aberrations is acceptable. It is important to mention that the advantages in size, speed, and lack of hysteresis give MDM an edge over conventional or traditional deformable mirrors.

9.7.1 Applications of MEMS Deformable Mirrors

A potential ophthalmic application of MDMs could correct aberrations of the human eye. Adaptive optics involving MDMs can provide adequate compensation for high-order aberrations of the eye, which is not possible using other devices or sensors. Other applications include low-cost adaptive solar telescopes and satellite tracking systems. But the most critical application of an MDM includes the high-power airborne laser capable of precision tracking of hostile intercontinental ballistic missiles during their launch, cruise, and terminal phases. These devices offer rapid compensation of thermal gradients in the laser beams traveling in various regions of the atmosphere. MDMs are best suited for applications where small size, light weight, and low cost are the principal design requirements. In other words, the future generations of miniaturized adaptive optics systems incorporating MDMs will be cheaper, faster, and more compact.

9.8   Carbon Nanotubes and CNT Arrays

CNTs and CNT arrays have received the greatest attention in NT-based sensors and devices for various applications. CNTs fall under two distinct categories: single-wall CNTs (SWCNT) and multiwall CNTs (MWCNT). Highly aligned MWCNTs up to 4 mm long can be fabricated and synthesized on silicon substrates. A CNT array can be fabricated involving either SWCNT or MWCNT elements or both. CNTs are considered smart materials because of their high mechanical strength, electrical conductivity, multifunctional capabilities, and piezoresistive and electrochemical sensing and actuation properties. Potential future applications have directed research activities toward the development of synthesis, material characterization, processing, and device fabrication techniques for CNT-based devices and sensors. CNTs can be grown in the form of powders or loose nanotubes. Thermally driven chemical vapor deposition (CVD) method uses hydrocarbon molecules as the carbon source, and iron, nickel, and cobalt as catalysts to grow high-density nanotubes or CNT arrays on silicon substrates. The growth mechanism is very complicated, which affects the length of the nanotubes and is dependent on the strength of the interaction between the nanoparticles and the support. In other words, the growth kinetics of a CNT is a complicated phenomenon, which is based on molecular dynamics. The electrochemical actuator can be formed by bonding an electrode to an appropriately grown nanotube array, which is typically 4 mm in length and 1×1 mm^2 in area. Electrochemical actuation for this particular CNT array requires a driving voltage between 2 and 3 V to produce 0.2 percent strain. The growth mechanism of CNT arrays involves patterned silicon substrates, but its characteristics can be investigated experimentally. Note the CNT array represents a new smart material with multifunctional capabilities, which can be tailored for various applications.

9.8.1   Potential Applications of CNT Arrays

9.8.1.1   Nanostructures/Nanocomposites Using CNT Arrays

MWCNTs can be used in the production of multifunctional, high-temperature nanocomposite materials because of their excellent electrical properties, superior melt stability, high thermal conductivity, and small diameters [9]. MWCNTs and matrix resins are generally used in injection, fabrication, and molding procedures to produce multifunctional, high-temperature nanocomposite materials. Matrix resin/MWCNTs mixtures can be prepared with concentrations ranging from 5 to 25 wt percent of the MWCNTs. However, the samples containing 20 wt percent of the MWCNTs are considered most desirable for this application. It is important to ensure that the MWCNTs are substantially aligned along the flow direction during the injection process. Note the electrical resistivity decreases with the increasing concentration of MWCNT. However, the thermal conductivity of the high-temperature nanocomposite material increases with the increasing concentration of MWCNT. These nanostrucures are best suited for development of defense-oriented products such as fighter aircraft, space weapons, rocket motors, warhead components, propellants, and missiles due to their inherent mechanical strength, high durability, light weight, and high thermal conductivity.

9.8.1.1.1 Nanocomposite Coating Contributes to Stealth Technology

NT materials scientists at the Institute of Technology in Roorke (India) have developed a microwave-absorbing nanocomposite coating that could make an aircraft or missile almost invisible to hostile radar, thereby significantly contributing to stealth technology. Microwave pulses emitted by radar are used to detect flying aircraft based on the radiation reflected by the metallic body of the aircraft. The (Indian Institute of Technology) IIT scientists claim that stealth technology involving nanocomposite technology will permit fighter or reconnaissance aircraft escape radar surveillance and protect its equipment from electronic jamming or enemy's missile attack. Nanocrystals of barium hexaferrite ($BaFe_{12}O_{19}$), also known as barium hexagonal ferrite, provide nanoparticles with sizes between 10 and 15 nm, which have the capability to absorb microwave energy. Note coating of nanocomposite material acts like a radar-absorbing material (RAM). Furthermore, the carbon-based nanocomposite material is a good absorber of microwave energy regardless of the radar frequency. However, it is more effective against radar operating at higher microwave frequencies and mm-wave frequencies. This material is known as a microwave ferrite material and it has a high complex permeability, with two components, namely, the real part and the imaginary part, which are primarily responsible for absorption losses at microwave frequencies. It is important to mention that in a magnetic crystal, the free energy of the system is strictly dependent on the orientation of the magnetization with respect to the crystallographic axes. The anisotropic energy has two components, namely, first-order (K_1) and second-order (K_2) components. Because the second-order anisotropic constant is very small compared to the first-order constant, the anisotropic energy is mostly dependent on the K_1 constant value. This constant is about 50,000 for nickel ferrites, 25,000 for manganese ferrites, and a whopping 4,000,000 for barium hexaferrites, thereby indicating its highest absorption energy capability at microwave frequencies. When an aluminum sheet with barium hexaferrite coating is exposed to 16 GHz microwave source, close to 90 percent of the microwave energy is absorbed, thereby leaving the reflected energy close to 10 percent. However, by optimizing the nanocomposite coating thickness and nanoparticle size, maximum absorption is possible for the incident radiation at aircraft or missile surfaces.

9.8.1.2 CNTs as Field Emitters or Electron Sources

Studies performed by NT scientists predict that planar arrays of CNTs can be used in the development of high-current density field emitters with cold cathodes. Arrays of CNT bundles might prove useful as miniaturized cold-cathode electron sources [10]. Note hot-cathode electron sources require operating voltages in excess of several 1000 V, whereas cold-cathode electron sources generating comparable current densities require only tens of volts to operate. In brief, arrays of CNTs can be used to develop cold-cathode electron sources, which will be miniaturized, light-weight electron-beam devices known as nanoklystrons. Preliminary estimates indicate that current densities are in the order of 100 mA/cm^2. However, reducing the screening effect, which is caused by the taller nanotubes screening the shorter ones, and increasing the spacing between the nanotubes to at least by a factor two times the height of the nanotubes, can significantly enhance the emission current density. Higher the number of CNTs higher will be the emission current density level. Such nanoklystyrons will provide enhanced frequency stability, which is a critical performance requirement for coherent pulse Doppler radar and CW missile illuminators.

A typical bundle of CNTs, each 70 μm in height and 5 μm in diameter and with a gap less than 5 μm between the adjacent bundles, can yield an acceptable level of electron density. However, to maximize the area-based current density, it is necessary to find an optimum combination of nanotube spacing and its height. It is important to point out that a single CNT might have a short lifetime, and hence arrays of bundle of CNTs are considered to achieve a reasonable current density, in case some CNT elements fail to contribute to overall density level. NASA scientists [10] indicate that planar arrays of bundles of CNTs with various bundle diameters, bundle heights, and bundle spacing must be developed to evaluate the overall performance of an electron source. It is the opinion of NASA scientists that bundles of MWCNTs involving 20 μm diameter CNTs and with average bundle height of 70 μm can yield a current density close to 2 A/cm^2 at an applied electric field intensity of 4.5 V/μm. This estimation of current density is based on array bundles of 2 μm diameter and spacing close to 5 μm.

9.8.1.3 CNT Technology for Biosensor Chemical and Environmental Applications

CNT arrays offer a promising host of smart nanosensors such as biosensors, chemical sensors, and environmental monitoring sensors. NT research scientists reveal that deployment of a nanoelectrode on a silicon substrate with a controlled site density of nickel nanoparticles can lead to development of a glucose monitoring device with higher accuracy and fast response [11]. The electrochemical properties of a nanoelectrode array also known as a tower electrode based on vertically aligned CNTs show great promise for a biosensor with low detection capability and ultrafast response. The fabrication steps needed to construct a nanotube-based tower-shaped electrode are shown in Figure 9.2. Coating of appropriate chemical materials on the nanoelectrode provides the important parameters on gaseous and liquid chemical, toxic agents in atmosphere and other environmental elements. The chemical sensing elements are capable of providing fast electron transfer capability and accurate cyclic voltammetric peak-separations needed for rapid heterogeneous/ homogeneous chemical reactions.

9.8.1.4 Nanotube Arrays for Electrochemical Actuators

NT researchers predict that patterned array towers using MWCNT arrays are best suited for the development of electrochemical actuators. MWCNT towers 1 mm^2 by 4 mm high can be grown on silicon substrates using CVD technology. One end of the tower can be connected to a wire impregnated with conductive epoxy on a glass substrate. Electrochemical actuation capability can be evaluated using a sodium chloride (NaCl) electrolyte solution. The actuation performance parameters of the nanotube tower actuator illustrated in Figure 9.2 can be evaluated in greater detail

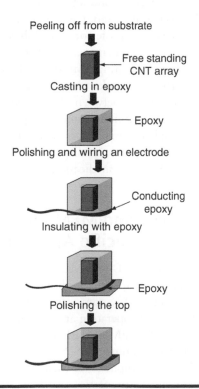

Figure 9.2   Fabrication of the tower electrode using CNT array technology.

based on the relationship between the applied actuation voltage and the tower displacement in the axial direction. The exposed conducting epoxy must be sealed and properly cured. Glass bead tape must be pasted on the top of the CNT tower structure to reflect the laser displacement sensor optical signal. The displacement of the CNT tower actuator can be measured as a function of applied voltage using a laser displacement sensor and a calibration chart can be made showing the displacement in micrometers as a function of actuation voltage. The tower performance test requires a nanotube electrode, a silver chloride reference electrode, and a platinum plate acting as a counter electrode.

9.8.1.5 Nanotube Probes and Dispensing Devices

NT scientists indicate the MWCNTs can be used as probes, needles, and dispensing elements. These NT-based devices would generally require two electrodes for each application. The nanotube probes can be used as a small biosensor to increase the sensitivity needed for cancer detection organs or tissues. The nanoelectrodes are required to probe the neuronal response of the cells in electrophysiology procedures and for locating the centers of epilepsy in neurological investigation. Nanotube technology offers strong, electrically conductive probes to provide measurement tools in a range of applications from nanomedicine to space exploration with high reliability and accuracy. Recently, a professor from Johns Hopkins University has presented a new NT-based technique that uses a burst of electricity to biomolecules from a gold launch pad, thereby creating a means of dispensing medication from tethers made from hydrocarbon molecules. A small pulse is sent through the gold electrodes, which breaks the bonds between the tethers and electrodes, thereby releasing the biomolecules. This technique could be applied to dispense medication from a biocompatible chip implanted in patients. The MB connected to the chip can be replaced without any interruption in dispensing of the medication or any discomfort to the patient receiving the medication.

9.9 Nanotechnology- and MEMS-Based Sensors and Devices for Specific Applications

This section exclusively deals with NT- and MEMS-based sensors and devices with unique capabilities best suited for specific applications. Such sensors and devices include acoustic sensors using NWs, mm-wave patch antennas using MEMS technology, CNT-based transistors operating at microwave and mm-wave frequencies, MEMS-based gyros, and MEMS-based accelerometers. Finally, materials required in the design and development of NT-and MEMS-based sensors and devices will be identified with an emphasis on their electrical, thermal, mechanical, electrochemical, and electro-optical properties.

9.9.1 Acoustic Sensors Using Nanotechnology for Underwater Detection Applications

The development and packaging of an underwater acoustic sensor is more difficult than typical microelectronic sensors because of the simultaneous need to protect the sensor or device from external environments, while performing its intended mission. The bio-inspired acoustic sensor must be modeled after the hearing mechanisms in fish and other underwater animals. The sound waves are transferred from the water through the tissues of the aquatic animals. The underwater acoustic sensor will allow sound to pass through only while interacting with the sensing element, which incorporates iron–gallium NWs. The fabrication, packaging, sealing, and encapsulation will require MEMS technology to achieve good signal quality and reliability over extended periods. The cavity containing the NWs must match the acoustic impedance of the seawater to preserve high sensor performance under various underwater environments. The sensor must be designed to withstand the pressures at the operating depth of typical sonar transducers. The package containing the sensor must protect the delicate sensing elements and interconnects from seawater-related chemical and corrosion effects. The sensor has potential applications in providing security to submarines, cruise ships, and coastal surveillance boats operating in various seawater environments.

9.9.2 MEMS Technology for mm-Wave Microstrip Patch Antennas

Microstrip patch antennas are widely deployed by communications systems, radar, and electronic warfare (EW) suites, where light weight, compact size, and low-profile are the critical design requirements. Silicon substrates are commonly used in the design and development of RF-MEMS devices. However, its high dielectric constant ($\varepsilon_r = 11.8$), its ability to excite surface waves that will limit the operating bandwidth, and higher loss tangent will limit the performance of microstrip patch antennas fabricated on silicon substrates. These problems are somewhat minimized by using high-resistivity silicon (HRS) substrates. RF sputtering is used to deposit silicon dioxide (SiO_2) films on HRS substrates and the same sputtering process can be used to deposit chromium–gold films on HRS substrates. Geometrical patterns can be made to form patch antenna elements with appropriate element dimensions and spacing between them. Although the patch antenna is realized on the suspended silicon dioxide dielectric membrane, the feed network can be formed on the bulk silicon substrate to minimize fabrication costs. It is critical to mention that the patch antenna fabrication process involves a low dielectric constant region (air cavity, $\varepsilon_r = 1$) surrounding the antenna resulting in improved bandwidth, low VSWR, and high return loss at mm-wave frequencies. Simulation data on a 35 GHz microstrip patch antenna using MEMS technology indicates a return loss better than −25 dB, directivity greater than 9.5 dB, E-plane and H-plane bandwidths of 60°, and an operation bandwidth better than 3.6 percent. These performance parameters at 35 GHz indicate that MEMS-based microstrip patch antennas can be designed at higher microwave and mm-wave frequencies for applications where minimum size, optimum RF performance, and low-profile are the principal design requirements.

9.9.3   Carbon Nanotube-Based Transistors and Solar Cells

NT research scientists at the San Diego (California) and Clemson universities (South Carolina) have discovered that specially synthesized CNT structures exhibit electronic properties, which are significantly improved over conventional silicon transistors currently used in computers. It is important to mention that CNTs are the rolled up sheets of carbon atoms and are more than 1000 times thinner than human hair. This discovery reveals a new era of ultraminiaturized electronic components with much smaller dimensions that will replace standard silicon transistors. The design and development of CNT-based transistors will involve a nanofabrication process, low-temperature solid-state physics for quantum computing and silicon nanophotonic technology. CNT-based transistors are still under research and development and, therefore, their performance parameters are not available. CNT technology offers the potential for high efficiency solar cells (Figure 9.3).

9.9.4   Nanotechnology-Based Sensors for Weapon Health and Battlefield Environmental Monitoring Applications

This section deals with NT-based sensors and devices capable of monitoring weapon health and battlefield environmental parameters. Both these monitoring schemes are considered very vital to ensure that weapons designed for specific applications are in a state of readiness and are ready to undertake the missions they designed for. Monitoring of battlefield environmental conditions or parameters is equally important. Prediction of sandstorms, rainfall, or snow in a battlefield is equally important so that weapons' operating modes can be changed to maintain minimum acceptable performance of the weapons.

9.9.4.1   Nanotechnology-Based Sensors to Monitor Weapon Health

Intermittent monitoring of the health of critical defensive and offensive weapon systems is of importance to ensure the weapons are ready to perform at optimum performance levels. The weapons could be air-to-surface missiles; air-to-air IR

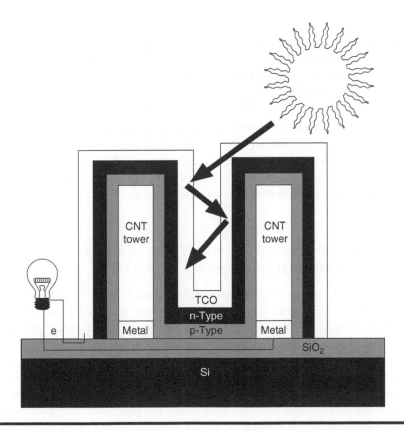

Figure 9.3 High efficiency solar cell architecture using an array of CNTs. (Courtesy of *Photonics Spectra*, July 2007, 7, Figure 4.)

missiles; submarine-based torpedoes; airborne, seaborne, and ground-based radar; airborne lasers; EW equipment; air-to-ground, air-to-air RF links; communication receivers; high-power RF sources; RF/IR detectors; T/R modules in active phase array antennas; unattended remotely located weapon systems; and other military systems and subsystems.

Some of the weapon system elements such as batteries, detectors, magnetic cores, and electrolytes have limited shelf lives and specific dormant lives. These elements need intermittent monitoring of their operational status. Weapon system engineers must quantify the monitoring requirements for remotely operated NT-based sensors for a host of candidate weapon systems. Monitoring of the outgassing of weapons propellant by chemical sensors, battery voltage/current ratings, receiver sensitivity, detector bias levels, magnetic field intensity, hazardous gas leakage, and other critical parameters is necessary to ensure the readiness of the weapons involved [11]. NT sensors using thin films of functionalized carbon annotate materials (CNTs) will be available in the near future to monitor weapon parameters in the field such as power consumption, reliability, maintainability, survivability, and maintenance cost. These sensors will be able to identify out-of-specification weapons, to predict shelf life, improve the reliability, confirm the readiness of the weaponry stockpile, and support weapon shelf life extension via innovative field maintenance procedures [11]. Health of complex weapon systems such as radar, missiles, lasers, electronic jamming equipment, and CW missile illuminators must be monitored at intervals to ensure reliability and weapon readiness status. Time is of great essence. Therefore, the application of MEMS and NT sensor research for rapid development of prognosis/ diagnostic systems is very critical in monitoring the health of missiles, high power-phased array radar, and airborne missile-tracking lasers.

9.9.4.2 Nanotechnology-Based Sensors to Monitor Battlefield Environmental Conditions

NT-based remote miniaturized sensors are required to alert soldiers of harmful chemical agents, biological agents, and toxic gases on the battlefield and its immediate vicinity. Biohazardous materials can be detected using passive and active wireless sensors based on monitoring the reflected phase. Battlefield monitoring sensors must be designed to monitor the hazardous environmental conditions. Microsensors must be deployed to monitor the temperature, pressure, humidity, vibration, and wind data. The sensors must provide field operation commanders accurate and real-time data. Note the prognostic/diagnostic concept can be used to access the shock or impact the missile has encountered and to provide the battlefield conditions needed for situation awareness [11]. In the case of complex weapon systems, NT-based embedded sensors can play a key role in monitoring environmental parameters such as temperature, pressure, humidity, etc., except for silent environments inherent to ground, air, and underwater military operations [11]. Unattended monitoring sensors on the battlefield must be fully operational for a minimum of 30 days. The remote-monitored sensors located at ranges greater than 60 ft or so must operate in a standby mode to preserve electrical power. These sensors can be actuated from acoustic, thermal, IR, and magnetic sources when a noticeable change is observed in the ambient level.

9.9.5 MEMS-Based Gyros and Applications

Precision navigation aids are extremely important for safe navigation, smart munitions, land vehicles used in military operations, robotic applications, airborne radar, and personnel locators. The highest precision inertial measurement units (imu) such as ring-laser gyros and fiber-optic gyros have been in use for several years.

These gyros provide a good overall performance, but at the expense of high cost, large size, and excessive power consumption. MEMS designers have developed MEMS-based gyros, which are best suited for guidance and navigation system applications. Gyros using MEMS technology not only offer excellent performance but also significant savings in cost, power consumption, and size. Because of these advantages, MEMS-based gyros offer precision guidance for military and aerospace electronic systems [12].

Critical components of a MEMS gyro, various functional elements, self-test port, and calibration circuit are shown in Figure 9.4. MEMS-based gyros are considered essential elements in automobile safety systems, because of their high reliability and minimum cost. These gyros are widely used in vehicle stability control and automobile navigation because of high precision performance. Automobile navigation systems combine accelerometers and gyros for backup of the global positioning system (GPS) function. The approach of combining several sensors is considered the most reliable and efficient, particularly for precision military applications.

Gyro stability is the most critical performance parameter and is given the most attention during the design phase of a gyro. Most military equipment requires gyro

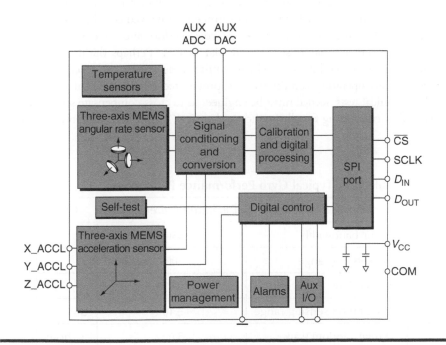

Figure 9.4 Block diagram of a MEMS-based gyro showing the various interface and functional elements. (Courtesy of *Military and Aerospace Electronics*, May 2007, 26, Figure 1.)

stability in few tenths of a degree per hour, which will undoubtedly increase the sensor cost. Many gyros have inherent biases and sensitivities to temperature, supply voltage, vibration, and other parameters. Implementation of self-test and calibration features can be incorporated in the gyro design flow with minimum cost, if the production quantities vary from 100,000 to 1,000,000 as in the case of automobile manufacturers. However, for others operating at low production volumes, the infrastructures to implement these features cannot be justified. Some gyro manufacturers provide built-in calibration circuits through the system-ready components. These components can significantly improve the in-system bias stability, depending on the sophistication of the factory calibration and the level of semiconductor integration around the gyro assembly. Selection of a precision-calibrated gyro is very crucial with short-term stability, which can be accomplished with a minimum of components. However, the long-term stability may require sophisticated filters and other sensor combinations. Typical gyro performance parameters are summarized in Table 9.3.

Design of a three-degrees-of-freedom inertial sensor is straightforward. But the design and development of an inertial sensor with six degrees of freedom will be a challenging problem for gyro engineers, for which semiconductor technology alone is not sufficient at the present moment. Achievement of high gyro accuracy across several sensor axes requires module-level integration, which is very costly. However, proper selection of materials and various processes involved could decrease the unit cost. A good balance between the sensor cost, performance, and ease-of-use is of critical importance in addressing a potential market. Perhaps the most significant challenge is in the module tests, which require measurements of several parameters under various operating conditions. A sophisticated combination of the electrical and mechanical tests needed must be engineered to cross-compensate several sensors involved in connecting for all significant electrical, positional, and motion effects.

Table 9.3   Typical Gyro Performance Parameters

Performance Parameters	Estimated Current Value	Unit
Sensor bandwidth	350	Hz
Rate sensing range	300	°/s
Bias stability	0.016	°/s
Acceleration sensing range	±10	G
Angular random walk	3.6	°/$\sqrt{\text{hour}}$
Axis nonorthogonality	±0.02	°

9.9.6   MEMS-Based Accelerometers and Applications

Precision accelerometers are widely used in commercial, industrial, and aerospace applications. Low-cost accelerometers are best suited for the consumer electronics market, because an extremely large number of sensors are involved. Two- and three-axis MEMS-based accelerometers ranging from1.5 to10 g are readily available in the market. The low-gravity sensors are most ideal for electronic systems that require the detection of small changes in force resulting from motion, fall, tilt, positioning, shock, and vibration. The two-axis or XZ-axis accelerometer is capable of sensing motion in both lateral and perpendicular planes and is considered the most cost-effective sensor, and hence a three-axis or XYZ-axis sensor is not needed. Potential applications include advanced pedometers, which can record the foot motion, distance, and speed of the runner's foot. Accelerometers with various sensitivities are available capable of satisfying the needs of multiple applications and functions. MEMS-based accelerometers are widely used in many consumer products, including industrial, healthcare monitoring, and embedded systems using six sensing parameters to meet applications requirement. The majority of these sensors are deployed in GPSs, hard drive protection, camcorder stability, and scrolling interfaces in gaming devices. Key features of MEMS accelerometers include full-range of directional motion, low-g capability, low-current consumption (500 μA), sleep-mode for extended battery use (3 μA), low-voltage operation (2.2–3.6 V), microelectronic interface compatibility, and fast power-up response at 1 ms.

9.9.7   Material Requirements and Properties for MEMS- and NT-Based Sensors and Devices

9.9.7.1   Introduction

In recent years, enormous progress has been made in the design, development, fabrication, and commercialization of MEMS- and NT-based sensors and devices. To meet the requirements of these advances, a new class of materials will be required for the design and development of MEMS and NT sensors [12]. Currently, traditional materials such as single crystal and polycrystalline silicon, silicon oxide, silicon nitride, and aluminum are widely used in silicon microelectronic devices. Now it is possible to introduce and integrate a variety of new metals and alloys such as copper, nickel, gold, palladium, tungsten, and nickel–titanium, ceramics and glasses, namely silicon carbide, quartz, zinc oxide, zirconium oxide, diamond, and lead–zirconium–titanate (PZT) and potential polymers. On the basis of the current information on new materials, the MEMS designer has multiple choices for the materials to optimize the performance and reliability for the sensors or devices operating under severe thermal and mechanical environments. MEMS designers have identified three distinct requirements for the materials: compatibility with silicon technology, appropriate electromechanical properties, and low residual stresses under changing temperatures. Judicial selection of materials is absolutely necessary for shock-resistant microbeams or cantilever structures, force sensors, micromachined flexures, and micromechanical (MM) filters to retain their performance levels under variable operating environments.

9.9.7.2 Material Requirements for Fabrication of MEMS Sensors and Devices

Different material requirements for fabrication and packaging may be required depending on the MEMS sensor type and operating environments. Fabrication of a MEMS or NT sensor or device requires different sets of materials and various processes, because numerous elements are involved in the design of the sensor. Each critical element of the device may require a specific material to optimize the overall MEMS device performance. In brief, a combination of various processing techniques such as bulk micromachining, surface micromachining, electroplating, and lithography are needed to introduce and integrate a large number of engineering materials in the design of MM elements [12]. Microsystems material can be classified into four categories: metals and alloys, ceramics and glasses, polymers and elastomers, and composites using nanomaterials or smart materials. It is important to mention that structural materials are those that constitute load-bearing beams, membranes, base plates, and support elements, whereas the transducer materials constitute devices, which convert energy from one physical domain to another such as RF energy into acoustic energy. Materials required in the fabrication of a transducer will have different properties than those required in the fabrication of the device.

9.9.7.3 Properties of Materials Required for Mechanical Design of MEMS Devices

The critical properties of materials commonly required in the mechanical design of MEMS and NT-based sensors and devices are Young's modulus, yield strength, fracture strength, torsional strength, residual strength modulus of rigidity, Poisson's ratio, shock resistance, coefficient of thermal expansion, specific heat per unit mass, fracture toughness, and fracture strength. It is important to point out that fracture strength is the most critical property of a structure, whereas fracture toughness is strictly a material property. Because it is extremely difficult to obtain the fracture toughness value, the tensile fracture strength is commonly used. It is also difficult to define the material requirements for microscale structures. However, by focusing on sensor or device structure with minimum feature sizes greater than 1 μm, it is possible to quantitatively relate MM properties to bulk properties in many cases. Some properties of certain materials such as fracture strength, yield strength, residual strength, and Young's modulus have values determined by interactions at multiple length scale (typically in the range of 0.01–1 μm), which can vary significantly in microscale structures compared to bulk values as shown in Table 9.4.

9.9.7.4 Properties of Materials Required for Thermal Design of MEMS Devices

Thermal design of MEMS and NT devices requires materials with specific values of critical thermal parameters such as coefficient of thermal expansion (α), mass density (ρ), specific heat capacity (C_{sh}), and thermal conductivity (K) to optimize the device thermal performance under thermal environments. Proper material integration is of critical importance, because it can affect the adhesion, seal, phase stability, thermal reaction, and thermal stress under variable thermal environments. The thermal properties of widely used materials in the design of MEMS and NT devices and sensors are summarized in Table 9.5.

Table 9.4 Comparison of Bulk and Microscale (μ) Properties of Materials Best Suited for MEMS and NT Devices

Metal	E_{bulk} (GPa)	E_μ (GPa)	$\sigma_{F,bulk}$ (MPa)	$\sigma_{F,\mu}$ (MPa)
Aluminum	69	70	205	152
Copper	125	120	398	350
Gold	80	70	220	310
Nickel	214	185	410	515
Polysilicon	130	160	2000	1250
Single crystal	130	125	2000	1350
Silicon carbide	430	400	4000	—
Silicon nitride	280	255	5200	6150
Silicon oxide	50–80	80	800–1100	1000

Note: The subscript μ stands for microscale values and bulk stands for bulk values for the materials widely used in the mechanical design of MEMS and NT devices.

Table 9.5 Thermal Properties of Materials Widely Used in MEMS/NT Sensors and Devices

Material	Coefficient of Thermal Expansion α (10^{-6}/K)	Mass Density ρ (g/cm^3)	Specific Heat Capacity C_{sh} (J/kg K)	Thermal Conductivity K (W/m K)
Nickel (bulk)	12.72	8.91	443	91
Gold	14.22	19.35	129	317
Chromium	10.02	7.15	449	94
Silicon nitride	2.91	3.14	712	19
Silicon [100]	2.31	2.34	713	156

9.10 Summary

This chapter summarizes the design aspects and performance capabilities of MEMS and NT-based sensors and devices. The properties and applications of photonic crystals are summarized. Capabilities of PBG devices and PCFs are described. Performance parameters of OEOs are discussed. PCFs offer higher light-coupling efficiency, large numerical aperture, ultralow dispersion, most robust and stable structure, and uniform index profiles. An OPO offers the lowest phase noise even at higher microwave frequencies. No other oscillator can match the short-term frequency stability (0.1 ppm/°C) and phase noise (better than −143 dBc at 10 kHz offset in a 10 GHz oscillator) of an OEO. These superior performance parameters are due to elimination of mode-hopping effects. Potential applications of nanocrystals, NWs, nanobubbles, and nanorods are briefly described. High spatial resolution and sensitivity of NWs are best suited for avalanche photodetectors in integrated nanophotonic systems. Performance capabilities of MDMs are identified, highlighting their deployment in low-cost adaptive optics. Adaptive optics incorporating MEMS mirrors are best suited to correct higher-order aberrations in the human eye and to eliminate wave front-aberrations experienced by airborne laser-tracking systems. Applications of CNTs and CNT arrays, in high-resolution displays, high-current density cold-cathode electron emitters, optical emission spectroscopes, MEMS biosensors, electrochemical actuators, electrodes for MBs, and advanced composite structures best suited for rocket motors and warheads are discussed in great detail. Applications of MWCNTs in multifunctional, high-temperature nanocomposites are identified. Application of barium hexaferrite nanocomposite coatings leading to stealth technology is briefly discussed. The coatings of this nanocomposite material can eliminate the detection of fighter or reconnaissance aircraft or missiles by enemy radar operating at higher microwave frequencies. NT-based sensors and devices such as acoustic sensors, mm-wave microstrip patch antennas with low side lobes and return loss, CNT-based transistors, weapon health and battlefield environmental monitoring sensors, and sensors for detection of chemical, biological, and toxic agents are described, identifying their unique capabilities. Potential industrial and military applications of MEMS-based gyros and accelerometers are identified with particular emphasis on reliability and performance. Material requirements and their properties best suited for the design and development of various MEMS- and NT-based sensors or devices are summarized. Critical thermal and mechanical properties of bulk and microscale materials required in the fabrication and packaging of MEMS- and NT-based microsystems and devices are summarized. Materials and properties required for structures are identified with emphasis on Young's modulus and fracture strength values for bulk and microscale materials.

References

1. Comments by Sr. editors, Fluorescent nanoparticles and cancer imaging, *Photonics Spectra*, March 2006, 5.
2. A.R. Jha, Applications of photonic crystal fiber in opto-electronic-oscillator, Technical presentation at the Royal Institute of Technology, Stockholm, Sweden, September 2006.
3. A.R. Jha, *Fiber Optic Technology*, Noble Publishing, Thomasville, Georgia, 2004, 7.
4. H. Hogan, The photoconduction of ZnO nanowires, *Photonics Spectra*, May 2007, 81.
5. M.J. Lander, Zinc phosphide nanowires for photon detection, *Photonics Spectra*, February 2007, 5.
6. Comments by Sr. editors, Technology in cancer and treatment, *Photonics Spectra*, March 2006, 5.
7. A.L. Fisher Sr. (Ed.), Laser heat forms nanobubbles, *Photonics Spectra*, May 2007, 48.
8. R. Gauhan, MEMS deformable mirrors bring adaptive optics within reach, *Photonics Spectra*, February 2005, 84.
9. H. Manohara and M. Bronkowsti, Arrays of bundles of carbon nanotubes as field emitters, *NASA Tech Briefs*, February 2007.
10. Sr. editors, Multifunctional, high-temperature nanocomposites, *NASA Tech Briefs*, May 2007, 48.
11. P. Ruffin, C. Brantley et al., Requirements for micro and nanotechnology-based sensors in weapon health and battlefield environmental monitoring applications, *Proceedings of SPIE Optical Conference on Security and Defense in Stockholm*, Sweden, Vol. 6172.
12. V.T. Srikar and M. Shearhy, Materials selection in micromechanical design: An application of the Ashby approach, *Journal of MEMS*, 12(1), February 2003, 3–4.

Index

A